Progress in Polymer Composites for Different Applications

Progress in Polymer Composites for Different Applications

Editor

Kamila Sałasińska

Basel • Beijing • Wuhan • Barcelona • Belgrade • Novi Sad • Cluj • Manchester

Editor
Kamila Sałasińska
Faculty of Materials Science
and Engineering
Warsaw University of
Technology
Warsaw
Poland

Editorial Office
MDPI
St. Alban-Anlage 66
4052 Basel, Switzerland

This is a reprint of articles from the Special Issue published online in the open access journal *Polymers* (ISSN 2073-4360) (available at: www.mdpi.com/journal/polymers/special_issues/polymer_compo_applications).

For citation purposes, cite each article independently as indicated on the article page online and as indicated below:

Lastname, A.A.; Lastname, B.B. Article Title. *Journal Name* **Year**, *Volume Number*, Page Range.

ISBN 978-3-7258-0246-3 (Hbk)
ISBN 978-3-7258-0245-6 (PDF)
doi.org/10.3390/books978-3-7258-0245-6

© 2024 by the authors. Articles in this book are Open Access and distributed under the Creative Commons Attribution (CC BY) license. The book as a whole is distributed by MDPI under the terms and conditions of the Creative Commons Attribution-NonCommercial-NoDerivs (CC BY-NC-ND) license.

Contents

About the Editor vii

Preface ix

Mateusz Barczewski, Aleksander Hejna, Kamila Sałasińska, Joanna Aniśko, Adam Piasecki, Katarzyna Skórczewska and Jacek Andrzejewski
Thermomechanical and Fire Properties of Polyethylene-Composite-Filled Ammonium Polyphosphate and Inorganic Fillers: An Evaluation of Their Modification Efficiency
Reprinted from: *Polymers* **2022**, *14*, 2501, doi:10.3390/polym14122501 1

Jacek Andrzejewski and Sławomir Michałowski
Development of a New Type of Flame Retarded Biocomposite Reinforced with a Biocarbon/Basalt Fiber System: A Comparative Study between Poly(lactic Acid) and Polypropylene
Reprinted from: *Polymers* **2022**, *14*, 4086, doi:10.3390/polym14194086 24

Magdalena Zdanowicz and Kamila Sałasińska
Characterization of Thermoplastic Starch Plasticized with Ternary Urea-Polyols Deep Eutectic Solvent with Two Selected Fillers: Microcrystalline Cellulose and Montmorillonite
Reprinted from: *Polymers* **2023**, *15*, 972, doi:10.3390/polym15040972 48

Mateusz Barczewski, Maria Kurańska, Kamila Sałasińska, Joanna Aniśko, Joanna Szulc, Izabela Szafraniak-Wiza, et al.
Comprehensive Analysis of the Influence of Expanded Vermiculite on the Foaming Process and Selected Properties of Composite Rigid Polyurethane Foams
Reprinted from: *Polymers* **2022**, *14*, 4967, doi:10.3390/polym14224967 60

Adam Olszewski, Paulina Kosmela, Wiktoria Żukowska, Paweł Wojtasz, Mariusz Szczepański, Mateusz Barczewski, et al.
Insights into Stoichiometry Adjustments Governing the Performance of Flexible Foamed Polyurethane/Ground Tire Rubber Composites
Reprinted from: *Polymers* **2022**, *14*, 3838, doi:10.3390/polym14183838 78

Worawat Poltabtim, Arkarapol Thumwong, Ekachai Wimolmala, Chanis Rattanapongs, Shinji Tokonami, Tetsuo Ishikawa and Kiadtisak Saenboonruang
Dual X-ray- and Neutron-Shielding Properties of Gd_2O_3/NR Composites with Autonomous Self-Healing Capabilities
Reprinted from: *Polymers* **2022**, *14*, 4481, doi:10.3390/polym14214481 99

Matea Lapaš Barišić, Hrvoje Sarajlija, Eva Klarić, Alena Knežević, Ivan Sabol and Vlatko Panđurić
Detection of Leachable Components from Conventional and Dental Bulk-Fill Resin Composites (High and Low Viscosity) Using Liquid Chromatography-Tandem Mass Spectrometry (LC-MS/MS) Method
Reprinted from: *Polymers* **2023**, *15*, 627, doi:10.3390/polym15030627 115

Mirela Văduva, Teodora Burlănescu and Mihaela Baibarac
Functionalization of Carbon Nanotubes and Graphene Derivatives with Conducting Polymers and Their Applications in Dye-Sensitized Solar Cells and Supercapacitors
Reprinted from: *Polymers* **2024**, *16*, 53, doi:10.3390/polym16010053 131

Paulina Latko-Durałek, Michał Misiak and Anna Boczkowska
Electrically Conductive Adhesive Based on Thermoplastic Hot Melt Copolyamide and Multi-Walled Carbon Nanotubes
Reprinted from: *Polymers* **2022**, *14*, 4371, doi:10.3390/polym14204371 . . . . **168**

Claudia Sergi, Libera Vitiello, Patrick Dang, Pietro Russo, Jacopo Tirillò and Fabrizio Sarasini
Low Molecular Weight Bio-Polyamide 11 Composites Reinforced with Flax and Intraply Flax/Basalt Hybrid Fabrics for Eco-Friendlier Transportation Components
Reprinted from: *Polymers* **2022**, *14*, 5053, doi:10.3390/polym14225053 . . . . **181**

Martina Polaskova, Tomas Sedlacek, Zdenek Polasek and Petr Filip
Modification of Polyvinyl Chloride Composites for Radiographic Detection of Polyvinyl Chloride Retained Surgical Items
Reprinted from: *Polymers* **2023**, *15*, 587, doi:10.3390/polym15030587 . . . . **200**

Marius Murariu, Yoann Paint, Oltea Murariu, Fouad Laoutid and Philippe Dubois
Engineering Polypropylene–Calcium Sulfate (Anhydrite II) Composites: The Key Role of Zinc Ionomers via Reactive Extrusion
Reprinted from: *Polymers* **2023**, *15*, 799, doi:10.3390/polym15040799 . . . . **213**

Krzysztof Moraczewski, Tomasz Karasiewicz, Alicja Suwała, Bartosz Bolewski, Krzysztof Szabliński and Magdalena Zaborowska
Versatile Polypropylene Composite Containing Post-Printing Waste
Reprinted from: *Polymers* **2022**, *14*, 5335, doi:10.3390/polym14245335 . . . . **239**

About the Editor

Kamila Sałasińska

Kamila Salasinska, PhD, DSc, graduated from the Faculty of Environmental Engineering of the Warsaw University of Technology. In 2015, she obtained her PhD from the Faculty of Materials Engineering at the same university. From 2015 to 2021, she was an employee of the Central Institute for Labour Protection—National Research Institute, and she now works at the Faculty of Materials Science and Engineering at WUT. In 2023, she obtained her Doctor of Science degree in engineering and technical sciences in the materials engineering discipline. In her research work, she designs and assesses the properties of polymer materials and their composites. Much of the work she has carried out and published so far concerns the burning behavior of polymers and the search for new, environmentally friendly fire retardants. Dr. Salasinska is a co-author of more than 60 publications from the JRC list, 4 chapters, 6 patents, and 3 patent applications. She has managed three projects and two research tasks, and she has been the contractor of another fourteen. Moreover, she has served as a Guest Editor of several Special Issues and is a member of the scientific committees of conferences. Dr. Sałasińska is a member of the Polish Society of Composite Materials, the Polish Carbon Society, and the Polish Society of Calorimetry and Thermal Analysis. The most important awards she has received include the Gold Medal of the International Invention Competition Concours Lepine 2020 and two Golden Laurels of Innovation of the Supreme Technical Organization (2018 and 2020).

Preface

Composite materials result from searching for more effective design solutions or creating new material characteristics. The industrial use of fiber-reinforced polymers was initiated in 1935 by Owens Corning, when the first glass-reinforced bell was produced. The mass use of composite materials in industry, construction, and transport began in 1990-2000. In 2022, the global market value reached more than USD 117 billion, and it is still growing. This Special Issue aimed to assemble works associated with the different aspects of polymeric composites. This Special Issue is a collection of 13 original papers in the areas of the design, production, and assessment of the properties of new materials. In summary, the articles presented in this Special Issue perfectly demonstrate the multitude of applications and potential research directions that there are for composites.

I wish to express gratitude to the Editors-in-Chief of *Polymers* for the opportunity to create and run this Special Issue, as well as to all the authors who contributed to this Special Issue's success with their high-quality papers. Moreover, I would like to thank the reviewers who helped to enhance the level of the manuscripts published here. Finally, the Section Managing Editor should be acknowledged for their excellent management of the editorial process.

Kamila Sałasińska
Editor

Article

Thermomechanical and Fire Properties of Polyethylene-Composite-Filled Ammonium Polyphosphate and Inorganic Fillers: An Evaluation of Their Modification Efficiency

Mateusz Barczewski [1,*], Aleksander Hejna [2,*], Kamila Sałasińska [3,4,*], Joanna Aniśko [1], Adam Piasecki [5], Katarzyna Skórczewska [6] and Jacek Andrzejewski [1]

1. Institute of Materials Technology, Faculty of Mechanical Engineering, Poznan University of Technology, Piotrowo 3, 61-138 Poznan, Poland; joanna.anisko@put.poznan.pl (J.A.); jacek.andrzejewski@put.poznan.pl (J.A.)
2. Department of Polymer Technology, Gdansk University of Technology, Narutowicza 11/12, 80-233 Gdansk, Poland
3. Faculty of Materials Science and Engineering, Warsaw University of Technology, Wołoska 141, 02-507 Warsaw, Poland
4. Department of Chemical, Biological and Aerosol Hazards, Central Institute for Labour Protection—National Research Institute, Czerniakowska 16, 00-701 Warsaw, Poland
5. Institute of Materials Engineering, Faculty of Materials Engineering and Technical Physics, Poznan University of Technology, Jana Pawła II 24, 60-965 Poznan, Poland; adam.piasecki@put.poznan.pl
6. Faculty of Chemical Technology and Engineering, Bydgoszcz University of Science and Technology, Seminaryjna 3, 85-326 Bydgoszcz, Poland; katarzyna.skorczewska@pbs.edu.pl
* Correspondence: mateusz.barczewski@put.poznan.pl (M.B.); aleksander.hejna@pg.edu.pl (A.H.); kamila.salasinska@pw.edu.pl (K.S.); Tel.: +48-61-647-58-58 (M.B.)

Citation: Barczewski, M.; Hejna, A.; Sałasińska, K.; Aniśko, J.; Piasecki, A.; Skórczewska, K.; Andrzejewski, J. Thermomechanical and Fire Properties of Polyethylene-Composite-Filled Ammonium Polyphosphate and Inorganic Fillers: An Evaluation of Their Modification Efficiency. Polymers 2022, 14, 2501. https://doi.org/10.3390/polym14122501

Academic Editor: Roberto Scaffaro

Received: 24 May 2022
Accepted: 15 June 2022
Published: 20 June 2022

Publisher's Note: MDPI stays neutral with regard to jurisdictional claims in published maps and institutional affiliations.

Copyright: © 2022 by the authors. Licensee MDPI, Basel, Switzerland. This article is an open access article distributed under the terms and conditions of the Creative Commons Attribution (CC BY) license (https://creativecommons.org/licenses/by/4.0/).

Abstract: The development of new polymer compositions characterized by a reduced environmental impact while lowering the price for applications in large-scale production requires the search for solutions based on the reduction in the polymer content in composites' structure, as well as the use of fillers from sustainable sources. The study aimed to comprehensively evaluate introducing low-cost inorganic fillers, such as copper slag (CS), basalt powder (BP), and expanded vermiculite (VM), into the flame-retarded ammonium polyphosphate polyethylene composition (PE/APP). The addition of fillers (5–20 wt%) increased the stiffness and hardness of PE/APP, both at room and at elevated temperatures, which may increase the applicability range of the flame retardant polyethylene. The deterioration of composites' tensile strength and impact strength induced by the presence of inorganic fillers compared to the unmodified polymer is described in detail. The addition of BP, CS, and VM with the simultaneous participation of APP with a total share of 40 wt% caused only a 3.1, 4.6, and 3 MPa decrease in the tensile strength compared to the reference value of 23 MPa found for PE. In turn, the cone calorimeter measurements allowed for the observation of a synergistic effect between APP and VM, reducing the peak heat rate release (pHRR) by 60% compared to unmodified PE. Incorporating fillers with a similar thermal stability but differing particle size distribution and shape led to additional information on their effectiveness in changing the properties of polyethylene. Critical examinations of changes in the mechanical and thermomechanical properties related to the structure analysis enabled the definition of the potential application perspectives analyzed in terms of burning behavior in a cone calorimetry test. Adding inorganic fillers derived from waste significantly reduces the flammability of composites with a matrix of thermoplastic polymers while increasing their sustainability and lowering their price without considerably reducing their mechanical properties, which allows for assigning developed materials as a replacement for flame-retarded polyethylene in large-scale non-loaded parts.

Keywords: polyethylene; composite; fire behavior; fire retardant; copper slag; basalt powder; expanded vermiculite

1. Introduction

Polyethylene (PE), despite being many decades since its first application, due to its unique properties, resulting from the possibilities for broad adjustments of its structural parameters, including its polydispersity, molecular weight, or copolymerization with other polymers, is still gaining ground in the field of scientific research [1–3]. The low price of this polymer, high chemical resistance, and excellent processability are the most frequently mentioned features that justify the fact that it is the most commonly processed polymer material. On the other hand, PE is a low-melting polymer characterized by high flammability. One of the thermoplastic polymers' most frequently used modification methods is their application as a matrix in polymer composites. The scope of the research carried out and the introduction of powder or fibrous fillers included both the use of inorganic materials and plant-based fillers [4–8]. While quite a lot of attention is paid to the use of waste fillers of plant origin, at the same time, one should bear in mind the need to utilize inorganic compounds generated, among others, during production and technological processes in the metallurgical and mining industries. As the research [9,10] has shown, the introduction of low-processed inorganic fillers in the form of powders, in most cases, leads to the deterioration of the composites' strength. Nevertheless, the enhancement of stiffness at elevated temperatures is noticeable, which is quite auspicious.

Much effort in the previous considerations has been devoted to increasing the temperature and thermomechanical stability and improving the fire resistance of polyethylene. Therefore, considering the high thermal resistance of inorganic fossil fillers, their use for these purposes was justified. However, it should be noted that the effects of introducing inorganic fillers, including nanometric ones, often lead to the achievement of entirely different thermal stability effects in terms of the increase mentioned above [11]. The result of the modification depends strictly on the filler's chemical structure, size, roughness, and the specific surface area of its particles. For example, montmorillonite (MMT) often induces different effects on the polymers' thermal stability depending on its dispersion in the matrix. The intercalated montmorillonite, in many reports, showed a negative impact on the thermal stability of the polymeric matrix, which was attributed to the presence of cationic compounds used for its modification.

On the other hand, the exfoliation of MMT yielded thermal stabilization of various polymers due to finer particles' dispersion [11]. Many published studies have shown that adding inorganic fillers can inhibit the burning rate, reduce smoke emission, lower the heat release rate, and increase char formation [12]. Often, the impact of temperature-stable powder fillers, such as silica [13], basalt [14], or metal oxides [15], which are not reactive with multiple thermoplastic polymer matrices, is mainly related to the reduction in the amount of organic polymer constituting the fuel during the burning process. Considering the results presented by Motahari et al. [13], it should also be considered that the addition of highly porous inorganic fillers may affect the thermal conductivity of polymers, which may directly influence the change in the combustion process.

Unfortunately, as demonstrated by several studies [12,14,16], the addition of sole fillers without flame retardants that causes additional, complex behavior during the fire, including intumescent effects, was insufficient. Recent research has shown that the effective modification of polymers is often achieved through synergy between multiple modifiers and fillers introduced into them. This phenomenon is successfully used to produce complex flame-retardant systems [17–19]. In the reported works developed, it has been shown that the simultaneous use of inorganic fillers with dedicated flame retardants can bring beneficial effects with regard to the overall flammability of polymer composites [17,20,21]. In such cases, increasing the efficiency of the powder fillers in the composition is not only based on reducing the amount of polymer. Additional benefits may be attributed to the barrier effects toward free radicals due to the presence of thermally stable fillers with a plate structure, such as exfoliated phyllosilicates [21]. The process of using modified mesoporous silica was also described, which led to the physical protection and hindered the volatilization of the oligomers due to the filler's agglomeration in the composite's surface layer during exposure to the flame [17].

Apart from the beneficial flammability reduction obtained by the simultaneous introduction of flame retardants and inorganic fillers, which is required by the most demanding industries such as the automotive industry [22], it is also essential to analyze obtained materials while considering the full spectrum of their properties, including their mechanical and thermomechanical performance. Many systems containing significant amounts of fire retardants (20–30 wt%) are not considered composites [23]. The additional introduction of powder or plate-shaped inorganic fillers with a low aspect ratio results in generating a hybrid structure containing insoluble particles with different properties and limited adhesion dispersed in the polymer matrix. Due to the insufficient interfacial interactions, the reinforcing effect on the polymer matrix is often limited. Moreover, a high share of the flame retardants required to achieve the expected flammability class intensifies the reduction in mechanical properties. Therefore, the substantive analysis of the correlation between the incorporation of additional inorganic, highly temperature-stable fillers into fire-retarded polymers is still an actual research topic. The significant deterioration of the mechanical performance is often ignored or omitted. The analysis of mechanical properties does not emphasize the limitations resulting from the total share of additives and fillers (up to 70 wt%) [24].

The study aimed to comprehensively evaluate introducing low-cost inorganic, fossil, and waste fillers into the flame-retarded polyethylene composition. Critical examinations of changes in the mechanical and thermomechanical properties related to the structure analysis enabled the definition of the potential application perspectives analyzed in terms of burning behavior in a cone calorimetry test. Incorporating three fillers with similar thermal stability but differing particle size distributions and shapes yielded additional information on their effectiveness in changing the properties of polyethylene.

2. Experimental

2.1. Materials

The commercial-injection-molding grade of high-density polyethylene (HDPE), type M300054, delivered by SABIC (Netherlands), was applied as a matrix for preparing composites. According to producer data, its density is 0.954 g/cm^3, and it has a melt flow rate (MFR) of 30 g/10 min (190 °C, 2.16 kg).

The intumescent flame retardant used for modification of the polyethylene was commercial Exolit AP 422, delivered by Clariant. It is a composition based on an ammonium polyphosphate (APP).

Three different inorganic fillers were used for manufacturing the composites: copper slag (CS), basalt powder (BP), and expanded vermiculite (VM). Copper slag (CS), a by-product generated from a suspension furnace, was derived in the form of fine powder with a grain density of 3.04 g/cm^3 from Polish copper-rich deposits. The chemical composition of CS, as declared by the supplier, consists of: 41.2 wt% SiO_2, 19.1 wt% Al_2O_3, 13.1 wt% CaO, 12.0 wt% Fe_2O_3 + FeO, 4.9 wt% MgO, 1.1 wt% Cu. The material showed a moisture content of 0.13 wt%. Natural BP, with a density of 2.95 g/cm^3, was a waste product obtained from the production of asphalt aggregate in Poland. Dominant chemical ingredients of BP given by the supplier in the technical data sheets were: 47.89 wt% SiO_2, 15.17 wt% Al_2O_3, 10.92 wt% Fe_2O_3, 9.49 wt% CaO, 7.57 wt% MgO, 3.33 wt% Na_2O, 2.04 wt% TiO_2, 0.90 wt% K_2O, 0.55 wt% P_2O_5, 0.20 wt% MnO, 0.01 wt% SO_3, and 0.01 wt% F. Thermally expanded vermiculite (VM) before pre-processing was characterized with a density of 2.61 g/cm^3 and a particle size up to 1.6 mm. It was provided by Perlit Polska (Poland). Before use, VM was subjected to milling with the knife mill Retsch GM200 with a knife rotational speed of 5000 rpm and 5 min, and was sieved by a Fritsch Analysette 3 mechanical siever using 100 μm mesh. The annealing process was carried out at a temperature of 1260 °C, and the chemical composition according to the manufacturer's data is 38.0–49.0% SiO_2, 20.0–23.5% MgO, 12.0–17.5% Al_2O_3, 0.3–5.4% Fe_2O_3, 5.2–7.9% K_2O, 0.0–1.2% FeO, 0.7–1.5%

CaO, 0.0–0.8% Na$_2$O, 0.0–1.5% TiO$_2$, 0.0–0.5 Cr$_2$O$_3$, 0.1–0.3% MnO, 0.0–0.6% Cl, 0.0–0.6% CO$_2$, 0.0–0.2% S. Broader information about the used fillers was presented in a previous works [25–27].

2.2. Sample Preparation

The composites were prepared by mixing them in a molten state. The HDPE pellets were pulverized into a fine powder using a Tria 25-16/TC-SL high-speed knife grinder to facilitate a more efficient physical mixing process with powdered organic filler. The polymeric powder was then preliminary mixed with 20 wt% of APP and 5, 10, and 20 wt% of filler using a Retsch GM200 knife mixer (5 min, 3000 rpm). Before being mixed in the molten state, the compositions were dried in a laboratory cabined dryer Memmert ULE 500 for 12 h at 70 °C. The mixtures were processed using a ZAMAK EH16.2D co-rotating twin-screw extruder operating at 100 rpm, with a maximum temperature for the process of 190 °C. For tensile and impact strength tests, the specimens with dimensions 100 × 100 × 4 mm^3 were manufactured with an Engel HS 80/20 HLS injection molding machine operating at 210 °C. The injection molding process was conducted with the following parameters: mold temperature T_{mold} = 30 °C, injection speed V = 100 mm/s, forming pressure P_f = 5 MPa, and cooling time t = 60 s. Standardized specimens for mechanical testing were mechanically processed.

2.3. Methods

The particle size distribution of inorganic fillers was characterized using a laser particle sizer Fritsch ANALYSETTE 22 apparatus (Weimar, Germany) operating in the range of 0.08–2000 µm.

Scanning electron microscopy (SEM) was performed using the model Tescan MIRA3 microscope (Brno-Kohoutovice, Czech Republic). The measurements were conducted with an accelerated voltage of 5 kV and magnifications of 200× and 2000×. The measurements were conducted with an accelerated voltage of 12 kV in the backscattered electrons (BSE) and secondary electron (SE) modes. The thin carbon coating (~20 nm) was deposited on samples using the Jeol JEE 4B vacuum evaporator.

The Fourier transform infrared spectroscopy (FT-IR) measurements were realized using a spectrometer Jasco FT/IR-4600 (Tokyo, Japan) at room temperature (23 °C) in the Attenuated Total Reflectance (ATR - FT-IR) mode. A total of 32 scans at a resolution of 4 cm^{-1} were used in all cases to record the spectra.

The specific weight of the applied fillers and resulting composites was determined using a gas pycnometer Pycnomatic from Thermo Fisher Scientific Inc. (Waltham, MA, USA). The following measurement settings were applied: gas—helium; target pressure—2.0 bar (29.0 psi); flow direction—reference first; temperature control—on; temperature set—20.0 °C; cell size—medium, 40 cm^3; the number of cleaning cycles—3; the number of measurements—10.

The results obtained from the pycnometric measurements were used to determine the porosity of the composites as the difference between the theoretical and experimental density values. The theoretical values were calculated according to Equation (1):

$$\rho_{theo} = \rho_m \cdot (1 - \varphi) + \rho_f \cdot \varphi \tag{1}$$

where: ρ_{theo}—theoretical density of the composite, g/cm^3; ρ_m—density of the matrix, g/cm^3; ρ_f—density of the filler, g/cm^3; and φ—a volume fraction of the filler.

To quantitatively determine the composite's porosity, Equation (2) was applied as follows:

$$p = \frac{\rho_{theo} - \rho_{exp}}{\rho_{theo}} \cdot 100\% \tag{2}$$

where: p—porosity of the material, %; and ρ_{exp}—an experimental value of composite density, g/cm^3.

The thermal diffusivity measurements were prepared using a modified Ångström method with a Maximus (Poland, Poznan) apparatus. A more comprehensive description of the experiments was described in detail in the literature [28,29]. During investigations, the microheater was charged by 23 V to heat the samples in a time of 400 s.

The thermal properties of the studied materials were analyzed using the differential scanning calorimetry (DSC) method. Samples of 5 ± 0.2 mg were placed in aluminum crucibles with pierced lids and were heated from 20 °C to 200 °C with a rate of 10 °C/min, held at this temperature for 10 min, and then cooled back to room temperature with a cooling rate of 10 °C/min. The procedure was realized twice. A Netzsch DSC 204F1 Phoenix (Selb, Germany) apparatus and an inert nitrogen atmosphere were used. The crystallinity degree X_{cr} was calculated according to Formula (3):

$$X_{cr} = \frac{\Delta H_m}{(1-f)\cdot \Delta H_{100\% PE}} \cdot 100\% \qquad (3)$$

where: ΔH_m—melting enthalpy of a sample, $\Delta H_{100\% PE}$—melting enthalpy of 100% crystalline PE, $\Delta H_{100\% PE}$ = 288 J/g [30], and f is the filler content.

Thermogravimetric analysis (TGA) was used to study the thermal decomposition of polyethylene and its composites. The 10 ± 0.2 mg samples were heated in the temperature range of 25–900 °C with a 10 °C/min heating rate using a Netzsch TG209 F1 (Selb, Germany) apparatus. The measurements were realized using Al_2O_3 crucibles in an inert atmosphere (nitrogen). The first mass derivative (DTG) was calculated in reference to the obtained mass vs. temperature curves. The 5% mass loss ($T_{5\%}$) and residual mass at 900 °C were determined.

The cone calorimeter measurements were conducted using a Fire Testing Technology Limited (UK, East Grinstead) apparatus according to the ISO 5660 standard to identify the burning behavior under forced-flaming conditions. The samples of 100 × 100 × 4 mm^3 were placed horizontally at 25 mm below a conical heater and tested at a heat flux of 35 kW/m^2 with piloted ignition. All samples were tested three times. The residues were photographed using an EOS 400 D digital camera from Canon Inc. (Tokyo, Japan).

The elastic modulus, elongation at break, and yield strength were tested through tensile testing. The tensile tests were performed per ISO 527 with a Zwick/Roell Z020 tensile tester model 5101 (Ulm, Germany) at room temperature. The elastic modulus measurements were conducted at a cross-head speed of 1 mm/min, while a different part of the experiment was carried out at 50 mm/min. Nine samples of each kind were tested.

The impact strength of the unnotched samples was examined by the Charpy method according to the ISO 179 standard at 25 °C. The Zwick/Roell HIT 25P (Ulm, Germany) impact tester with a 5 J hammer was applied for the measurement, and the peak load was determined as the maximum force (*Fmax*). For each series, seven specimens were tested.

The hardness was evaluated using a KB Prüftechnik (Hochdorf-Assenheim, Germany)apparatus with a ball indentation hardness test according to the ISO 2039 standard. The presented averaged values were based on a minimum of 15 tests from each series.

Vicat softening point temperature (VST) and heat deflection temperature (HDT) investigations were prepared with the use of a CEAST HV3 apparatus (Pianezza, Italy). The measurements were carried out in an oil bath following the ISO 306 standard in the A50 measurement configuration (50 N, 50 °C/h) and ISO 75 (0.455 MPa), respectively. The experiments were conducted for six specimens from each series.

3. Results and Discussion

3.1. Fillers' Characterization

The cumulative size distribution Q3(x) and adequate histograms dQ3(x) made for the three inorganic fillers used in this study are presented in Figure 1. The analysis of the graphs shows that the copper slag has the largest particle size, while in the case of the other fillers, most of the filler particles are of a comparable size. The VM exhibits two modes of particle size distribution due to the fraction of finely divided filler plates formed during the grinding of the expanded filler.

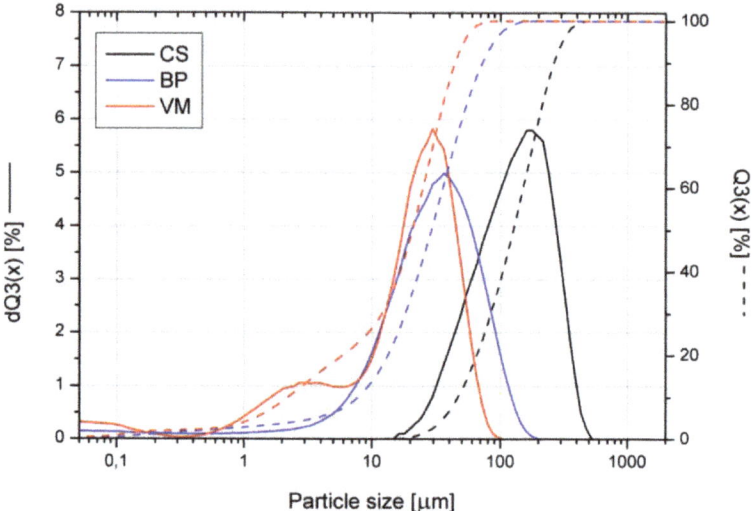

Figure 1. Particle size distribution of inorganic fillers.

3.2. Structural Analysis

Figures 2 and 3 compile brittle fracture SEM images of polyethylene and its composites. Figure 2 summarizes the images taken with low magnification to evaluate the fillers' dispersion in the polyethylene matrix. The use of the BSE mode allowed for the differentiation of the applied fillers' particles from the APP; thanks to the various density of materials, they could be distinguished in the obtained SEM images. In the case of all materials, a homogeneous distribution of the flame retardant in the polymer matrix can be observed. In the case of composites containing BP and VM, the distribution of the filler particles also does not raise any significant concerns. It allows the compositions to be defined as homogeneous. Modified CS composites with a much larger particle size in the tested area of the series with a lower filler concentration (5 and 10 wt%) revealed the presence of larger filler fragments, while only in the case of the PE/APP/20CS composite were different sizes of copper slag particles observed to be distributed evenly on the whole analyzed area. However, based on the analysis performed, it can be concluded that none of the fillers used tended to create agglomerated structures in the PE matrix. Moreover, the introduction of inorganic fillers in various concentrations did not deteriorate the APP dispersion. Despite the long process of drying, for the composites containing in their structures natural composite inorganic BP and VM fillers, an increased number of micropores were noted, which may come from the residual moisture from the filler.

Additionally, Figure 3 summarizes the SEM images taken in two modes, SE and BSE, for the PE/APP composition and composites, demonstrating the highest concentration of the filler (20 wt%) that can both distinguish the presence of filler particles and APP as well as assess the nature of the breakthrough. These images were taken at a higher magnification, making it possible to evaluate the adhesion at the polymer–filler interface indirectly. The break-out sites of APP particles of a regular shape are observed for all compositions. It can be concluded that the particles of inorganic fillers are characterized by better adhesion; in their case, there were no irregular pull-out holes and gaps in the interfacial region, which could suggest a loss of cohesion between the composite material and the combined materials. In the case of composites containing expanded vermiculite, broken fragments of the plate filler distributed in the sample volume are visible, as are the structures of non-comminuted filler packages that PE has not intercalated. The partial disintegration of the filler into the micrometric form of well-dispersed plates may be beneficial from the point of view of obtaining a limited flame effect [31,32]. Due to using a low processing temperature, degradation of the APP during processing may be omitted, which is also confirmed by the regular shape of the fire-retardant particles and the lack of voids at the PE/APP interface.

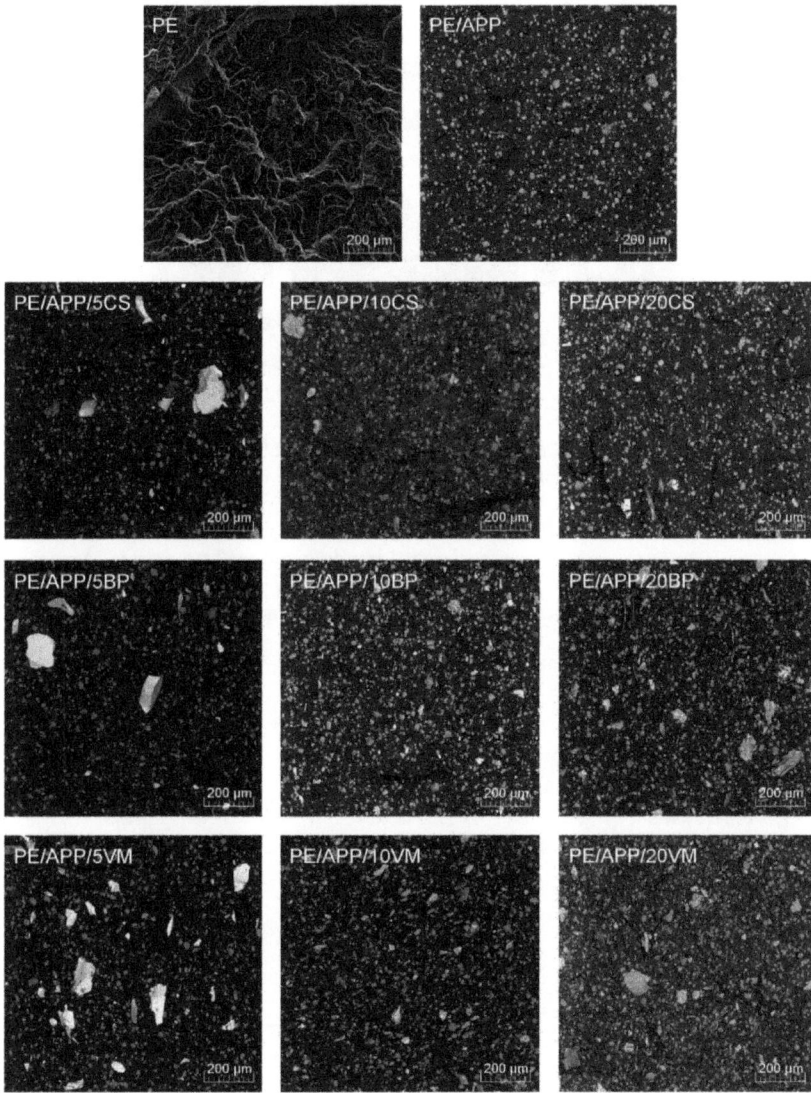

Figure 2. SEM images of PE and PE composites' brittle fractures (mag. 200×, BSE mode).

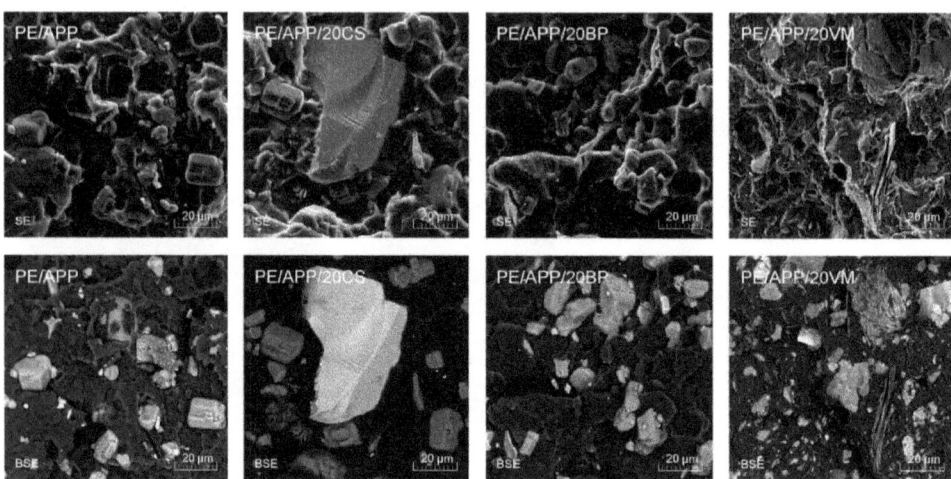

Figure 3. SEM images of PE and PE composites' brittle fractures (mag. 2000×, SE and BSE mode).

Table 1 summarizes the results of the physical properties measurements. Based on the analysis of the density of the fillers and the injection-molded samples, it was possible to determine the volumetric fraction of the filler in the composites and their porosity following Equation (3). Since the CS and BP fillers had a comparable density, it can be seen that the volumetric fraction of the filler is similar to those composite series. The higher volumetric content of the filler for the series containing VMs is due to the lower density and the extensive surface of expanded vermiculite, which was not entirely mechanically degraded during mechanical grinding and melt processing. According to the classification presented in [33], all samples reveal a low porosity, which excludes its strong effect on the mechanical properties of the samples and makes manufactured parts of good quality.

Table 1. Physical parameters of injection-molded PE and PE–composite samples.

Sample	Density [g/cm^3]	Volumetric Content of the Filler [%]	Porosity [%]
PE	0.949 ± 0.003	-	-
PE/APP	1.043 ± 0.001	10.95	1.33
PE/APP/5CS	1.083 ± 0.001	13.2	1.45
PE/APP/10CS	1.131 ± 0.001	15.6	1.09
PE/APP/20CS	1.219 ± 0.001	21.2	2.27
PE/APP/5BP	1.081 ± 0.001	13.2	1.59
PE/APP/10BP	1.139 ± 0.002	15.7	0.38
PE/APP/20BP	1.239 ± 0.001	21.4	0.48
PE/APP/5VM	1.087 ± 0.001	13.4	0.91
PE/APP/10VM	1.129 ± 0.001	16.0	1.04
PE/APP/20VM	1.222 ± 0.001	22.0	1.44

Figure 4a,b presents the spectra of the unmodified polyethylene matrix and composites containing applied fillers. Spectra of unfilled PE show an appearance typical for polyolefins [34]. The most significant absorption bands were noted around 2847 and 2914 cm^{-1} and were associated with the symmetric and asymmetric stretching vibrations of carbon–hydrogen bonds in the backbone of polyethylene. Signals attributed to these bonds' bending and rocking vibrations were also noted at 1460, 1470, 718, and 729 cm^{-1}. The positions of these signals are in line with the literature data on polyethylene materials [35]. A more detailed analysis of PE spectra confirms its type—high-density polyethylene (HDPE). According to Jung et al. [36], magnification of the 1300–1400 cm^{-1} region may provide essential insights related to the exact type of PE. Figure 4b (zoom 1300–1400 cm^{-1}) points to the absence of the 1377 cm^{-1} absorption band and visible signals at 1367 and 1352 cm^{-1} that indicate HDPE. The filler incorporation hardly affected the position and magnitude of bands characteristic of HDPE, which points to the lack of matrix decomposition despite the use of hard and rigid mineral fillers.

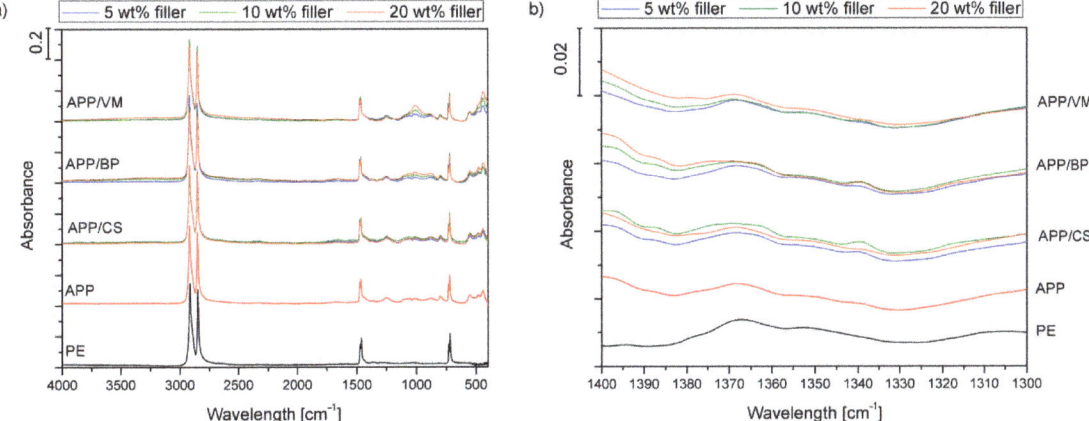

Figure 4. FTIR spectra of PE and PE-based composites in the range of 4000–400 cm^{-1} (a) and 1400–1300 cm^{-1} (b).

Spectra of the prepared composites show additional absorption bands related to the composition of the applied fillers. All spectra contain a minor peak around 1256 cm^{-1}, characteristic of the stretching vibrations of P=O bands present in polyphosphate structures [37]. Bands related to the vibrations of single phosphorous–oxygen bonds were also noted around 800 and 880 cm^{-1} [38]. Moreover, materials containing basalt and vermiculite fillers show small signals in the range 990–1010 cm^{-1}, typical for stretching Si–O bonds, as reported in previous work [39]. Bending vibrations of these bands were expressed by the signals around 450 cm^{-1}. They were more pronounced for the composites filled with vermiculite, which is in line with our previous studies, indicating a powerful absorption band around 1000 cm^{-1} [26].

3.3. Thermal Properties

Thermal diffusivity (D) quantifies materials' ability to conduct heat relative to their ability to store heat. This parameter may be determined using the Angström method, which is a steady-state measurement using an alternating-current heating plate [40]. This method in various modifications has been successfully used in multiple studies of polymers and their composites [28,29,41–43]. As previously discussed by Wenelska et al. [44], the determination of this property can help compare the thermal properties of PE-based composites and their flammability. Results obtained for considered materials showed a reciprocal tendency to those discussed in earlier studies because the presence of flame retardant and fillers lowered D values in comparison to unmodified polyethylene.

Moreover, observed results showed that the overall change in thermal diffusivity, considering standard deviations, caused by the additional incorporation of inorganic fillers may be omitted. Figure 5 shows the averaged thermal diffusivity values obtained for PE and its composites. According to indirect density-based porosity measurements and SEM observations, the lowered thermal diffusivity may be connected with a high amount of well-dispersed additives and fillers, as well as the microporosity occurring in the composite structure. As discussed by Prociak et al. [29], the cell size, in the case of porous materials, may significantly influence the D value; even the smallest amount of pores may affect the thermal behavior of the polymeric materials. Simultaneously, it should be underlined that the measured thermal diffusivities for all materials are at a comparable level. Despite the highest porosity of the samples containing CS, the thermal diffusivity of the prepared composites did not differentiate itself from the other samples. The lowest D values noted for BP-filled composites should be connected with the better conductivity of basalt powder itself in comparison to VM and CS rather than to structural changes, including the presence of the voids in the injection-molded samples.

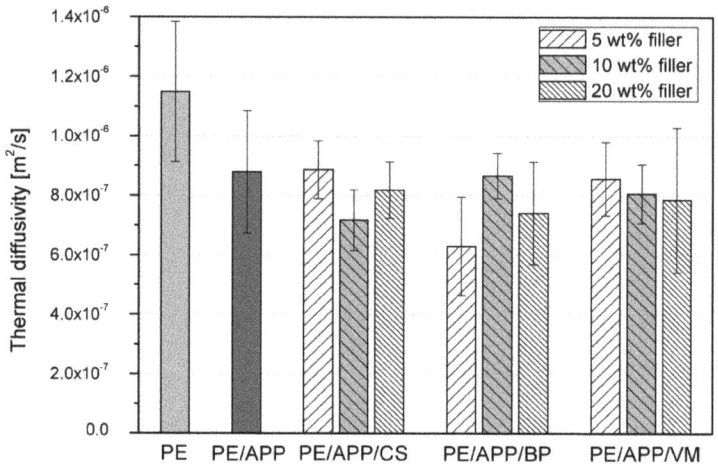

Figure 5. Thermal diffusivity of PE and PE-based composites.

The structure-related evaluation was based on a thermal analysis assessed by employing differential scanning calorimetry. Figure 6 shows the DSC curves from signals recorded during the second heating and first cooling. Additional thermal parameters such as the crystallization temperature (T_C), second melting temperature (T_{M2}), heat of fusion (ΔH_M), and crystallinity level, calculated according to Equation (1), are collectively presented in Table 2. The nucleation density and size of the spherulites depend on the crystallization temperature, degree of undercooling, and molecular weight of the polymeric matrix [45]. Therefore, the incorporation of fillers may cause changes in the crystallization behavior by the heterogeneous nucleation and change the thermal diffusivity of the polymeric melt. Changes in nucleation are related to promoting spherulite generation on the filler particles' surfaces, decreasing crystallites' thickness, and causing the epitaxial growth of the spherulites. Considering the DSC method's sensitivity, the observed melting and crystallization temperature changes of all APP-modified and inorganic filler composites can be negligible. At the same time, evident differences between individual material series are visible based on changes in the heat of fusion measured during the second heating process and the crystallinity calculated on its basis. Considering the low susceptibility of polyethylene to heterogeneous nucleation, the observed increase in crystallinity in the case of a 20 wt% addition of APP may be considered as having a substantial effect on the change in the PE structure. It should be mentioned that the achieved results are contrary to former studies [45], where incorporating APP into the HDPE matrix resulted in almost no effect on composite crystallinity. The difference may result from the different molecular weights of HDPE grades, which affect susceptibility to heterogeneous nucleation [46].

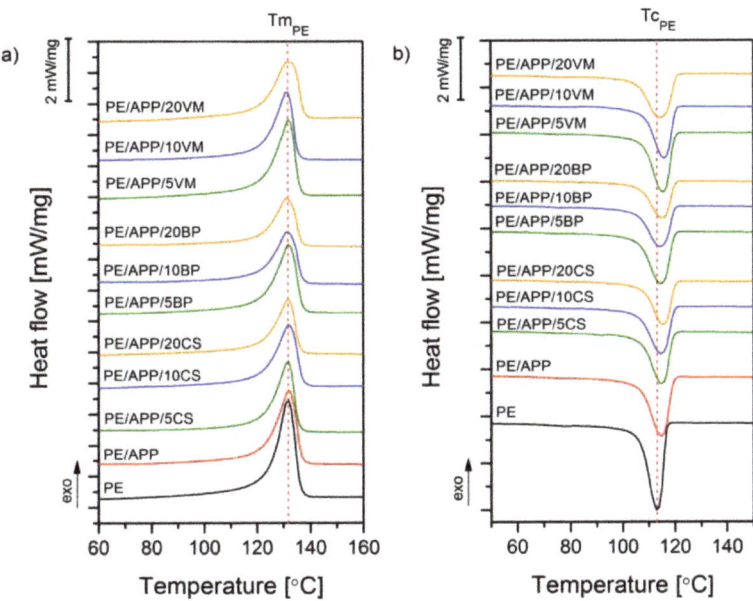

Figure 6. DSC curves obtained during the second heating (**a**) and first cooling (**b**).

Table 2. Thermal parameter obtained from DSC experiments.

Sample	T_C [°C]	T_{M2}	ΔH_m [J/g]	X_c [%]
PE	113.1	131.7	198.2	68.8
PE/APP	114.7	131.9	162.7	70.6
PE/APP/5CS	114.8	131.6	141.7	65.6
PE/APP/10CS	114.6	131.9	140.6	69.7
PE/APP/20CS	115.4	131.7	118.4	68.5
PE/APP/5BP	114.5	132.0	155.1	71.8
PE/APP/10BP	114.3	131.4	126.1	62.5
PE/APP/20BP	115.0	131.2	110.5	63.9
PE/APP/5VM	115.3	132.2	162.5	75.2
PE/APP/10VM	115.6	131.7	155.5	77.1
PE/APP/20VM	114.2	132.1	148.0	85.6

Interestingly, the addition of the lowest amounts of the inorganic fillers (VM and BP) resulted in the intensification of the nucleation effect, leading to an improved crystallinity level for the composites. However, for VM-filled composites, the crystallinity has been increasing gradually with the filler content; for basalt-filled composites, the opposite effect was noted. It should be mentioned that both micrometric and nanosized vermiculite were previously described as fillers with a confirmed nucleating ability on HDPE [45]; the exfoliated silicate layers may act as nucleation sites for the secondary nucleus of the composites during crystallization.

3.4. Thermal Stability and Fire Behavior under Forced-Flaming Conditions

Results obtained with thermogravimetric analysis are summarized in Table 3 and Figure 7. The presented data highlight that incorporated flame retardants and inorganic fillers influence the thermal stability of the polymer. Unmodified HDPE degraded completely at approx. 500 °C. The most noticeable difference between polyethylene and PE/APP with fillers is that the polymer decomposed in a single step, whereas the composites present a two-step degradation. From Figure 7, it can be seen that APP/CS, APP/BP, and APP/VM have a similar course of DTG curves and show the primary weight loss at 348–376 °C and 469–477 °C. In the first one, the main products were H_2O and NH_3, resulting from the thermal decomposition of polyphosphate. In turn, the second was related, apart from the decomposition of HDPE, to the release of phosphoric, polyphosphoric, and metaphosphoric acids from APP [47–49].

Table 3. Thermal properties of PE and PE-based composites tested in a nitrogen atmosphere.

Sample	$T_{5\%}$ [°C]	1st DTG Peak [°C; %/min]	2nd DTG Peak [°C; %/min]	Residual Mass at 900°C [%]
PE	424	-	471; −31.30	0
PE/APP	417	370; −0.71	404; −27.90	7.6
PE/APP/5CS	423	371; −0.59	473; −30.00	16.7
PE/APP/10CS	426	376; −0.50	470; −25.96	21.5
PE/APP/20CS	423	374; −0.45	474; −20.82	28.9
PE/APP/5BP	422	361; −0.74	473; −23.39	14.5
PE/APP/10BP	418	361; −0.51	469; −25.40	22.9
PE/APP/20BP	418	363; −0.51	474; −18.24	33.0
PE/APP/5VM	425	355; −0.72	473; −25.96	17.4
PE/APP/10VM	423	362; −0.61	476; −24.32	21.9
PE/APP/20VM	422	348; −0.48	477; −21.07	31.1

The second stage of decomposition was delayed compared to polyethylene, and the decomposition rate was much lower (maximal reduction by 42% for PE/APP/20BP). Inorganic carbonaceous residues remained after the major decomposition step, between 14.5 and 33.0 wt%. According to the TG result, the APP combined with BP, excluding the system with 5 wt% of inorganic filler, had more residue than the other systems under the same decomposition condition. In the case of APP and APP with the lowest amount of filler, the carbonaceous char was consumed in the subsequent minor decomposition step above 500 °C. The main mass loss stage and the residue it generates may influence the fire behavior of materials [47,50].

In turn, the $T_{5\%}$ weight loss temperature corresponding to the onset temperature of each flame-retarded HDPE occurred earlier than that of unmodified polymer. This is due to the relatively low temperatures, in which APP begins to decompose and forms phosphorus or phosphoric acid, promoting chain stripping, cross-linking, and char formation [51]. The TGA results indicated that the developed systems might have potential as a flame retardant for PE.

The cone calorimeter provides parameters such as time to ignition (TTI), heat release rate (HRR), including the peak heat release rate (pHRR) and total heat release (THR), effective heat of combustion (EHC), maximum average rate of heat emission (MARHE), and specific extinction area (SEA), as shown in Table 3. The HRR and THR curves for unmodified HDPE and its composites are shown in Figure 8.

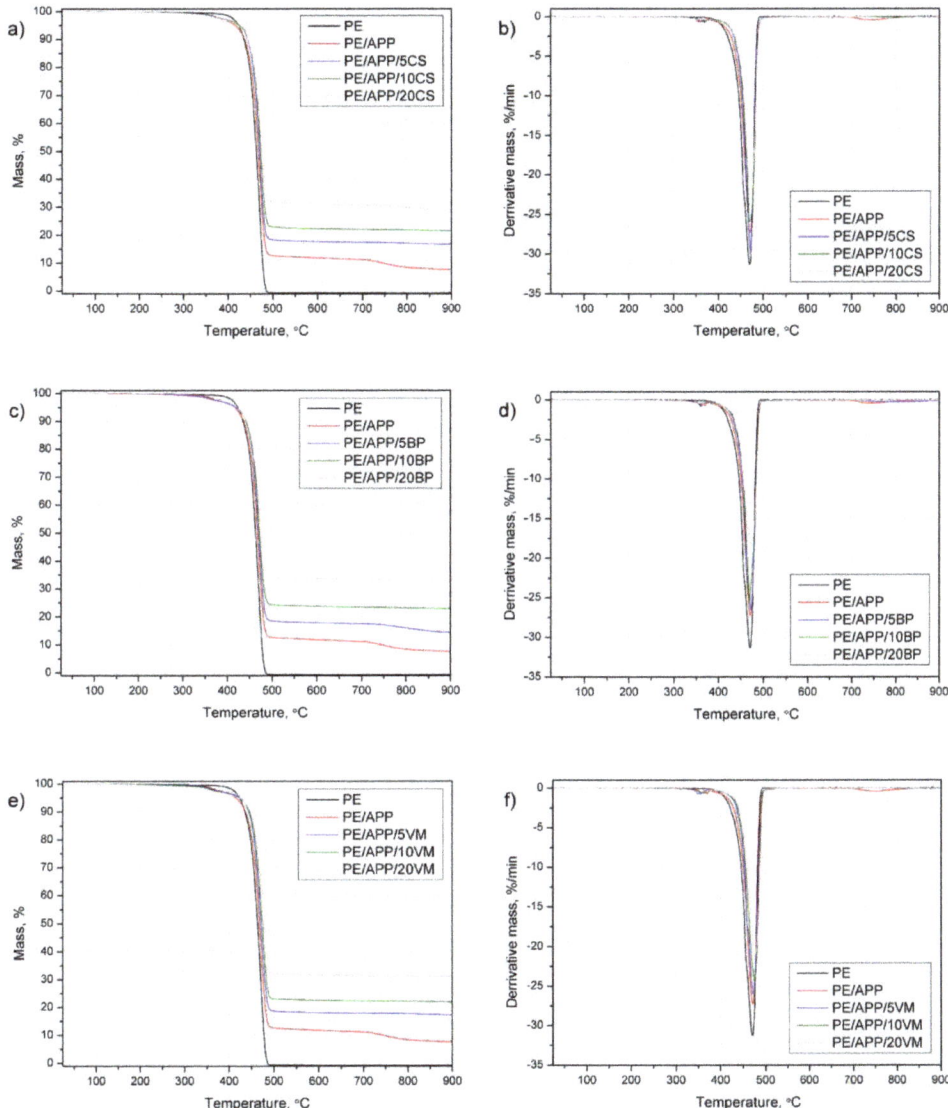

Figure 7. TG (**a,c,e**) and DTG (**b,d,f**) curves of PE and PE-based composites tested in a nitrogen atmosphere.

Polyethylene burned intensively after ignition in 128 s with a pHRR of 414 kW/m². The HRR curve course of HDPE is characteristic of a non-charring material, dominated by a pronounced pHRR. The addition of commercial fire retardant elongated the TTI without significant changes in the pHRR. Along with incorporating APP and CS, BP, or VM, in most cases, the time to ignition and burning time was prolonged, while the pHRR values decreased. Adding 5 wt% of inorganic components and 20 wt% of APP to HDPE resulted in a pHRR reduction from 6% (PE/APP/5BP) to 27% (PE/APP/5VM), whereas the samples with the highest additive amount showed reductions from 25% (PE/APP/20CS) to as high as 60% (PE/APP/20VM). The APP/VM was the most effective in reducing the burning intensity, and its use had changed the HRR curve's course to the type of a charring or residue-forming polymer [47,52]. VM is known for its flame-retardant effects [18,53,54]. The decrease in HRR values also reduced indices illustrating the flame spread or fire growth rates, such as MARHE and FIGRA. The reduction in FIGRA was variable, whereas the decrease in MARHE, excluding HDPE modified with CS, showed gradual decline dependence according to the increasing number of fillers. The highest reduction in FIGRA and MARHE, 2.5 and more than 3 times, respectively, was noted for PE/APP/20VM.

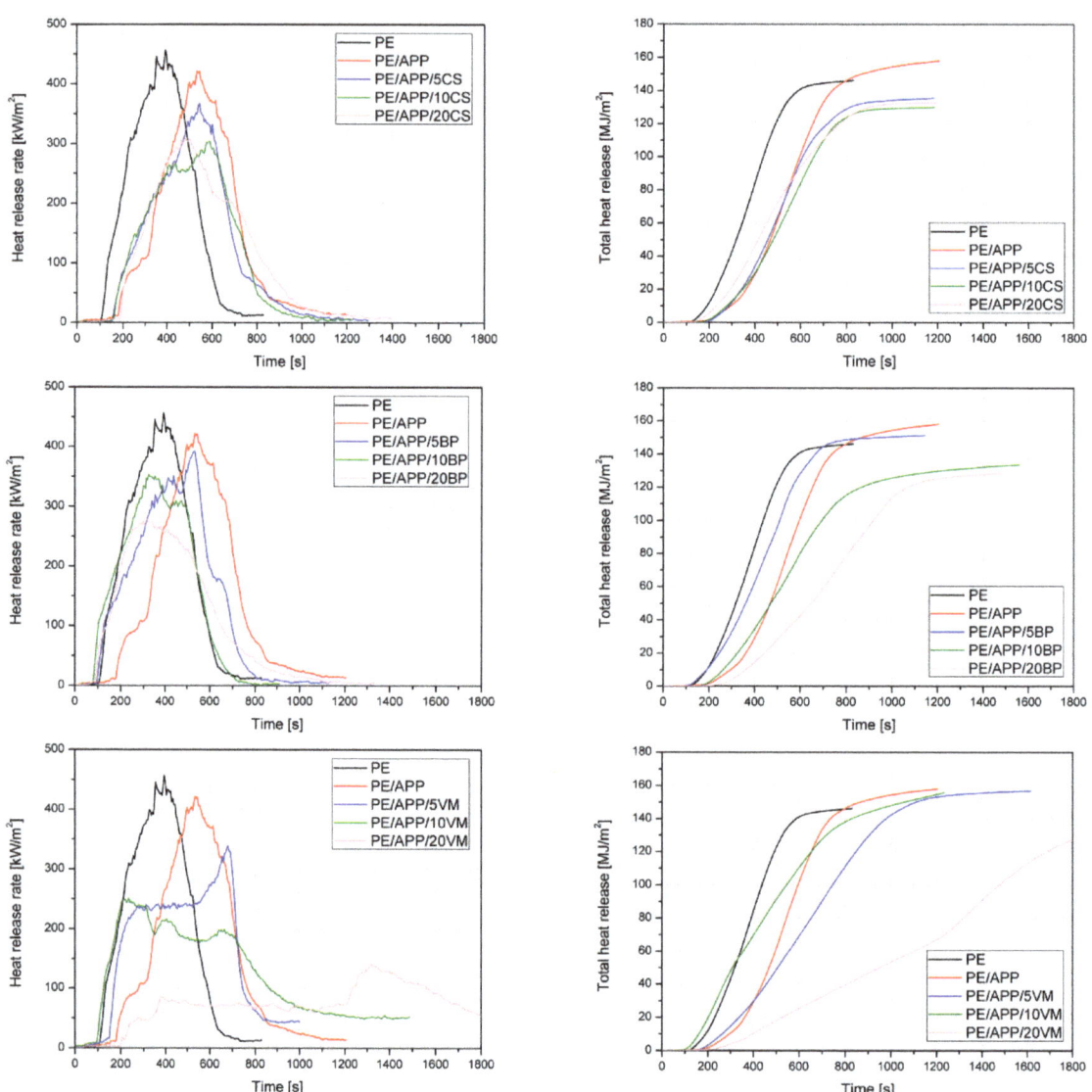

Figure 8. Heat release rate and total heat release curves of PE and PE-based composites.

THR is a measure of the fire load, indicating incomplete combustion by reducing combustion efficiency and/or char creation [55]. APP and APP combined with 5, or in some cases 10 wt% inorganic components caused an increase in the total heat release (Table 4). From Figure 9, showing the total heat output versus time, it can be observed that HDPE modified with the investigated systems in most cases did not reach higher values than polyethylene during the first 800 s; however, the materials burned much longer. Samples with CS and 20 wt% VM did not achieve THR values as high as HDPE throughout the test, while the rest reached it just at the end of the flame burning. The highest reduction, equal to approx. 15%, was noted for PE/APP/20BP and PE/APP/10CS. In turn, the highest THR and the standard deviation were obtained for composites with VM. EHC of unmodified polyethylene is relatively high and similar to non-flame-retarded polyolefins [47,56]. The additions led to the change in the gas-phase activity, and excluding the samples from 10 and 20 wt% of VM, the decrease in EHC was observed. This indicates that fuel dilution effects due to the release of incombustible products, or flame inhibition due to the release of phosphorus species acting as radical scavengers, may have occurred [57]. Replacing some amount of the HDPE with inorganic components cannot be excluded, reducing the emission of volatile decomposition products into the combustion zone. The increased flame retardancy was accompanied by a moderate increase in the CO yield of between 1% and 29%. Moreover, the average yield of residue after the burning of 20APP was 18% and increased to 33–36% with increases in the content of the inorganic component. The change tendency of the residue yield in the CC test is similar to that in the TG analysis.

Table 4. Cone calorimeter data of PE and PE modified with fire-retardant systems.

Materials	TTI, s	pHRR, kW/m^2	MARHE, kW/m^2	FIGRA, kW/m^2	THR, MJ/m^2	EHC, MJ/kg	Residue, %	CO Yield, kg/kg	SEA, m^2/kg
PE	128 (10)	414 (37)	231 (15)	1.0 (0.1)	148 (5)	42 (2)	9 (0)	0.0249 (0.0)	332 (14)
PE/APP	174 (8)	423 (35)	195 (4)	0.8 (0.0)	153 (12)	41 (2)	18 (1)	0.0295 (0.0)	411 (25)
PE/APP/5BP	123 (20)	391 (5)	210 (15)	0.7 (0.1)	154 (15)	41 (2)	18 (3)	0.0281 (0.01)	456 (40)
PE/APP/10BP	119 (52)	327 (58)	203 (51)	0.9 (0.4)	137 (7)	39 (2)	23 (5)	0.0272 (0.0)	373 (62)
PE/APP/20BP	131 (67)	277 (71)	166 (47)	0.7 (0.4)	128 (3)	38 (1)	33 (5)	0.0260 (0.0)	419 (82)
PE/APP/5CS	159 (27)	366 (16)	176 (7)	0.7 (0.1)	135 (5)	39 (2)	23 (2)	0.0314 (0.0)	469 (10)
PE/APP/10CS	154 (19)	286 (26)	162 (3)	0.5 (0.0)	127 (7)	38 (1)	26 (0)	0.0321 (0.0)	491 (25)
PE/APP/20CS	165 (55)	310 (34)	166 (13)	0.8 (0.2)	132 (12)	38 (2)	33 (1)	0.0320 (0.0)	479 (27)
PE/APP/5VM	158 (13)	301 (69)	162 (23)	0.4 (0.1)	156 (9)	41 (1)	20 (0)	0.0311 (0.)	521 (33)
PE/APP/10VM	114 (32)	252 (47)	157 (33)	0.8 (0.5)	158 (19)	44 (5)	36 (2)	0.0310 (0.0)	404 (47)
PE/APP/20VM	185 (84)	165 (40)	93 (42)	0.3 (0.4)	140 (23)	43 (5)	25 (1)	0.0251 (0.0)	422 (138)

The values in parentheses are the standard deviations.

The total smoke release from the forced flaming combustion represents the cumulative smoke amount generated per unit area of the tested material [58,59]. The addition of a commercial flame retardant led to a considerable increase in TSR from 1211 m^2/m^2 to 1606 m^2/m^2. From the curves' profiles in Figure 9, it is observed that most of the samples showed an increase in values over unmodified HDPE after 600 s of the test. The exception is PE/APP/20VM, which burned for more than 1800 s, and values higher than polyethylene appeared only for tests of about 1300 s. Notably, the relation between the data and the number of additives can be observed only for series with VM. A decrease in value with an increase in vermiculite content may be due to the release of a higher amount of water. Similarly, a specific extinction area, which corresponds to the surface of light-absorbing particles present in the smoke generated from 1 kg of material, increased due to the addition of developed systems. SEA values ranged from 373 to 521 m^2/kg and were independent of the amount or even type of additives.

An effective protective layer suppresses the release of combustible volatiles as well as heat transfer into the materials, leading to a reduced HRR and prolonged burning time [47]. The residues of HDPE, polymer-modified with APP, and systems with BP, CS, or VM after cone calorimetric tests are shown in Figure 10. Unlike the composites, in the case of PE and PE/APP, there is no residue or only a little left. For the flame-retarded polyethylene, a continuous black char layer was formed on top of the burning materials. The increase in thickness was limited so that intumescence was practically ruled out. However, the formed residual layer was limiting heat transfer to the pyrolysis front and mass to the flame.

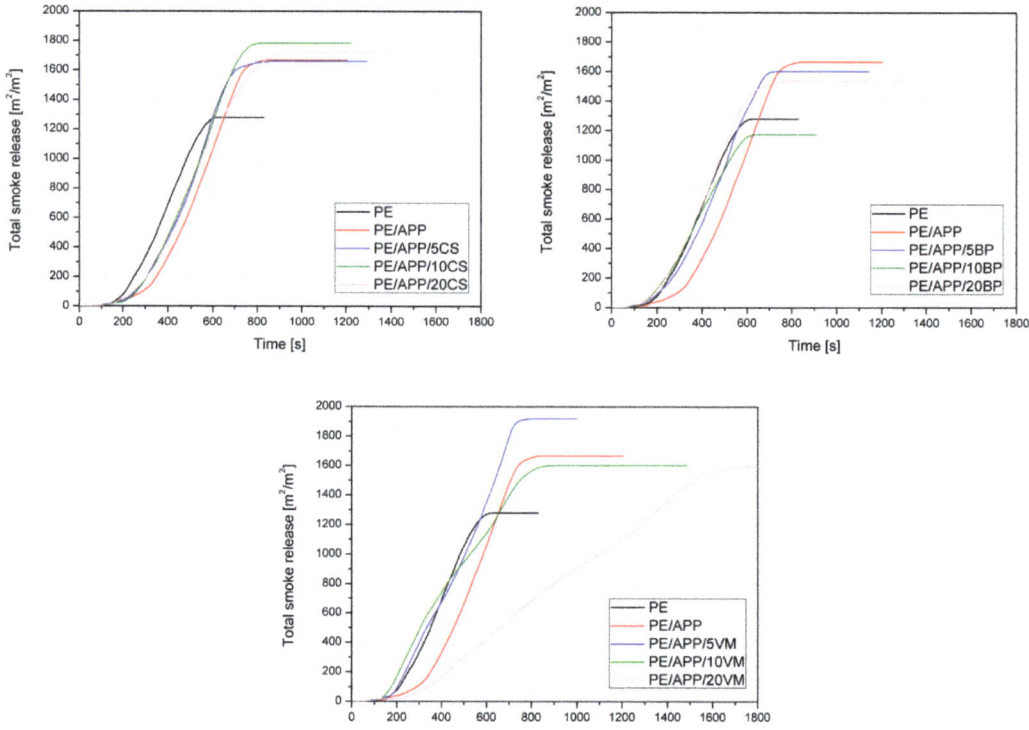

Figure 9. Total smoke release curves of PE and PE-based composites.

The change in the course of the curve with the reduction in pHRR and the considerable growth in residue exhibited a flame-retardant mode of action in the condensed phase. The residue yielded up to 0.3, demonstrating the formation of an inorganic carbonaceous char accompanied by an appropriate decrease in the fuel involved. In turn, the decrease in EHC and the increased CO yield and smoke production may indicate flame retardant effects in the gas phase. The increase in smoke emission and an EHC decrease by approx. 10% may suggest flame inhibition. Both flame retardant effects, the increase in residue due to charring and the decrease in EHC due to fuel dilution/flame inhibition, were detected. However, due to the linear increase in residue, charring started to outperform the gas-phase mechanisms with a high amount of additives [47].

Figure 10. Sample appearance after the cone calorimetry test.

3.5. Mechanical and Thermomechanical Properties

Table 5 summarizes the results of the mechanical and thermomechanical tests. The evaluation of the mechanical performance of polyethylene and its composites took into account the tensile test, Charpy impact strength, and hardness. The introduction of APP and inorganic fillers increased the elasticity modulus of the specimens. The addition of a flame retardant resulted in an increase in the Young modulus by 28%. In contrast, stiff domains of powder fillers caused a further improvement in this mechanical parameter. The PE/APP/20VM sample reveals the highest stiffness, which showed more than a two-fold increase in the stiffness compared to the reference material (unmodified PE). The increase in Young's modulus caused by APP and fillers was reported earlier in the literature and is an expected effect resulting from stiff structures that block macromolecular mobility in traces of deformation [60–62]. At the same time, the most beneficial impact of the increase in stiffness caused by the introduction of VM may result from the increase in the degree of crystallinity for the polymer matrix, which usually leads to improvement of this mechanical parameter [63]. On the other hand, as was observed in SEM images, vermiculite-based composites, due to their complex structure, in the form of ground well-dispersed small plates as well as multilayered non-intercalated packets, reveal much higher volumetric content of the filler than CS- and BP-filled composite series, which also affect the composite stiffness [60,64]. Therefore, it can be supposed that while the final improvement of the elasticity modulus is caused by the presence of particulate fillers (CS and BP) acting as rigid stiff domains dispersed in the PE matrix, similar E values of VM-filled composites result from the higher crystallinity and volumetric content of the plate-shaped filler.

Table 5. Mechanical and thermomechanical properties of PE and PE-based composites.

Sample	Tensile Strength, σ_M [MPa]	Elasticity Modulus, E [MPa]	Elongation at Break, ε [%]	Charpy Impact Strength, a_k [kJ/m^2]	Shore D Hardness [°ShD]	Vicat Softening Temperature, VST [°C]	Heat Deflection Temperature, HDT [°C]
PE	23.0 (1.4)	673 (32.5)	98 (12)	3.64 (0.38)	61.2 (0.6)	74.5 (0.8)	61.9 (4.1)
PE/APP	19.9 (0.66)	862 (13.9)	12 (3.2)	1.84 (0.15)	64.3 (1.0)	73.6 (0.3)	73.0 (4.5)
PE/APP/5CS	19.4 (0.24)	924 (39.8)	7.7 (0.96)	1.79 (0.14)	65.4 (0.6)	73.5 (1.0)	63.2 (0.5)
PE/APP/10CS	20.6 (1.75)	1260 (100)	5.3 (1.80)	1.20 (0.31)	64.8 (0.8)	76.0 (0.6)	73.6 (6.0)
PE/APP/20CS	19.9 (0.65)	1335 (49.5)	3.7 (0.08)	2.22 (0.09)	66.1 (0.5)	77.6 (0.4)	80.4 (6.4)
PE/APP/5BP	19.7 (0.66)	978 (48.8)	9.3 (0.87)	2.05 (0.67)	65.1 (0.8)	73.5 (1.0)	73.3 (0.3)
PE/APP/10BP	18.6 (0.43)	1070 (55.0)	8.6 (0.43)	1.62 (0.62)	66.1 (0.8)	74.8 (1.3)	77.9 (5.6)
PE/APP/20BP	18.4 (0.31)	1200 (102)	2.1 (0.09)	1.04 (0.22)	66.5 (1.6)	81.6 (0.9)	106.5 (5.4)
PE/APP/5VM	19.3 (0.53)	976 (42.1)	6.3 (1.1)	1.43 (0.25)	65.2 (0.7)	75.6 (0.4)	74.0 (1.2)
PE/APP/10VM	19.1 (0.45)	1070 (61.1)	4.7 (0.34)	1.21 (0.06)	66.4 (0.8)	77.9 (0.9)	77.7 (5.2)
PE/APP/20VM	19.0 (1.37)	1450 (115)	1.9 (0.32)	1.00 (0.03)	67.9 (1.2)	82.3 (1.0)	79.4 (2.3)

The tensile strength of all the materials containing flame retardant and fillers decreased compared to the reference sample. However, it should be noted that even in the case of composites with the highest concentration, which contained a total amount of additives of 40 wt% (20 wt% APP and 20 wt% of filler), the reduction in the tensile strength was not so significant that it could constitute a considerable limitation in its use. The lowest σ_M value (18.4 MP) was recorded for the PE/APP/20BP series, as it only has a deterioration of 20% compared to the reference sample. The particles of any of the additives used (FR and fillers) did not have a large shape factor to constitute a filler, enabling effective stress transfer and resulting in reinforcement for the polymer. Their presence led to the creation of points of stress accumulation during strain, causing the destruction of materials at lower strength values. It should be emphasized that the obtained tensile strength results are favorable, taking into account the high filling melt with inorganic materials introduced into the non-polar polymer without the use of a compatibilizer and surface modification, which, according to previously published studies [64], are crucial from the point of view of obtaining the mechanical properties of particulate-shaped composites.

A phenomenon of decreasing elongation at the break of polymers due to the addition of fillers and modifiers is widely described and reported in the literature [65–67] and is connected with the accumulation of stresses at the polymer-filler interphase. According to Pukanszky et al. [68], the dominant effect affecting changes in the mechanisms of micromechanical deformations is debonding, understood as the loss of adhesion between the polymer matrix and the filler. Considering the lack of additional compatibilizers enhancing interfacial adhesion, the dominant factor influencing the limitations of elongation at reaction is not so much the particles' size and shape as it is the filler's volumetric content [68,69]. The minimum modifier content (APP) in the considered case was 20 wt%.

The composites were made by adding 5 to 20 wt% of the inorganic filler. All materials showed a drastic drop in elongation at break compared to unmodified PE. It should be emphasized that the individual composite series showed different ε_b values at a comparable mass concentration of the filler, which resulted from the different volumetric content and the degree of dispersion in the matrix. However, these values, from the point of view of the functional properties of the final products, can be considered comparable.

All the impact tests of polyethylene and its composites were performed on the notched specimens; therefore, all of them were fully broken. The addition of powder fillers and a flame retardant reduced the impact resistance. Interestingly, incorporating an inorganic filler to the polyethylene modified with APP did not cause any additional reduction in the impact strength of the composite compared to the PE/APP series. The PE/APP/20CS and PE/APP/5BP samples were characterized by their higher impact strength. The first series mentioned above had the highest impact strength among the modified material series. The most important, that is, more than threefold, deterioration of the impact toughness was noted for composites modified with vermiculite. The impact strength of composites reinforced with dispersion fillers decreases with the increasing volume of the filler [70,71]. VM appeared both in the form of single dispersed plates in the polymer matrix and exfoliated flakes with concertina-shaped domains. Composites manufactured with their use were characterized by a more significant volumetric share of filler in the matrix than other series and an increased concentration of stresses around the unsaturated spaces of the exfoliated filler not intercalated by the polymer. The decrease in the impact strength of the thermoplastic composites is often referred to as the increase in the composite stiffness caused by rigid filler structures dispersed in a polymeric matrix. At the same time, the stress concentration occurring around particles may result in the appearance of additional crack propagation points during the dynamic loading of the material [61,71]. Paradoxically, as demonstrated by Sewda and Maiti [71], the use of a compatibilizer to increase adhesion between the filler and the polymer may reduce the normalized relative impact strength value. Thus, the noticeably greater brittleness of the VM-filled composites observed in the case under consideration should be assigned with a different adhesion to the polyethylene matrix but with an increased volumetric fraction of the plate shape of the filler. Moreover, the drastically smaller size of the filler and the resulting shortened interparticle distance in the case of VM-filled composites also plays a role in worsening the response of the composites manufactured with their share to the impact load [70]. To summarize, compared to the lowered elongation at break of the modified polyethylene series, the materials should be classified as brittle.

The presence of fossil or waste-originated inorganic fillers, characterized by an increased hardness compared to the polymer matrix, leads to an increase in the parameter hardness of composites produced with their use [35,62,69,72]. The incorporation of APP improved the Shore D hardness by 5%, while the additional introduction of inorganic fillers affects a further gradual increase in this mechanical property, along with a rising filler content. The highest hardness was shown for the PE/APP/20VM series samples. However, it should be emphasized that the differences in hardness between the composite batches made using different fillers negligible. It is justified considering both the similar chemical composition and the dominant share of SiO_2 in the fillers [14,26,35].

Table 5 also summarizes the results of thermomechanical tests determined in static point load (*VST*) and three-point bending (*HDT*) conditions. By analyzing the thermomechanical properties of the composites, it can be concluded that introducing all types of inorganic fillers increased their thermomechanical stability in the case of the highest concentration of fillers in the composite. The modification effectiveness of individual fillers was varied. Interestingly, for all fillers, the increased hardness values were comparable, so it can be concluded that the changes in *VST* were additionally associated with changes in the macromolecular structure of the polymer matrix. At the same time, taking into account the results of DSC studies, changes in the thermomechanical properties of certain static conditions can only be associated with an increased degree of crystallinity in the case of a series of materials reinforced with vermiculite. In the case of the remaining series, there was no significant increase in crystallinity caused by the fillers and APP. Therefore, in the case of CS- and BP-filled composites, the dominant role in increasing the load-carrying capacity and increased resistance to point loading was played by the rigid filler domains in the polymer matrix.

In contrast, in the case of VM-containing composites, the effectiveness of the interaction was additionally increased by modifying the crystal structure of the polymer. It should also be noted that for PE/APP/VM composites, higher VST values than HDT were recorded in all cases, which may be related to the plate-shaped structure of the filler, which creates targeted structures in the wall layers of the injected samples during the technological process, resulting in an increased resistance to operation of the indenter during mechanical (hardness tests) and thermomechanical (VST) measurements. Referring to the research presented by Rusu et al. [73], exceeding the concentration of the filler accompanying the appearance of agglomerated zinc powder filler structures in HDPE resulted in the lack of dependence of the influence of the increasing amount of filler on the flexural modulus, although it did not reduce the effectiveness of the beneficial effect on hardness and VST. Given the SEM observations of PE-based composites and the noticed differences in the distribution and particle size of the filler, the lack of an exact correlation between VST and HDT is justified.

4. Conclusions

Correlative analysis of polymer composites produced based on ammonium-polyphosphate-modified polyethylene using three types of thermally stable inorganic dispersion fillers has been realized. According to the assumptions, adding fillers increased the stiffness and hardness of composites both at room and at elevated temperatures, which may significantly increase the applicability range of the flame retardant polyethylene. Realized studies using a cone calorimeter showed that adding a micrometric filler with high thermal stability causes some beneficial effects in terms of reducing flammability. Introducing a filler with a complex plate shape (ground expandable vermiculite) allowed for the obtainment of a synergistic effect, significantly reducing the heat rate release. This may be related to the presence of a fine fraction of the filler in the form of plates and the presence of unbroken packets of exfoliated filler without interactions with polymer. Their presence in the swelled surface of APP-modified polymer during combustion caused the effect of the formation of a ceramic surface layer and increased barrier properties.

The introduction of all types of fillers resulted in the deterioration of the overall mechanical performance of the composites compared to polyethylene. However, it should be added that the changes in the tensile strength and impact strength were not at the level that would significantly limit the use of developed compositions.

Author Contributions: Conceptualization, M.B., A.H. and K.S. (Kamila Sałasińska); methodology, M.B., A.H., K.S. (Kamila Sałasińska) and J.A. (Jacek Andrzejewski); formal analysis, M.B., A.H. and K.S. (Kamila Sałasińska); investigation, M.B., A.H., K.S. (Kamila Sałasińska), J.A. (Joanna Aniśko), A.P., K.S. (Katarzyna Skórczewska) and J.A. (Jacek Andrzejewski); resources, M.B., A.H. and J.A. (Jacek Andrzejewski); writing—original draft preparation, M.B., K.S. (Kamila Sałasińska), A.H.; writing—review and editing, M.B., A.H. and K.S. (Kamila Sałasińska); visualization, M.B. and K.S. (Kamila Sałasińska); supervision, M.B.; project administration, M.B.; funding acquisition, M.B. All authors have read and agreed to the published version of the manuscript.

Funding: The results presented in this paper were funded with grants for education allocated by the Ministry of Science and Higher Education in Poland, executed under project no. 0513/SBAD/4741.

Institutional Review Board Statement: Not applicable.

Informed Consent Statement: Not applicable.

Data Availability Statement: Not applicable.

Conflicts of Interest: The authors declare no conflict of interest.

References

1. Ronca, S. Polyethylene. In *Brydson's Plastics Materials*; Elsevier: Amsterdam, The Netherlands, 2017; pp. 247–278.
2. Fan, Y.; Xue, Y.; Nie, W.; Ji, X.; Bo, S. Characterization of the Microstructure of Bimodal HDPE Resin. *Polym. J.* **2009**, *41*, 622–628. [CrossRef]
3. Knuuttila, H.; Lehtinen, A.; Nummila-Pakarinen, A. Advanced Polyethylene Technologies—Controlled Material Properties. In *Long Term Properties of Polyolefins. Advances in Polymer Science*; Springer: Berlin/Heidelberg, Germany, 2004; pp. 13–28.
4. Barton-Pudlik, J.; Czaja, K. Conifer Needles as Thermoplastic Composite Fillers: Structure and Properties. *BioResources* **2016**, *11*, 6211–6231. [CrossRef]
5. Aguliar, H.; Yazdani-Pedram, M.; Toro, P.; Quijada, R.; Lopez-Manchado, M.Á. Synergic Effect of Two Inorganic Fillers on the Mechanical and Thermal Properties of Hybrid Polypropylene Composites. *J. Chil. Chem. Soc.* **2014**, *59*, 2468–2473. [CrossRef]
6. Li, M.; Chen, Y.; Wu, L.; Zhang, Z.; Mai, K. A Novel Polypropylene Composite Filled by Kaolin Particles with β-Nucleation. *J. Therm. Anal. Calorim.* **2019**, *135*, 2137–2145. [CrossRef]
7. Fu, S.-Y.; Lauke, B.; Mäder, E.; Yue, C.-Y.; Hu, X. Tensile Properties of Short-Glass-Fiber- and Short-Carbon-Fiber-Reinforced Polypropylene Composites. *Compos. Part A Appl. Sci. Manuf.* **2000**, *31*, 1117–1125. [CrossRef]
8. Olesik, P.; Kozioł, M.; Jała, J. Processing and Structure of HDPE/Glassy Carbon Composite Suitable for 3D Printing. *Compos. Theory Pract.* **2020**, *20*, 72–77.
9. Jakubowska, P.; Borkowski, G.; Brząkalski, D.; Sztorch, B.; Kloziński, A.; Przekop, R.E. The Accelerated Aging Impact on Mechanical and Thermal Properties of Polypropylene Composites with Sedimentary Rock Opoka-Hybrid Natural Filler. *Materials* **2022**, *15*, 338. [CrossRef]
10. McGauran, T.; Dunne, N.; Smyth, B.M.; Cunningham, E. Incorporation of Poultry Eggshell and Litter Ash as High Loading Polymer Fillers in Polypropylene. *Compos. Part C Open Access* **2020**, *3*, 100080. [CrossRef]
11. Chrissafis, K.; Bikiaris, D. Can Nanoparticles Really Enhance Thermal Stability of Polymers? Part I: An Overview on Thermal Decomposition of Addition Polymers. *Thermochim. Acta* **2011**, *523*, 1–24. [CrossRef]
12. Razak, J.A.; Akil, H.M.; Ong, H. Effect of Inorganic Fillers on the Flammability Behavior of Polypropylene Composites. *J. Thermoplast. Compos. Mater.* **2007**, *20*, 195–205. [CrossRef]
13. Motahari, S.; Motlagh, G.H.; Moharramzadeh, A. Thermal and Flammability Properties of Polypropylene/Silica Aerogel Composites. *J. Macromol. Sci. Part B* **2015**, *54*, 1081–1091. [CrossRef]
14. Barczewski, M.; Sałasińska, K.; Kloziński, A.; Skórczewska, K.; Szulc, J.; Piasecki, A. Application of the Basalt Powder as a Filler for Polypropylene Composites With Improved Thermo-Mechanical Stability and Reduced Flammability. *Polym. Eng. Sci.* **2019**, *59*, E71–E79. [CrossRef]
15. Hirschler, M.M. Reduction of Smoke Formation from and Flammability of Thermoplastic Polymers by Metal Oxides. *Polymer* **1984**, *25*, 405–411. [CrossRef]
16. Mostovoy, A.; Bekeshev, A.; Tastanova, L.; Akhmetova, M.; Bredihin, P.; Kadykova, Y. The Effect of Dispersed Filler on Mechanical and Physicochemical Properties of Polymer Composites. *Polym. Polym. Compos.* **2021**, *29*, 583–590. [CrossRef]
17. Matar, M.; Azambre, B.; Cochez, M.; Vahabi, H.; Fradet, F. Influence of Modified Mesoporous Silica SBA-15 on the Flammability of Intumescent High-Density Polyethylene. *Polym. Adv. Technol.* **2016**, *27*, 1363–1375. [CrossRef]
18. Ren, Q.; Zhang, Y.; Li, J.; Li, J.C. Synergistic Effect of Vermiculite on the Intumescent Flame Retardance of Polypropylene. *J. Appl. Polym. Sci.* **2011**, *120*, 1225–1233. [CrossRef]
19. Salasinska, K.; Celiński, M.; Mizera, K.; Kozikowski, P.; Leszczyński, M.K.; Gajek, A. Synergistic Effect between Histidine Phosphate Complex and Hazelnut Shell for Flammability Reduction of Low-Smoke Emission Epoxy Resin. *Polym. Degrad. Stab.* **2020**, *181*, 109292. [CrossRef]
20. Wu, Z.-H.; Qu, J.-P.; Zhao, Y.-Q.; Tang, H.-L.; Wen, J.-S. Flammable and Mechanical Effects of Silica on Intumescent Flame Retardant/Ethylene–Octene Copolymer/Polypropylene Composites. *J. Thermoplast. Compos. Mater.* **2015**, *28*, 981–994. [CrossRef]
21. Huang, G.; Gao, J.; Li, Y.; Han, L.; Wang, X. Functionalizing Nano-Montmorillonites by Modified with Intumescent Flame Retardant: Preparation and Application in Polyurethane. *Polym. Degrad. Stab.* **2010**, *95*, 245–253. [CrossRef]
22. Isa, I.A.A.; Jodeh, S.W. Thermal Properties of Automotive Polymers III—Thermal Characteristics and Flammability of Fire Retardant Polymers. *Mater. Res. Innov.* **2001**, *4*, 135–143. [CrossRef]
23. Hu, X.-P.; Li, Y.-L.; Wang, Y.-Z. Synergistic Effect of the Charring Agent on the Thermal and Flame Retardant Properties of Polyethylene. *Macromol. Mater. Eng.* **2004**, *289*, 208–212. [CrossRef]
24. Hamid, M.R.Y.; Ab Ghani, M.H.; Ahmad, S. Effect of Antioxidants and Fire Retardants as Mineral Fillers on the Physical and Mechanical Properties of High Loading Hybrid Biocomposites Reinforced with Rice Husks and Sawdust. *Ind. Crops Prod.* **2012**, *40*, 96–102. [CrossRef]
25. Hejna, A.; Piszcz-Karaś, K.; Filipowicz, N.; Cieśliński, H.; Namieśnik, J.; Marć, M.; Klein, M.; Formela, K. Structure and Performance Properties of Environmentally-Friendly Biocomposites Based on Poly(ε-Caprolactone) Modified with Copper Slag and Shale Drill Cuttings Wastes. *Sci. Total Environ.* **2018**, *640–641*, 1320–1331. [CrossRef] [PubMed]
26. Barczewski, M.; Mysiukiewicz, O.; Hejna, A.; Biskup, R.; Szulc, J.; Michałowski, S.; Piasecki, A.; Kloziński, A. The Effect of Surface Treatment with Isocyanate and Aromatic Carbodiimide of Thermally Expanded Vermiculite Used as a Functional Filler for Polylactide-Based Composites. *Polymers* **2021**, *13*, 890. [CrossRef] [PubMed]

27. Barczewski, M.; Mysiukiewicz, O.; Matykiewicz, D.; Skórczewska, K.; Lewandowski, K.; Andrzejewski, J.; Piasecki, A. Development of Polylactide Composites with Improved Thermomechanical Properties by Simultaneous Use of Basalt Powder and a Nucleating Agent. *Polym. Compos.* **2020**, *41*, 2947–2957. [CrossRef]
28. Jakubowska, P.; Sterzyński, T. Thermal Diffusivity of Polyolefin Composites Highly Filled with Calcium Carbonate. *Polimery* **2012**, *57*, 271–275. [CrossRef]
29. Prociak, A.; Pielichowski, J.; Sterzyñski, T. Thermal Diffusivity of Polyurethane Foams Measured by the Modified Ångström Method. *Polym. Eng. Sci.* **1999**, *39*, 1689–1695. [CrossRef]
30. Khonakdar, H.A.; Morshedian, J.; Wagenknecht, U.; Jafari, S.H. An Investigation of Chemical Crosslinking Effect on Properties of High-Density Polyethylene. *Polymer* **2003**, *44*, 4301–4309. [CrossRef]
31. Muiambo, H.F.; Focke, W.W.; Asante, J.K.O. Flame Retardant Properties of Polymer Composites of Urea Complex of Magnesium and Vermiculite. In Proceedings of the Europe/Africa Conference Dresden 2017–Polymer Processing Society PPS, Dresden, Germany, 27–29 June 2017; p. 050011.
32. Idumah, C.I.; Hassan, A.; Affam, A.C. A Review of Recent Developments in Flammability of Polymer Nanocomposites. *Rev. Chem. Eng.* **2015**, *31*, 149–177. [CrossRef]
33. Di Landro, L.; Montalto, A.; Bettini, P.; Guerra, S.; Montagnoli, F.; Rigamonti, M. Detection of Voids in Carbon/Epoxy Laminates and Their Influence on Mechanical Properties. *Polym. Polym. Compos.* **2017**, *25*, 371–380. [CrossRef]
34. Korol, J.; Hejna, A.; Wypiór, K.; Mijalski, K.; Chmielnicka, E. Wastes from Agricultural Silage Film Recycling Line as a Potential Polymer Materials. *Polymers* **2021**, *13*, 1383. [CrossRef] [PubMed]
35. Hejna, A.; Kosmela, P.; Barczewski, M.; Mysiukiewicz, O.; Piascki, A. Copper Slag as a Potential Waste Filler for Polyethylene-Based Composites Manufacturing. *Tanzan. J. Sci.* **2021**, *47*, 405–420. [CrossRef]
36. Jung, M.R.; Horgen, F.D.; Orski, S.V.; Rodriguez, C.V.; Beers, K.L.; Balazs, G.H.; Jones, T.T.; Work, T.M.; Brignac, K.C.; Royer, S.-J.; et al. Validation of ATR FT-IR to Identify Polymers of Plastic Marine Debris, Including Those Ingested by Marine Organisms. *Mar. Pollut. Bull.* **2018**, *127*, 704–716. [CrossRef] [PubMed]
37. Ni, J.; Chen, L.; Zhao, K.; Hu, Y.; Song, L. Preparation of Gel-Silica/Ammonium Polyphosphate Core-Shell Flame Retardant and Properties of Polyurethane Composites. *Polym. Adv. Technol.* **2011**, *22*, 1824–1831. [CrossRef]
38. Ma, T.-K.; Yang, Y.-M.; Jiang, J.-J.; Yang, M.; Jiang, J.-C. Synergistic Flame Retardancy of Microcapsules Based on Ammonium Polyphosphate and Aluminum Hydroxide for Lithium-Ion Batteries. *ACS Omega* **2021**, *6*, 21227–21234. [CrossRef]
39. Salasinska, K.; Barczewski, M.; Aniśko, J.; Hejna, A.; Celiński, M. Comparative Study of the Reinforcement Type Effect on the Thermomechanical Properties and Burning of Epoxy-Based Composites. *J. Compos. Sci.* **2021**, *5*, 89. [CrossRef]
40. Zhu, Y. Heat-Loss Modified Angstrom Method for Simultaneous Measurements of Thermal Diffusivity and Conductivity of Graphite Sheets: The Origins of Heat Loss in Angstrom Method. *Int. J. Heat Mass Transf.* **2016**, *92*, 784–791. [CrossRef]
41. Kloziński, A.; Jakubowska, P.; Ambrożewicz, D.; Jesionowski, T. Thermal properties of polyolefin composites with copper silicate. *AIP Conf. Proc.* **2015**, *1664*, 060016. [CrossRef]
42. dos Santos, W.N.; dos Santos, J.N.; Mummery, P.; Wallwork, A. Thermal Diffusivity of Polymers by Modified Angström Method. *Polym. Test.* **2010**, *29*, 107–112. [CrossRef]
43. Sadej, M.; Gierz, L.; Naumowicz, M. Polyurethane Composites with Enhanced Thermal Conductivity Containing Boron Nitrides. *Polimery* **2019**, *64*, 592–595. [CrossRef]
44. Wenelska, K.; Maślana, K.; Mijowska, E. Study on the Flammability, Thermal Stability and Diffusivity of Polyethylene Nanocomposites Containing Few Layered Tungsten Disulfide (WS 2) Functionalized with Metal Oxides. *RSC Adv.* **2018**, *8*, 12999–13007. [CrossRef] [PubMed]
45. Tjong, S.C.; Bao, S.P. Crystallization Regime Characteristics of Exfoliated Polyethylene/Vermiculite Nanocomposites. *J. Polym. Sci. Part B Polym. Phys.* **2005**, *43*, 253–263. [CrossRef]
46. Seven, K.M.; Cogen, J.M.; Gilchrist, J.F. Nucleating Agents for High-Density Polyethylene-A Review. *Polym. Eng. Sci.* **2016**, *56*, 541–554. [CrossRef]
47. Deng, C.; Yin, H.; Li, R.-M.; Huang, S.-C.; Schartel, B.; Wang, Y.-Z. Modes of Action of a Mono-Component Intumescent Flame Retardant MAPP in Polyethylene-Octene Elastomer. *Polym. Degrad. Stab.* **2017**, *138*, 142–150. [CrossRef]
48. Liu, G.; Chen, W.; Yu, J. A Novel Process to Prepare Ammonium Polyphosphate with Crystalline Form II and Its Comparison with Melamine Polyphosphate. *Ind. Eng. Chem. Res.* **2010**, *49*, 12148–12155. [CrossRef]
49. Camino, G.; Costa, L.; Trossarelli, L. Study of the Mechanism of Intumescence in Fire Retardant Polymers: Part V—Mechanism of Formation of Gaseous Products in the Thermal Degradation of Ammonium Polyphosphate. *Polym. Degrad. Stab.* **1985**, *12*, 203–211. [CrossRef]
50. Schartel, B. Phosphorus-Based Flame Retardancy Mechanisms—Old Hat or a Starting Point for Future Development? *Materials* **2010**, *3*, 4710–4745. [CrossRef]
51. Wang, D.-Y.; Liu, Y.; Wang, Y.-Z.; Artiles, C.P.; Hull, T.R.; Price, D. Fire Retardancy of a Reactively Extruded Intumescent Flame Retardant Polyethylene System Enhanced by Metal Chelates. *Polym. Degrad. Stab.* **2007**, *92*, 1592–1598. [CrossRef]
52. Schartel, B.; Hull, T.R. Development of Fire-Retarded Materials—Interpretation of Cone Calorimeter Data. *Fire Mater.* **2007**, *31*, 327–354. [CrossRef]
53. Naveen, J.; Jawaid, M.; Zainudin, E.S.; Sultan, M.T.H.; Yahaya, R.; Majid, M.S.A. Thermal Degradation and Viscoelastic Properties of Kevlar/Cocos Nucifera Sheath Reinforced Epoxy Hybrid Composites. *Compos. Struct.* **2019**, *219*, 194–202. [CrossRef]

54. Wang, F.; Gao, Z.; Zheng, M.; Sun, J. Thermal Degradation and Fire Performance of Plywood Treated with Expanded Vermiculite. *Fire Mater.* **2016**, *40*, 427–433. [CrossRef]
55. Günther, M.; Levchik, S.V.; Schartel, B. Bubbles and Collapses: Fire Phenomena of Flame-retarded Flexible Polyurethane Foams. *Polym. Adv. Technol.* **2020**, *31*, 2185–2198. [CrossRef]
56. Schartel, B.; Braun, U.; Schwarz, U.; Reinemann, S. Fire Retardancy of Polypropylene/Flax Blends. *Polymer* **2003**, *44*, 6241–6250. [CrossRef]
57. Salmeia, K.; Fage, J.; Liang, S.; Gaan, S. An Overview of Mode of Action and Analytical Methods for Evaluation of Gas Phase Activities of Flame Retardants. *Polymers* **2015**, *7*, 504–526. [CrossRef]
58. Schartel, B.; Perret, B.; Dittrich, B.; Ciesielski, M.; Krämer, J.; Müller, P.; Altstädt, V.; Zang, L.; Döring, M. Flame Retardancy of Polymers: The Role of Specific Reactions in the Condensed Phase. *Macromol. Mater. Eng.* **2016**, *301*, 9–35. [CrossRef]
59. Battig, A.; Fadul, N.A.-R.; Frasca, D.; Schulze, D.; Schartel, B. Multifunctional Graphene Nanofiller in Flame Retarded Polybutadiene/Chloroprene/Carbon Black Composites. *e-Polymers* **2021**, *21*, 244–262. [CrossRef]
60. Ozmusul, M.S.; Picu, R.C. Elastic Moduli of Particulate Composites with Graded Filler-Matrix Interfaces. *Polym. Compos.* **2002**, *23*, 110–119. [CrossRef]
61. Argon, A.S.; Cohen, R.E. Toughenability of Polymers. *Polymer* **2003**, *44*, 6013–6032. [CrossRef]
62. Kosciuszko, A.; Czyzewski, P.; Wajer, Ł.; Osciak, A.; Bielinski, M. Properties of Polypropylene Composites Filled with Microsilica Waste. *Polimery* **2020**, *65*, 99–104. [CrossRef]
63. Humbert, S.; Lame, O.; Séguéla, R.; Vigier, G. A Re-Examination of the Elastic Modulus Dependence on Crystallinity in Semi-Crystalline Polymers. *Polymer* **2011**, *52*, 4899–4909. [CrossRef]
64. Walter, R.; Friedrich, K.; Privalko, V.; Savadori, A. On Modulus and Fracture Toughness of Rigid Particulate Filled High Density Polyethylene. *J. Adhes.* **1997**, *64*, 87–109. [CrossRef]
65. Móczó, J.; Pukánszky, B. Polymer Micro and Nanocomposites: Structure, Interactions, Properties. *J. Ind. Eng. Chem.* **2008**, *14*, 535–563. [CrossRef]
66. Zare, Y. The Roles of Nanoparticles Accumulation and Interphase Properties in Properties of Polymer Particulate Nanocomposites by a Multi-Step Methodology. *Compos. Part A Appl. Sci. Manuf.* **2016**, *91*, 127–132. [CrossRef]
67. Khalaf, M.N. Mechanical Properties of Filled High Density Polyethylene. *J. Saudi Chem. Soc.* **2015**, *19*, 88–91. [CrossRef]
68. Pukánszky, B.; Van Es, M.; Maurer, F.H.J.; Vörös, G. Micromechanical Deformations in Particulate Filled Thermoplastics: Volume Strain Measurements. *J. Mater. Sci.* **1994**, *29*, 2350–2358. [CrossRef]
69. Pukánszky, B. Particulate Filled Polypropylene: Structure and Properties. In *Polypropylene Structure, Blends and Composites*; Springer: Dordrecht, The Netherlands, 1995; pp. 1–70.
70. Zhang, S.; Cao, X.Y.; Ma, Y.M.; Ke, Y.C.; Zhang, J.K.; Wang, F.S. The Effects of Particle Size and Content on the Thermal Conductivity and Mechanical Properties of Al2O3/High Density Polyethylene (HDPE) Composites. *Express Polym. Lett.* **2011**, *5*, 581–590. [CrossRef]
71. Sewda, K.; Maiti, S.N. Mechanical Properties of HDPE/Bark Flour Composites. *J. Appl. Polym. Sci.* **2007**, *105*, 2598–2604. [CrossRef]
72. Garbacz, T.; Dulebova, L. The Effect of Particulate Fillers on Hardness of Polymer Composite. *Adv. Sci. Technol. Res. J.* **2017**, *11*, 66–71. [CrossRef]
73. Rusu, M.; Sofian, N.; Rusu, D. Mechanical and Thermal Properties of Zinc Powder Filled High Density Polyethylene Composites. *Polym. Test.* **2001**, *20*, 409–417. [CrossRef]

Article

Development of a New Type of Flame Retarded Biocomposite Reinforced with a Biocarbon/Basalt Fiber System: A Comparative Study between Poly(lactic Acid) and Polypropylene

Jacek Andrzejewski [1,*] and Sławomir Michałowski [2]

[1] Institute of Materials Technology, Faculty of Mechanical Engineering, Poznan University of Technology, Piotrowo 3 Stree, 61-138 Poznan, Poland

[2] Department of Chemistry and Technology of Polymers, Cracow University of Technology, 24 Warszawska Street, 31-155 Kraków, Poland

* Correspondence: jacek.andrzejewski@put.poznan.pl; Tel.: +48-61-665-5858

Citation: Andrzejewski, J.; Michałowski, S. Development of a New Type of Flame Retarded Biocomposite Reinforced with a Biocarbon/Basalt Fiber System: A Comparative Study between Poly(lactic Acid) and Polypropylene. *Polymers* 2022, 14, 4086. https://doi.org/10.3390/polym14194086

Academic Editor: Paul Joseph

Received: 5 September 2022
Accepted: 26 September 2022
Published: 29 September 2022

Publisher's Note: MDPI stays neutral with regard to jurisdictional claims in published maps and institutional affiliations.

Copyright: © 2022 by the authors. Licensee MDPI, Basel, Switzerland. This article is an open access article distributed under the terms and conditions of the Creative Commons Attribution (CC BY) license (https://creativecommons.org/licenses/by/4.0/).

Abstract: A new type of partially biobased reinforcing filler system was developed in order to be used as a flame retardant for polylactic acid (PLA) and polypropylene (PP)-based composites. The prepared materials intended for injection technique processing were melt blended using the novel system containing ammonium polyphosphate (EX), biocarbon (BC), and basalt fibers (BF). All of the prepared samples were subjected to a detailed analysis. The main criterion was the flammability of composites. For PLA-based composites, the flammability was significantly reduced, up to V-0 class. The properties of PLA/EX/BC and PLA/EX/(BC-BF) composites were characterized by their improved mechanical properties. The conducted analysis indicates that the key factor supporting the effectiveness of EX flame retardants is the addition of BC, while the use of BF alone increases the flammability of the samples to the reference level. The results indicate that the developed materials can be easily applied in industrial practice as effective and sustainable flame retardants.

Keywords: poly(lactic acid); polypropylene; flame retardancy; mechanical performance; sustainable fillers; injection molding

1. Introduction

The flammability of polymer-based materials is one of the essential restrictions on the use of plastics. Plastic parts are usually easy to ignite and once ignited, the flammable gases, fumes, and decomposition products can fuel the process of further burning. For that reason, the modification of polymer-based materials is challenging; in practice, it is not possible to obtain fireproof plastics. However, modifications of the material are aimed to slow down or limit this combustion process and smoke emission, so as to reduce the hazardous effects of the flammability [1,2].

Even for composite materials, where thermoset or thermoplastic resin is used as a binder for mineral fillers, flammability is considered a serious problem. Due to the high thermo-mechanical resistance of reinforced composites, the area of application for this type of material frequently includes electrotechnical devices and machines operating at elevated temperatures. In such cases, the risks associated with fire ignition are very high. Despite these threats, manufacturers more often shy away from the use of metal composites in favor of composite products, especially those obtained by an injection molding technique. The use of flammability limitation techniques for this type of product is, therefore, a necessity. For many years, the most popular flame retardants used in thermoplastics have been halogen-based additives. For these compounds, the flammability is limited by the formation of free radicals. At high temperatures, these highly reactive compounds scavenge the polymer degradation products. Despite the increased effectiveness of halogenated-based agents, their use is now prohibited mainly due to the hazardous nature of these compounds, especially their toxicity and negative environmental impact [3–6]. Other flame retardants are still in common use, the most popular being metallic hydroxides (MH) and phosphorus-based additives. In the first category of MH materials, the most prevalent compounds are aluminum tri-hydroxide (ATH) and magnesium di-hydroxide (MDH). In both cases, the flame-retardant mechanism is based on water vapor release, which happens during combustion. Unfortunately, for this type of compound, the visible reduction in the flammability of polymers occurs with a minimum content of 50 wt%, which always leads to the deterioration of most of the essential material properties.

Currently, the most promising group of polymer flame-retardants are phosphorous-based compounds due to their lack of toxicity and relatively high effectiveness. For these same reasons, materials of this type were used in this research cycle. The two most popular phosphorous-based flame-retardants are red phosphorous and ammonium polyphosphate (APP). The flame-retardancy mechanism for this type of compound is based on the decomposition of the APP fillers into phosphoric acid and transformation into polyphosphoric acid at higher temperatures, leading to the formation of a rigid char layer. The initial decomposition reaction takes place in oxygen or nitrogen, which means that oxygen/nitrogen-containing plastics are more favorable for modification. In the research discussed in this article, flame retardants were used for a PLA and PP matrix; it was expected, due to the presence of oxygen in the PLA structure, that the effectiveness of flame retardancy for this polymer should be higher.

In the bio-based or biodegradable-plastics category, research on the use of flame retardants has mainly focused on PLA and its composites. At present, the study on PLA flame retardancy covers the use of metal hydroxides [7–9], carbon/graphite-based fillers [10–13], and nano additives [10,14–16]. The use of phosphorus-based compounds is also widely described in the literature and applies not only to pure PLA [17–20] but also to other types of biodegradable materials such as poly butylene succinate PBS [21,22], thermoplastic starch TPS [23,24], and different kinds of blends [25,26]. The necessity to use flame retardants is essential for composites with the addition of natural fillers. For this reason, the polymer matrix is very often made of traditional thermoplastics such as PVC, PE, or PP. The category of wood-polymer composite (WPC) materials is particularly important as the content of wood fillers usually exceeds 50 wt%. These composites are often used in construction; therefore, the appropriate selection of the matrix/filler/flame retardant system is the key to obtaining low flammability [27–30]. Since the total amount of fillers during the extrusion processing of WPC profiles is usually very high, it is possible to obtain the desired mechanical properties. However, the injection molding of high filler content composites is more difficult due to the need to maintain a sufficiently low viscosity of the material. This forces the use of more efficient reinforcement systems, usually based on glass fibers. For the presented study, basalt fibers were used, which, in addition to more sustainable manufacturing methods, are also characterized by higher thermal resistance than standard E-glass fibers [31–33].

For several years, the use of biochar/biocarbon particles as polymer filler has begun to be not only the subject of research work but also a competitive technology in many industrial applications [34,35]. There are already known examples of applications in the packaging and automotive industries, where, in many cases, biocarbon has replaced mineral fillers [36–38]. So far, the use of biocarbon fillers has mainly been considered a method of improving the ecological balance of polymer products. It has a logical justification in the case of products manufactured for the packaging industry or other types of products with low strength requirements [39,40]. For more demanding applications, in particular, in the technical applications of composites, the use of biocarbon has many limitations related to the low reinforcement factor for popular thermoplastics such as PP or PE in particular [41–45]; some exceptions are nylon plastics, where the high polarity of the polymer structure improves the effectiveness of matrix-filler interactions [46–50]. For the presented research, the use of bio-carbon fillers was aimed at limiting the flammability of the developed composites. Usually, materials of this type, due to the high addition of fillers, lose many desirable mechanical properties, therefore, the hybridization of the structure with synthetic fibers is usually a necessary procedure [51–53].

The idea of using biocarbon/biochar as a helpful material in reducing the flammability of polymer materials has not been a common research topic. Preliminary work on this subject was carried out by Snowdon et al. [54], however, flammability was not the core subject of the research. Another promising study was presented by Li et al. [55] where biocarbon modified with chitosan was used in the modification of PLA. Another study on the use of biocarbon/ATH mixture was presented by Wang et al. [56].

The aim of our study was to assess the possibility of using the hybrid filler system containing biocarbon (BC) particles, APP, and basalt fibers (BF) for comparative purposes. Due to the fact that both BC and BF belong to a group of sustainable materials, we decided to use PLA as the main investigated polymer, while polypropylene (PP) was used for comparative purposes. Since APP flame-retardants are most effective for oxygen-containing polymers, the expected results for PLA are expected to be more favorable. The developed materials are intended to be manufactured using injection molding technology; therefore, during the study, the materials were subjected to a two-stage processing procedure involving twin-screw extrusion blending and injection molding. The obtained samples were tested using several measuring methods. Mechanical properties were evaluated by static tensile/flexural tests and Charpy impact resistance measurements. Thermal analysis was carried out using the thermogravimetric (TGA) method and differential scanning calorimetry (DSC) analysis. Thermo-mechanical properties were compared using DMTA analysis and HDT/Vicat tests. Finally, flammability was evaluated using the microcalorimetry analysis and UL-94 method.

2. Materials and Methods

2.1. Materials

The type of poly(lactic acid) used during this study was Ingeo PLA 3001D, from NatureWorks (Minnetonka, MN, USA). The polypropylene resin was Moplen PP HP500N, from LyondellBasell (Plock, Poland). For all flame retarded samples, we used ammonium polyphosphate (APP); the filler was a supplier in powdery form. The type used was Exolit AP 422 (from Clariant, Muttenz, Switzerland). For reinforcement, we used chopped basalt fibers (BF) type BCS 13-1/4″-KV02M from Kamenny Vek company (Dubna, Russia). The biobased filler was biocarbon powder BC, obtained from the pyrolysis of the wood chips. The raw filler was supplied by the Fluid company (Sedziszow, Poland). Before processing, the BC filler was ball milled; the whole procedure and properties of the filler are presented in our previous studies [49,57]. The basic properties of all polymers and fillers are listed in Table 1. Data were collected from technical data sheets.

Table 1. The list of basic properties of the pure polymers and fillers.

	Polymers			
	Density (g/cm^3)	Melt Flow Rate (g/10 Min)	Tensile Strength (MPa)	Izod Impact Strength (J/m)
PLA	1.24	22	62	16
PP	0.9	12	34	25
	Fillers			
	Density (g/cm^3)	Particle size (μm)	Bulk density (g/cm^3)	Decomposition temperature (°C)
EX	1.9	15	0.7	>275
	Density (g/cm^3)	Particle size (μm)	Biomass type	Carbon content (%)
BC	1.5	1	Wood chips	65
	Density (g/cm^3)	Diameter (μm)	Fiber length (mm)	Sizing
BF	2.7	13	6.2	Silane (0.4%)

2.2. Sample Preparation

All materials formulations were melt mixed using a twin-screw extruder. The machine type used was Zamak EH-16.2D (Zamak Merkator company, Skawina, Poland), equipped with co-rotating 16 mm screws. Before processing, both polymers were dried in a cabinet oven to avoid unnecessary moisture (at 60 °C for 6 h). Powdery fillers (EX and BC) were also dried (80 °C/12 h), while BF fibers were used as received. The melt blending of all materials was conducted using the same conditions. All ingredients were dry-blended before extrusion. The extrusion temperature measured at the die-head was set to 195 °C, while the screw speed was 100 rpm. The extruded material was transferred to the pelletizer using the air-cooled belt conveyor. The obtained pellets were collected in sealed bags, while before injection molding, the drying procedure was again applied (60 °C, 6 h). For the purpose of sample shaping, we used the electric injection molding press model Engel E-MAC 50 (from Engel GmbH, Schwertberg, Austria). We used the same procedure for both types of materials (PLA and PP-based). The injection molding temperature was 210 °C (measured at the nozzle), and the mold temperature was set to 30 °C. The injection/holding pressure was 800/400 bar, respectively, while the holding/cooling time was 10/30 s. The obtained rectangular bars and dumbbell samples were collected for conditioning (at least 48 h). The full list of samples is highlighted in Table 2.

Table 2. The list of sample designation and composite formulations.

Sample	Matrix Polymer (wt%)	Ammonium Polyphosphate (EX) (wt%)	Biocarbon (BC) (wt%)	Basalt Fiber (BF) (wt%)
PLA	100	0	0	0
PLA/EX20	80	20	0	0
PLA/BC20	80	0	20	0
PLA/BF20	80	0	0	20
PLA/EX20/BC20	60	20	20	0
PLA/EX20/BF20	60	20	0	20
PLA/EX20/(BC-BF)20	60	20	10	10
PP	100	0	0	0
PP/EX20	80	20	0	0
PP/BC20	80	0	20	0
PP/BF20	80	0	0	20
PP/EX20/BC20	60	20	20	0
PP/EX20/BF20	60	20	0	20
PP/EX20/(BC-BF)20	60	20	10	10

2.3. Characterization

The mechanical properties of the obtained materials were evaluated using static tests (tensile, flexural) and a notched Izod impact test. The tensile test was performed in accordance with the ISO 527 standard, with measurements conducted with a cross-head speed of 10 mm/min. Flexural tests (ISO 178) were performed at the rate of 2 mm/min with a span distance of 64 m. Tensile/flexural measurements were conducted using the Zwick/Roell Z010 universal testing machine. Izod impact resistance tests (according to ISO 180 standard) were performed on the notched samples where the notch depth was 2 mm. We used the Zwick/Roell HIT 15 machine with a hammer equipped with a 5 J energy pendulum.

A scanning electron microscope (SEM) analysis was used to evaluate the structure's appearance. The observed samples were cryofractured; the specimen was immersed in liquid nitrogen before breaking. Scanning observations were conducted on the surface covered with a conductive layer of gold using a sputter coater. An EVO 40 SEM microscope (Carl Zeiss AG, Jena, Germany) was used to conduct the main analysis.

Differential scanning calorimetry (DSC) measurements were performed using the standard heating/cooling/heating procedure. Tests were carried out from 20 to 230 °C at the heating/cooling rate of 10 °C/min. During the measurements, samples ($\approx$5 mg) were placed inside the aluminum crucible; the oven chamber was purged with nitrogen (N2 flow = 20 mL/min), and we used the pierced lid. Samples were also cut from the injection-molded specimens. The apparatus used was the DSC F1 Phoenix from Netzsch company (Selb, Germany).

Thermogravimetric tests (TGA) were conducted using the Libra 209 F1 apparatus from the Netzsch company (Selb, Germany). The measurements were conducted from 30 to 800 °C at a heating rate of 10 °C/min. Similar to DSC, all samples were tested under a protective atmosphere of nitrogen. The average size of the samples was around 10 mg.

Thermal analysis was supplemented with dynamic mechanical and thermal analysis (DMTA) measurements. For the purpose of the study, the rotational rheometer Anton Paar MCR301 was used. The apparatus was equipped with solid sample clamps; we used the torsion mode deformation. All tests were conducted under the same conditions using rectangular samples (50 × 10 × 4 mm). The temperature range was set to 25–150 °C, while the heating rate was 2 °C/min. Constant strain amplitude was equal to 0.01%, while the deformation frequency was 1 Hz. The results in the form of storage modulus and tan δ plots were collected.

Flammability tests of the prepared samples were conducted using the pyrolysis combustion flow calorimeter (PCFC). Tests were carried out in accordance with the ASTM D7309-2007 standard. A microcalorimeter is a device designed to determine the rate of heat release by materials. The test method is based on the measurement of heat and oxygen loss during the thermal decomposition of a small sample of the material. The decomposition is carried out in an inert gas atmosphere in the temperature range of 150–750 °C with a heating rate of 1 °C/s. Then, the pyrolysis gases are oxidized in a high-temperature furnace at a temperature of 900 °C for complete oxidation. The results obtained during the tests allow for the determination of the HRR and THR parameters as well as the time and temperature of the HRR peaks.

Standard flammability tests were carried out using the UL-94 classification. Horizontal (HB) and vertical (V) combustion method measurements were performed using samples with dimensions of 125 × 10 × 4 mm, according to the PN-EN 60695-11-10 standard. Burn time was measured, and the dripping of burning material was monitored. According to the above observations, the combustion index, V, of the samples was calculated; it was on this basis that the materials were classified (according to the UL 94 HB methodology).

3. Results and Discussion

3.1. Structure Evaluation—Scanning Electron Microscopy Observations

The structure appearance of EX, BC, and BF-based samples is presented in Figure 1, while the pictures presenting the structure of the hybrid samples are shown in Figure 2. The direct comparison between the individual filler types revealed large differences in the structure appearance. The structure of composites with the addition of basalt fibers shows a rather typical picture of composites reinforced with short fibers [58,59]. It is worth noting that the structure appearance for both PP and PLA suggests a lack of strong matrix-fiber interactions since the composite interface can be clearly distinguished. A more interesting observation can be made when comparing the structure of powder filler composites EX and BC. For EX-based composites, the structure observations revealed a large number of irregular ammonium phosphate particles, with the average particle size around 20 μm, similar to other studies [60]. The particle size for the BC filler is much smaller ($\approx$1 μm), which is confirmed by SEM analysis [49]. The morphology of the sample surface revealed the brittle nature of the fracture mechanism. The irregular surface of the sample with the addition of EX is due to the presence of a relatively large filler particle rather than a more complex crack path. Since the structure of BC-based samples revealed that the dispersion of the BC particles was very uniform for both types of polymers, the expected mechanical properties should be more favorable than for EX-based samples, especially considering the significantly smaller size of the filler particle size for BC.

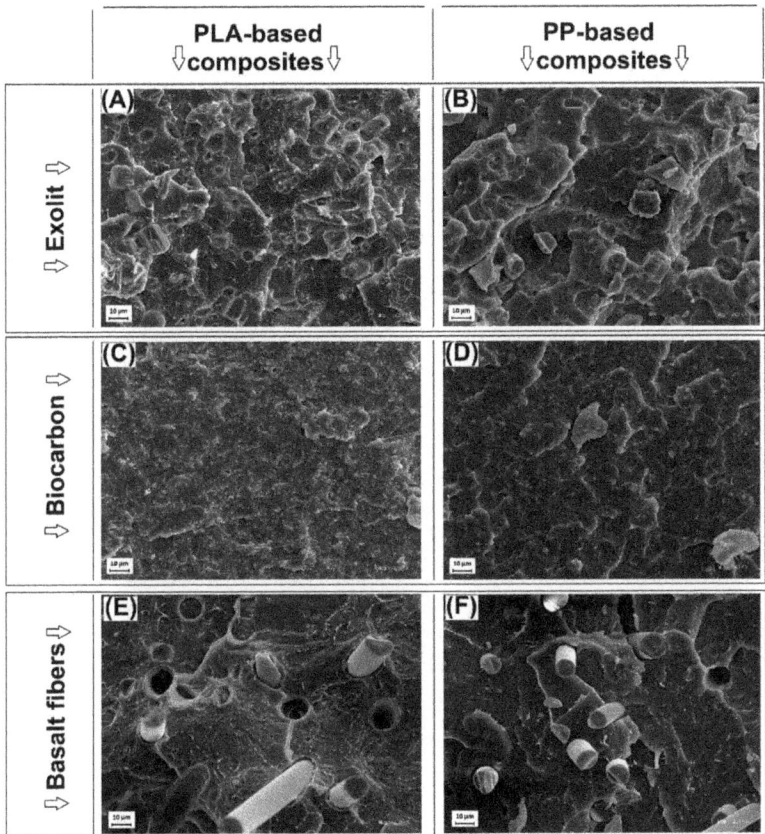

Figure 1. The structure appearance of the (**A,B**) EX-based composites, (**C,D**) BC-based, and (**E,F**) BF-based materials.

The main part of the research was focused on the examination of the composites with the mixed filler system, where, for all materials, the main filler was EX flame retardant (Figure 1A) while BC, BF, and BC/BF fillers were used as additional/supporting additives. The structure appearance for EX/BC composites was quite similar to pure EX materials since the large EX particles are the dominant element of the sample images. The presence of the BC filler was also visible; however, due to the smaller size of the BC particles, they are less distinctive. The addition of BF into the structure of EX/BF and EX/(BC–BF) samples led to the formation of regular holes, which are the residue of the fibers. The random orientation of the fibers suggests the reduction in anisotropy in the arrangement of the fibers during the injection process. That phenomenon was observed in our previous studies as well [57,61]. Spherical particles were still visible for all samples reinforced with short BF. Interestingly, the was no visible difference between the PP and PLA samples. For both polymers, the matrix structure is characterized by a slightly rough fracture surface, which is a typical feature for partly crystalline polymers.

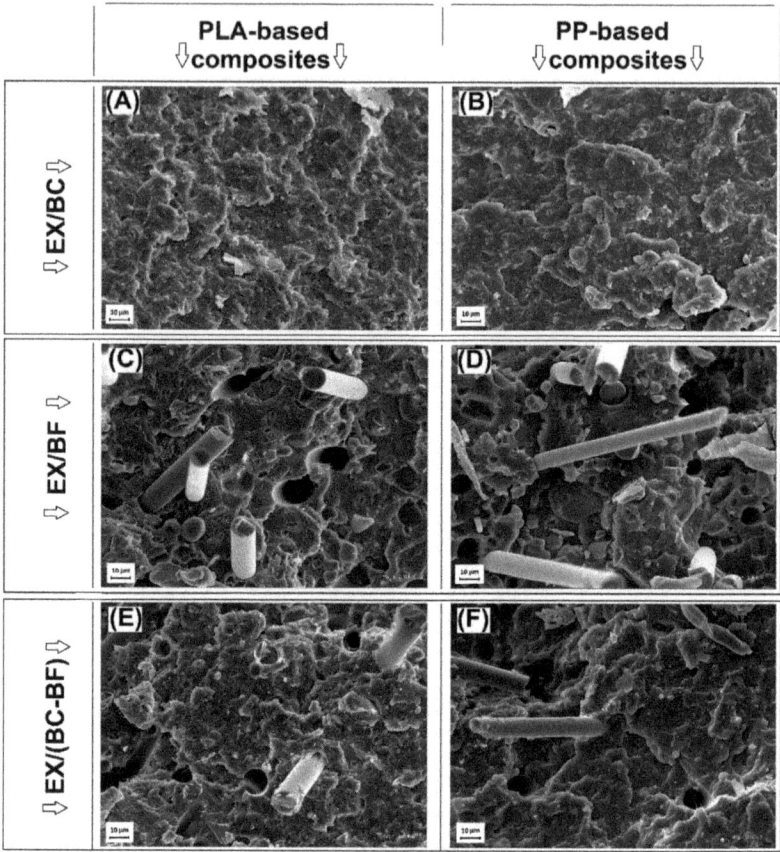

Figure 2. The structure appearance of the hybrid samples: (**A,B**) EX/BC, (**C,D**) EX/BF, and (**E,F**) EX/(BC-BF).

3.2. Thermal Stability—Thermogravimetric Analysis (TGA)

The results of the TGA analysis measurements are presented in the form of thermograms of weight loss (TG) and its first derivative (DTG). The plots for the input materials, pure polymers, and fillers are presented in Figure 3. Analogous thermograms for composite samples are presented in the graphs in Figures 4 and 5, respectively, for PLA and PP-based samples. Additional data (DTG peak position and char residue) are reported in Table S1 (Supplementary Information Section).

The analysis of the results of the pure matrix materials clearly shows a significantly higher thermal stability for PP. The weight loss caused by polymer decomposition of PP occurs at a temperature of about 400 °C, while for PLA, this process begins at around 350 °C. Similar results can be seen in the DTG curves where the shift of the PP and PLA peaks is about 90 °C in favor of PP. Due to the lack of additives in the form of fillers and a low tendency of residual char formation, the final weight of the PP and PLA samples is close to zero.

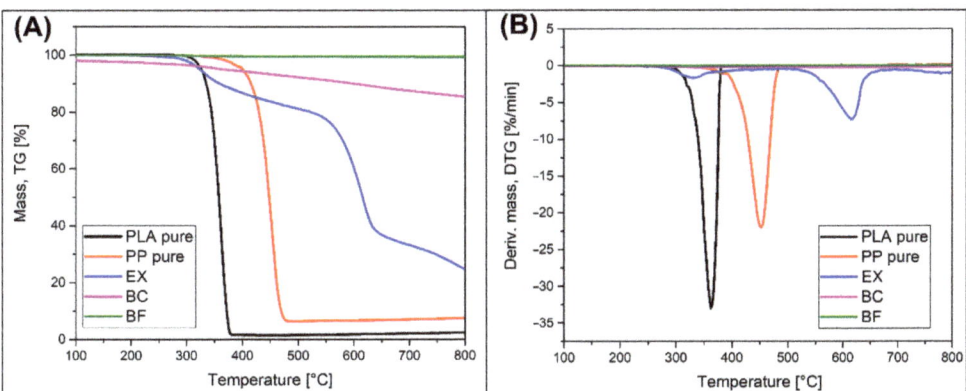

Figure 3. The thermogravimetric (TGA) analysis of pure polymers and filers: (**A**) TG and (**B**) DTG plots.

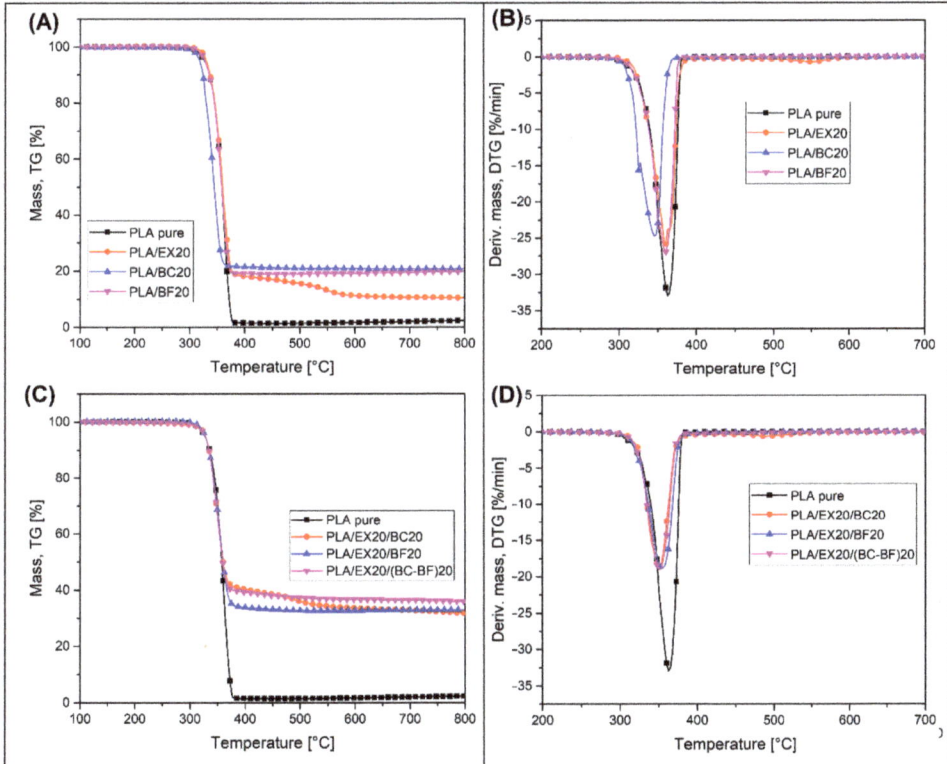

Figure 4. The thermogravimetric (TGA) analysis of PLA-based samples. Results are presented in the form of weight loss TG (**A**,**C**) and derivative weight loss plots DTG (**B**,**D**).

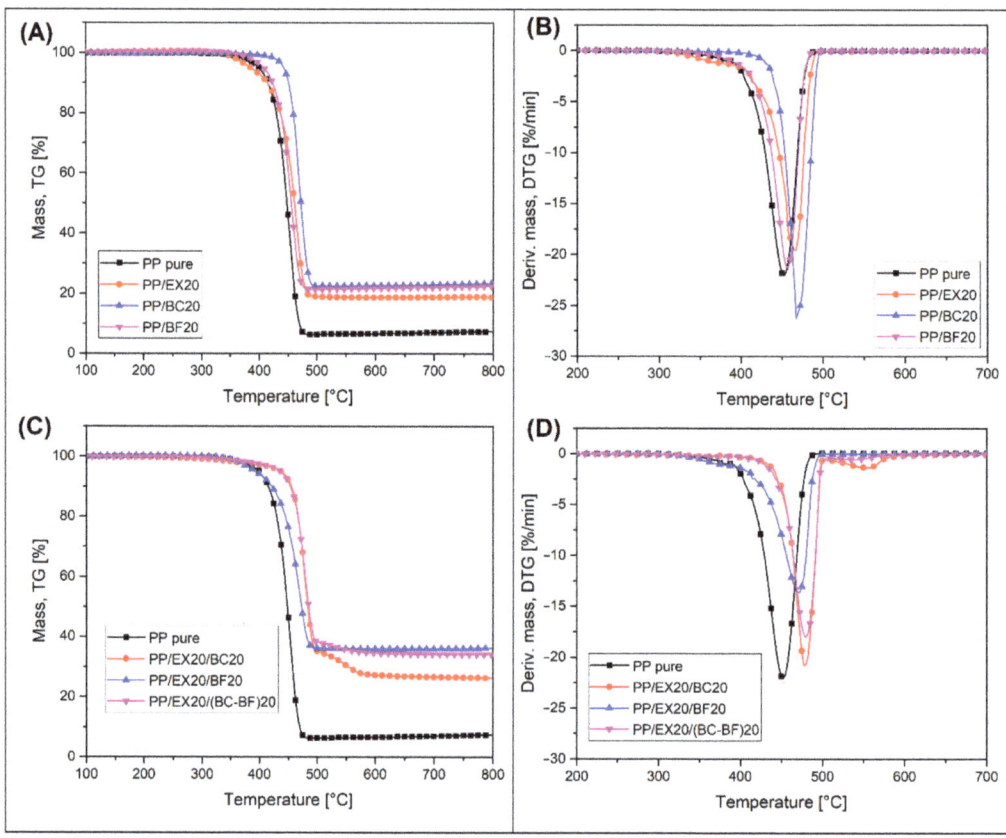

Figure 5. The thermogravimetric (TGA) analysis of PP-based samples. Results are presented in the form of weight loss TG (**A**,**C**) and derivative weight loss plots DTG (**B**,**D**).

The TG curve for the EX flame retardant agent indicates a two-stage process of decomposition of this compound. The first signs of decomposition in the form of visible weight loss are already observed at 300 °C, which is confirmed by the peak in the DTG curve at 340 °C. At higher temperatures, the decomposition process stabilizes to regain intensity above 500 °C. The temperature of the second DTG peak is 610 °C. The active compound used in the used variant of the additive (EXOLIT AP 422) is ammonium polyphosphate (APP) [62–64]. The first observed degradation step, ending at a temperature of about 450 °C, is related to the decomposition of APP into ammonia and phosphoric acid. The mass loss then does not exceed 20%; however, the high temperature allows the cross-linking reaction to occur and the formation of polyphosphoric acid, causing the formation of a char layer. In the second stage of decomposition, occurring at a temperature of 500–700 °C, polyphosphoric acid is decomposed into phosphorus oxides forming a stable char. At 800 °C, the weight of the sample is reduced to 20%.

In the case of the other used filler, basalt fibers (BF), very high-temperature stability can be observed in the entire tested measuring range. The initial weight of the sample was not changed even at 800 °C. The observed loss of 2% is related to the decomposition of the fiber sizing agent.

Biocarbon (BC) filler is characterized by a lower temperature stability. A weight loss of about 2% was already visible at the temperature of 100 °C; in the discussed case, it is related to the release of water vapor from the porous BC structure. A further stable weight loss, at the level of 15% at 800 °C, is due to the breakdown of the organic functional groups present in the BC structure. These are mainly the hydroxyl and carboxyl group residues of organic compounds that were decomposed during the biomass pyrolysis process. Interestingly, the presence of these compounds has, in many cases, a positive effect on the occurrence of increased adhesion at the polymer-filler interface BC [46,49,65].

In the case of the TGA analysis results for PLA-based composites (Figure 4), the thermal stability of the matrix is almost identical for all samples. The start of decomposition for composite samples coincides with the decomposition temperature of pure PLA. There is a slight deviation for the PLA/BC20 sample, where the onset of decomposition occurs at a temperature of about 10 °C lower than other materials. The reason for this phenomenon may be a partial degradation of the PLA structure during the processing procedure (extrusion and injection molding) caused by the moisture contained in the BC filler. Apart from a slight decrease in the PLA thermal stability of the sample from BC, it is worth noting that, for most of the samples, the residual char mass was very close to the filler content. This confirms the complete decomposition of the polymer matrix at high temperatures. Interestingly, for PLA/EX20 and PLA/EX20/BC20 samples, the presence of the second stage of weight loss (at 500 °C) confirmed the two-step mechanism of APP decomposition.

TGA measurements for PP samples are presented in Figure 5. Generally, the decomposition temperature of PP-based materials was significantly higher, and there is also a visible difference between the particular composites. The initial weight loss for most of the samples was similar to pure PP, which can be confirmed by comparing the temperature of a 5% weight loss. However, the addition of BC particles increased the thermal stability of composites by around 30 °C. This is an inverse change to that observed for the PLA samples. Similar to PLA, the residual char content for PP-based composites was very close to the initial content of the fillers. The only noticeable difference was observed for the PP/EX20/BC20 sample, where the EX filler decomposition was observed at 500 °C.

It is worth noting that, considering the general tendency of changes in thermal stability (5% weight loss), the addition of fillers to PLA does not significantly affect the decomposition temperature of the material, while for PP composites the decomposition temperature was slightly shifted. Similar results were obtained for the studies performed by Kadola et al. where PLA and PP-based samples were modified with three different fire retardants [66]. The introduction of modifiers to PP has always resulted in a significant change in the degradation temperature of the material, while the changes for PLA were relatively small.

The thermogravimetry results cannot be considered a valuable method for flame retardancy analysis; however, for selected samples containing the EX filler, it is possible to observe the additional weight loss step associated with the decomposition of polyphosphoric acid.

3.3. Thermal Properties, Phase Transitions—DSC Analysis

The results of the DSC thermal analysis are presented in the form of heating/cooling thermograms (see Figure 6). Some data and calculations are also presented in Table S2 (Supplementary Information Section). Since the melting temperature of PLA and PP are very close, the measurements were carried out in the same temperature range from room temperature up to 230 °C. The 1st heating signal for PLA-based samples (Figure 6A) and PP materials (Figure 6C) confirmed large differences in the phase transitions of the examined polymers. For PLA samples, the appearance of the cold crystallization phenomenon revealed a dominated amorphous nature of the matrix phase after the injection molding process. The presence of an increasing amount of the filler led to a small shift of the T_{cc} to a lower temperature ($\approx$10 °C); however, the magnitude of that change is not significant. The melting temperature for all materials is close to 170 °C. Compared to PLA, the DSC analysis of PP-based samples revealed a less complex shape of the thermograms, while for all samples, only a single melting peak was recorded. Similar to PLA materials, the peak position for PP samples was almost equal for all measurements ($\approx$165 °C). It seems that the addition of filler, even at a high amount of 40 wt%, did not lead to any significant change in composite characteristics during the heating stage of DSC analysis. The lack of differences between the plots' appearance was confirmed by the crystallinity calculations (see Table S2). For PLA samples, the reference results for pure polymer revealed a 10% crystallinity level, while for most of the composites, the calculations show around 20% of the crystalline phase. The calculations for PP-based samples revealed that even for pure PP the crystallinity reached around 49%, which means that the main expected differences in thermomechanical properties between PLA and PP-based materials are caused by the crystallinity level. The DSC cooling stage for PP-based samples did not reveal any unpredictable changes. The crystallization temperature for composites was mostly slightly higher compared to pure PP. However, the magnitude of the temperature shift was small and cannot be compared to highly efficient nucleating agents such as sorbitol-based modifiers [67,68].

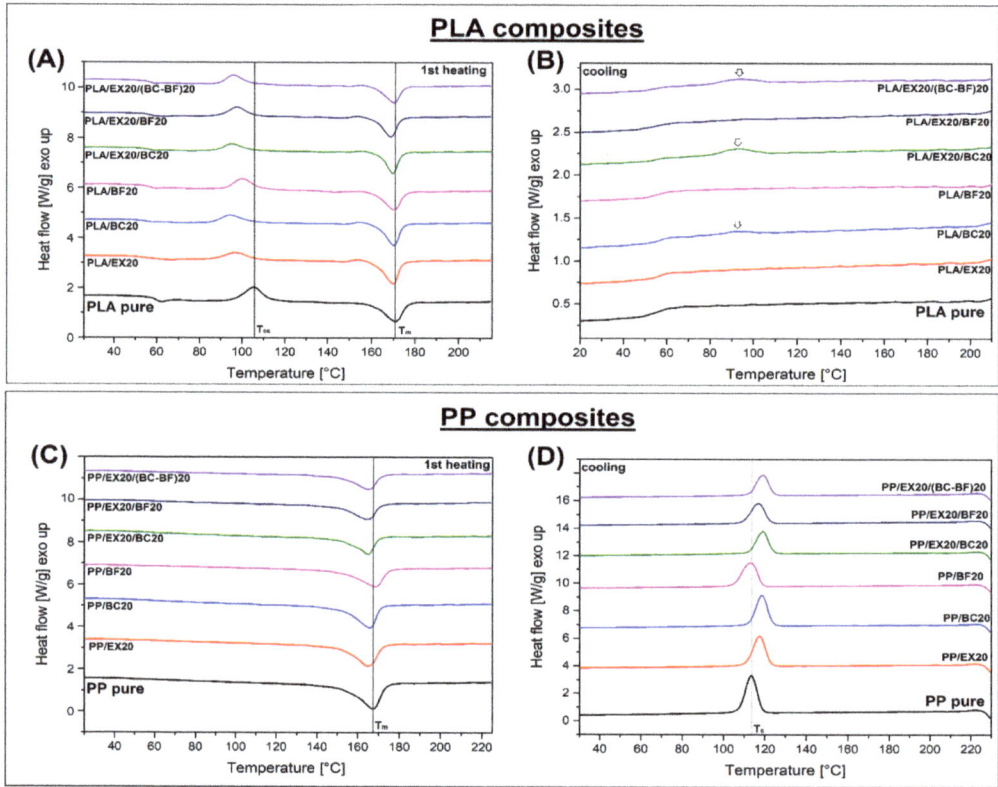

Figure 6. The DSC thermograms of (**A**,**B**) PLA and (**C**,**D**) PP-based materials. Plots present the 1st heating and cooling signals.

Interestingly, for both PLA and PP-based materials, the addition of BF led to a visible decrease in crystallinity. This behavior was recorded for PLA/BF20 and PP/BF20 samples; however, the hindering of the crystal phase growth at the observed level should not influence the important properties of composites. The described phenomenon, consisting of the reduction in the nucleation effect in PP as a result of the use of basalt-based materials, has already been observed in the literature [69]. Clearly, a more favorable nucleating effect was found for the BC additive, especially for PLA-based materials. For most of the tested materials, a cooling stage of the measurement did not reveal the distinct crystallization peak; it occurs only for samples containing BC particles. Unfortunately, the scale of this phenomenon is negligible, and it does not affect the level of crystallinity of the PLA/BC composites. However, it may be a suggestion for further work using PLA with a higher crystallization ability.

The results of the DSC analysis revealed some visible changes in crystallinity for the PLA samples. However, despite the fact that for most composites, the content of the crystalline phase was two times higher than for pure PLA, the amorphous phase that still predominates in the matrix structure significantly contributes to the reduction in thermal resistance. Due to the fact that, for PP, the content of the crystalline phase is significant (≈50%), the presence of fillers does not contribute to any noticeable changes in phase structure. The addition of fillers may contribute to some changes in crystalline structure morphology, however, which might improve the mechanical properties of PP composites.

3.4. Mechanical Properties Evaluation—Static and Impact Tests

The mechanical properties of the prepared materials are collected in the form of plots in Figure 7. The charts present the results of tensile strength/modulus, elongation at break, and notched Izod impact strength. The complete list of results is shown in Table S3 (Supplementary Information Section).

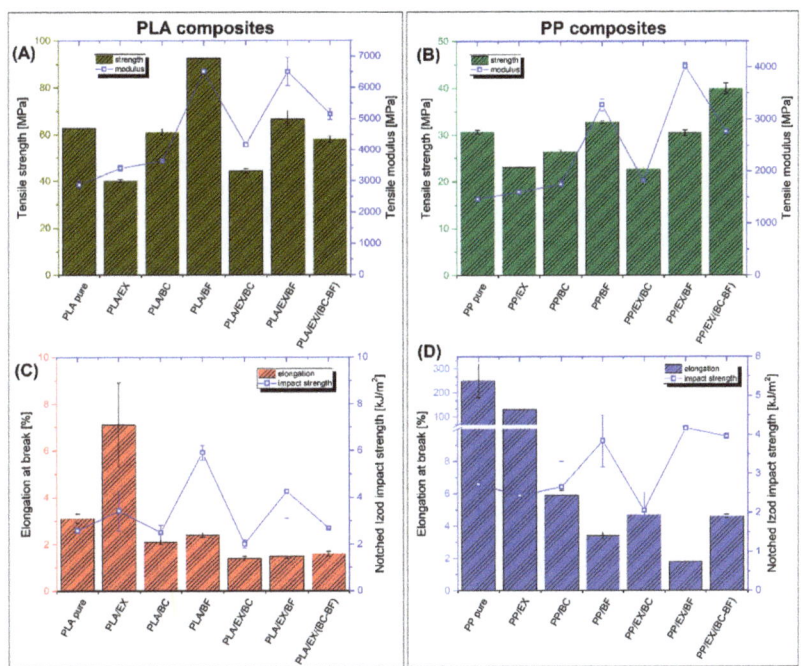

Figure 7. The results of the static tensile and Izod impact resistance tests. The tensile strength and modulus of (**A**) PLA-based and (**B**) PP-based composites. Elongation at break and impact strength, respectively, of (**C**) PLA and (**D**) PP-based materials.

The results for PLA-based samples are presented in Figure 7A,C. It is clear that the addition of 20 and 40 wt% of the fillers increases the stiffness of the material (tensile modulus). However, the reinforcing efficiency was very different when considering the absolute values of the modulus. For pure powder fillers (EX and BC), the stiffness improvement was only slight, which was also observed for PP-based samples (see Figure 7B). The most visible improvement was recorded for the 20 wt% BF samples, where the PLA/BF sample modulus reached around 6.5 GPa, compared to the initial 3 Gpa, for pure PLA. The reinforcing effect for the PP/BF sample was also significant since the modulus was 3.3 Gpa, compared to the reference 1.5 GPa. For hybrid composites, the total content of the fillers was 40 wt%.

However, the introduction of the EX/BC system did not improve the stiffness since the tensile modulus values were very close to the results of the BC-modified samples. This tendency was similar for PLA and PP-based samples as well. More favorable results were obtained for the hybrid samples containing reinforcing fibers. In comparison, the highest tensile modulus was recorded for EX/BF samples. Interesting conclusions can thus be made after the comparison of the tensile strength results.

For PLA-based composites, only the PLA/BF sample strength was visibly improved, up to 90 MPa from the initial 60 MPa for pure PLA. The lowest strength was recorded for the PLA/EX and PLA/EX/BC samples (≈40 MPa). This strength deterioration is quite an obvious phenomenon since the reinforcing efficiency for powdery fillers is low. For the other materials, the tensile strength results were close to the properties of pure PLA, from 60 to 70 MPa. Interestingly, for the PLA/BC samples, the measured strength was similar to pure PLA, which suggests that the matrix-filler interactions for the BC filler are stronger than for the EX particles. The results of the tensile strength measurement for PP-based composites revealed quite similar tendencies. Similar to PLA composites, the tensile strength was reduced for all samples with the addition of powdery fillers. Interestingly, the highest strength was observed for the PP/EX/(BC-BF) sample, while for the other BF reinforced composites (PP/BF and PP/EX/BF), the strength values are very close to the reference pure PP (≈35 MPa).

The elongation at break for all PLA-based composites was very low and usually did not reach 3%. However, some unexpected results were observed for the PLA/EX composite since the maximum strain increased up to 5%. The reason behind this improvement is not clear. However, part of the research indicates the occurrence of plasticizing effects with the use of small amounts of APP in PLA [19,70,71]. The partial solubility of APP in the polymer matrix is due to the low molecular weight of this compound. The micromechanism that improves the maximum elongation is caused by the change in the distance between the matrix polymer chains, which leads to the reduction in internal stresses during deformation. Unfortunately, in the case of the discussed samples, the effectiveness of this phenomenon is very limited and incomparably lower than the changes observed for more effective plasticizers [72–74]. Unlike pure PLA, the initial elongation at break for pure PP can reach even 250%. In comparison, the addition of fillers leads to a sharp reduction in sample strain. For most of the composite samples, the fracture appeared below 6% strain. Again, the EX-filled samples were characterized by different behavior, while the elongation at break for PP/EX composites was around 120%.

The impact resistance was relatively low for all of the prepared samples; however, some favorable improvement was noticed for BF-reinforced samples. For PLA-based samples, the reference impact strength was 2.5 kJ/m^2 (for pure PLA), while the highest value was recorded for PLA/BF samples, reaching 6 kJ/m^2. The visible improvement was also observed for the PLA/EX/BF sample (4.5 kJ/m^2), while for the rest of the composites, the impact strength was close to 3 kJ/m^2 or lower. Similar trends were observed for PP-based samples, where, for all BF-reinforced samples, the impact strength was close to 4 kJ/m^2. For the rest of the materials, the impact strength was below 3 kJ/m^2. The results of the impact tests show a favorable change in the fracture mechanism for the samples with the BF addition. The fiber pull-out mechanism, the main phenomenon leading to the toughness improvement in composite materials, is strongly limited for injection molded samples [75,76]; however, the impact strength is still visibly higher compared to the pure polymer. For the standard manufacturing procedure, where staple fibers are added to the matrix during the melt mixing procedure, the final filler length rarely exceeds 1 mm. That leads to a reduced interaction efficiency at the fiber-matrix interface [77,78].

Summarizing the mechanical test results, it can be noted that the addition of fibrous fillers is necessary to obtain relatively good mechanical performance. Without the BF reinforcement, the composites' brittleness led to mechanical property deterioration.

3.5. Thermo-Mechanical Properties—Comparison between DMTA Analysis and HDT/Vicat Tests

Since low thermal resistance is one of the most important limitations in the use of PLA, the thermomechanical properties of the prepared composites were investigated using DMTA analysis and a combination of heat deflection (HDT) and Vicat softening (VST) measurements. The results of DMTA tests are presented in the form of storage modulus and tan δ plots (see Figure 8). The HDT and VST test results are presented in Figure 9. The additional DMTA analysis for pure polymers is presented in Figure S1 (Supplementary Information Section).

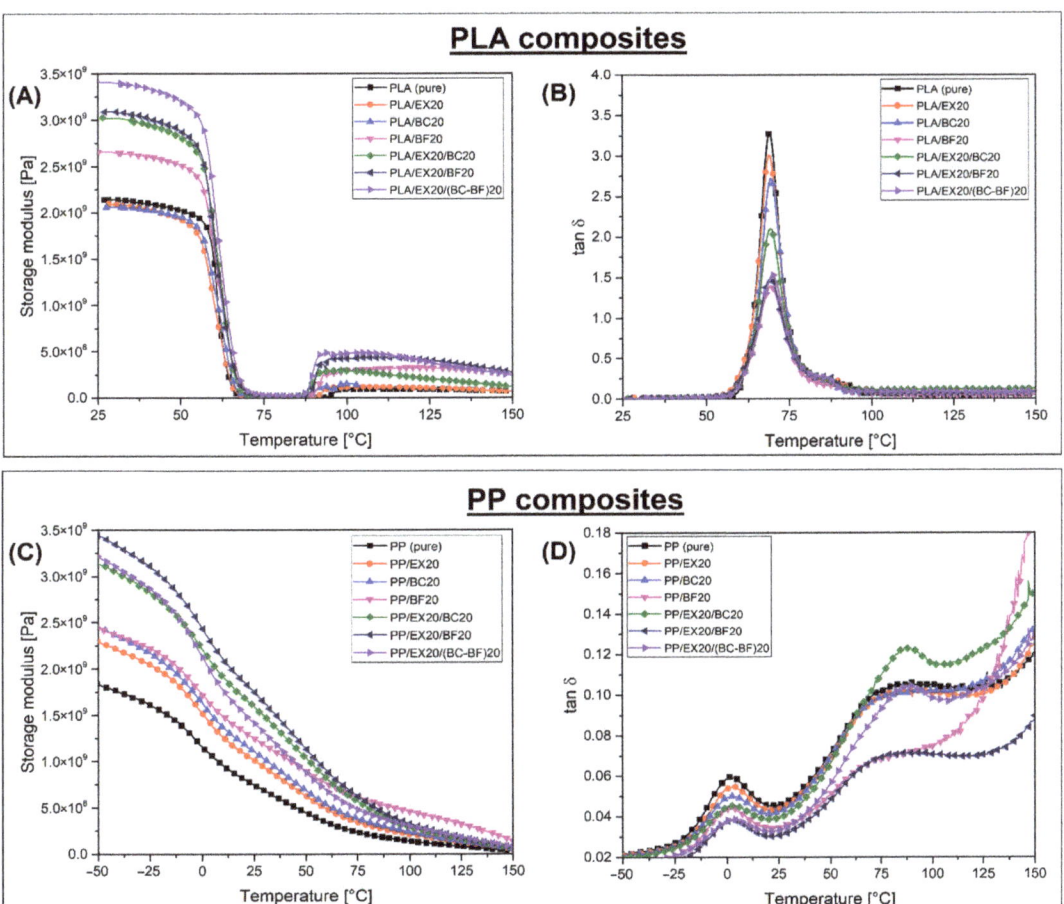

Figure 8. The results of the DMTA analysis of (**A**,**B**) PLA-based and (**C**,**D**) PP-based composites. The plots collect the results of the storage modulus and tan δ measurements.

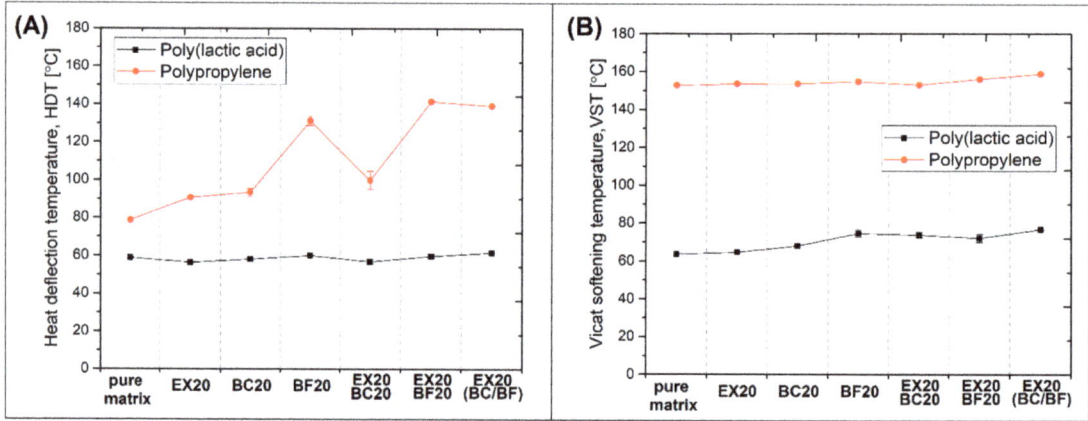

Figure 9. The results of (**A**) HDT and (**B**) VST measurements of all PLA and PP-based samples.

The comparison of the storage modulus for PLA-based samples (see Figure 8A) revealed large differences in material stiffness for different types of fillers. It can be expected that the reinforcing efficiency of the powdery fillers was negligible since the storage modulus values for PLA/EX20 and PLA/BC20 composites are almost similar to pure PLA. Visible changes were recorded for PLA/BF20 samples, which confirmed a more favorable strengthening mechanism. The highest stiffness was observed for composites with the addition of 40 wt% of the fillers, while the highest storage modulus was recorded for the PLA/EX20(BC-BF)20 sample. Despite the application of a large amount of filler, the glass transition phenomenon led to a large drop in storage modulus values. For all samples, the sharp stiffness reduction observed at around 60 °C led to material softening, resulting in a significant reduction in the use at higher temperatures. The tan δ plots confirmed that the Tg region for all composites was constant, which confirmed that strong filler/matrix interactions did not occur. The only noticeable difference refers to the area under the Tc peak, where, for highly filled samples, the peak value was strongly reduced. That phenomenon is typical for composite samples since the presence of the filler particles leads to a reduction in the polymer volume in the sample [79,80]. The appearance of the DMTA plots for PLA-based samples confirmed the amorphous character of the matrix phase. At higher temperatures (>90 °C), the storage modulus values increase, which is related to the cold crystallization phenomenon. However, it confirmed that during the standard injection molding process, where the mold temperature is below 100 °C, the possibility of obtaining a highly crystalline form of PLA is not possible.

The DMTA analysis for PP-based composites differs significantly from the results for PLA, mostly due to the highly crystalline structure of the matrix phase. Since the T_g of PP usually occurs close to 0 °C, the measurements were conducted from −50 °C in order to cover the whole glass transition region. The initial high storage modulus values for PP-based samples were similar to those obtained for PLA composites. That is quite natural since, at the beginning of the test, both polymers are in a glassy state. However, it is worth noticing that the starting temperature for PP-based materials was 75 °C lower than for PLA. The stiffness of PP recorded at 25 °C dropped to half its initial value. Such a decrease can be considered significant; however, the reduction in the modulus value throughout the whole measurement is relatively slow. The presence of the crystalline phase improves the thermal stability of the PP matrix. It is particularly evident in the glass transition region, where, unlike PLA, the PP composites do not lose stiffness so rapidly. Interestingly, for PP-based materials, the increase in the storage modulus was reported even for powder fillers, such as EX or BC. It is typical for highly crystalline polymers, where spherulites are formed around the filler particles [81,82]. The highest stiffness was reported for hybrid composites, with a 40 wt% addition of the fillers. Unlike PLA-based samples, where stiffness for all samples sharply drops at around 60 °C for all materials, the modulus for PP composites at elevated temperature (>50 °C) strongly depends on the filler content and type. This is why the expected thermal resistance will be highest for BF-reinforced hybrid samples. Similar to PLA, the position of the Tg peak was not shifted, which suggests a lack of strong filler-matrix interactions.

DMTA analysis makes it possible to observe, in a very accurate way, any changes in the stiffness of polymer composite materials on the temperature scale, but in industrial practice, thermomechanical resistance is usually evaluated using HDT or Vicat tests. Due to significant differences in measurement methodology, the results of these tests cannot be compared, therefore, it is not possible to use them interchangeably. In both cases, however, they are an indicator of the possible application temperature of the tested materials. The HDT results presented in Figure 9A confirm a large difference between PLA and PP-based materials. Even for pure polymer samples, there is around a 20 °C difference since HDT reached 60 °C and 80 °C, respectively, for pure PLA and pure PP. The difference for composites is even higher. This is easy to evaluate since, for all PLA-based composites, HDT was constant and close to the reference pure PLA. It is clear that for PP composites, the introduction of spherical fillers (EX and BC) led to only a small enhancement of the thermal resistance, a HDT of around 80–90 °C. A more significant improvement was recorded for BF, where for the PP/BF20 sample, the HDT reached 130 °C, while for the hybrid samples, PP/EX20/BF20 and PP/EX20/(BC-BF)20, the heat deflection was around 140 °C. The most important fact revealed during the analysis of HDT results is the lack of visible HDT improvement for PLA-based composites, even for highly filled hybrid samples. The advantage of using PP-based materials is even more visible when analyzing the VST results (see Figure 8B). The softening temperature for all PP materials was above 150 °C, which is very close to the melting point of the PP matrix. Unlike the HDT measurements, where, for all PLA-based samples, the results were similar, the VST measurements revealed some noticeable improvement after the addition of the fillers: from an initial 64 °C for pure PLA up to 77 °C for the PLA/EX20/(BC-BF)20 sample. However, that kind of difference cannot be considered a serious advantage. In such a situation, the beneficial role of the crystalline phase is revealed. Previous studies in this area confirmed that PLA-based composites are able to withstand prolonged exposure to high temperatures. However, the investigated samples were usually prepared by compression molding or annealing [83,84].

A thermomechanical analysis revealed significant drawbacks for all PLA-based injection-molded materials. For PLA, like other thermoplastic polyesters, such as PET or PBT, the crystallization process is much slower than for other semi-crystalline polymers such as PP, PE, or PA6. The manufacturing of specialized products requires the use of a high mold temperature, which unfortunately always increases production costs and process efficiency, and it often involves the use of dedicated nucleation agents [79,85–87].

3.6. Flame Retardancy—PCFC Microcalorimetric Analysis, UL-94 Tests

The results of the microcalorimetric analysis are collected in the form of separate plots for PLA-and PP-based samples in Figure 10. Some data and calculations are also presented in Table S4. The results of microcalorimetric measurements provide slightly more reliable information on the behavior of the tested materials during decomposition in the oxidizing environment, a distinction from TGA testing tests. The initial analysis of the HRR curve appearance indicates that the dynamics of the combustion process for most samples were very similar, which would suggest no significant differences in flammability. For other studies where phosphorous-based FR was used, the calorimetric measurements indicate a significant extension of the burning time of samples modified with APP at the same time the height of the HRR peak is reduced [66,71,88,89]. Regarding the PCFC measurements performed, the results are not conclusive and require additional calculations of the THR index, the changes of which slightly more clearly indicate the improvement of the results in the flammability tests.

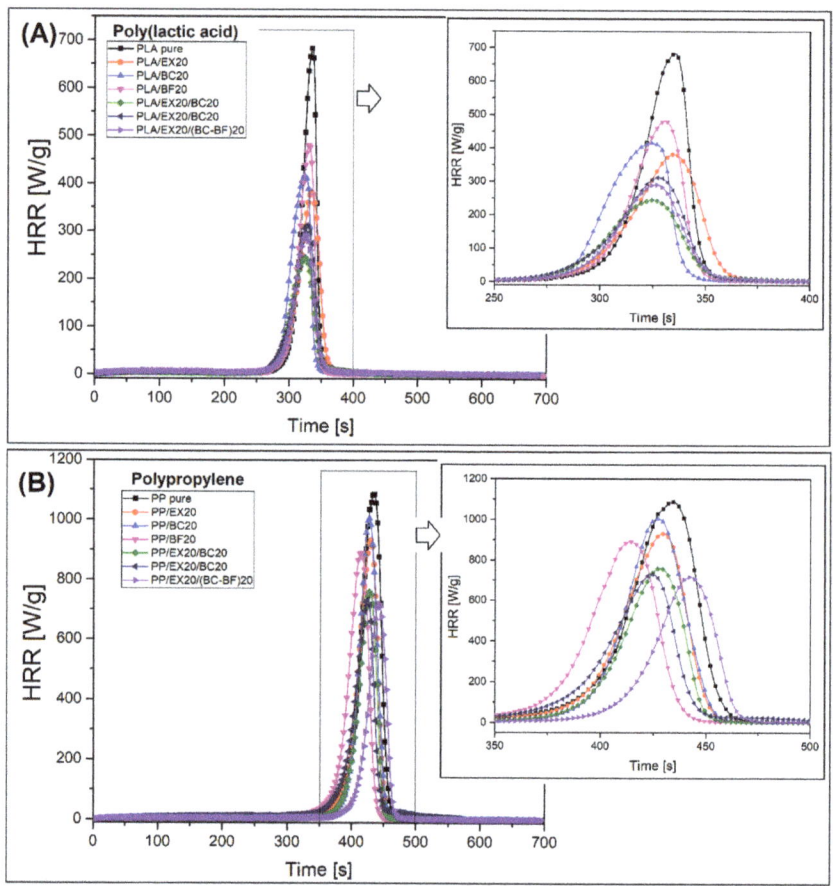

Figure 10. The relationship of the heat release rate (HRR) as a function of the measurement time of materials based on (**A**) PLA and (**B**) PP.

It is clear that, for all samples, the heat release rate plots (HRR) revealed a single-stage process. Heat release is considered the most critical factor during the assessment of polymer composite flammability, while the position of the HRR curve peak is usually treated as the key factor during combustion calorimetric measurements. For PLA-based samples, it is clear that the HRR peak was highest for pure PLA samples, while the addition of the fillers led to a reduction in the HRR maximum. Among samples with the addition of 20 wt% fillers, the lowest HRR peak was observed for the PLA/EX20 sample; while comparing the hybrid samples, the lowest peak value was recorded for the PLA/EX20/BC20 sample. The results of the HRR peak height are in line with other factors calculated during the measurement. The results of the total heat release (THR) are very close to the HRR peak comparison where the lowest THR value was recorded for hybrid composites containing EX and BC particles.

The HRR plots for PP composites revealed less efficient flammability reduction. Although the trend of HRR changes after the addition of fillers is favorable, the reduction in HRR and THR values indicates that materials' flammability will be relatively high. Even for the PP/EX20/(BC-BF)20 sample, where THR was reduced to 26.7 kJ/g, the result was worse than for pure PLA (THR = 20 kJ/g). This difference gives a clear idea of the flame retardant potential for both types of material, whereas, even for highly modified PP samples, the flammability reduction will be less effective, which was confirmed by other studies [90]. The differences in the results of the microcalorimetric analysis were confirmed during the standard UL-94 tests.

The burning classification for the prepared materials was evaluated using the UL-94 testing methodology. All samples were tested in horizontal and vertical positions. The sample appearance during testing is presented in Figure 11 for both horizontal and vertical tests. The results of the performed tests are collected in Table 3.

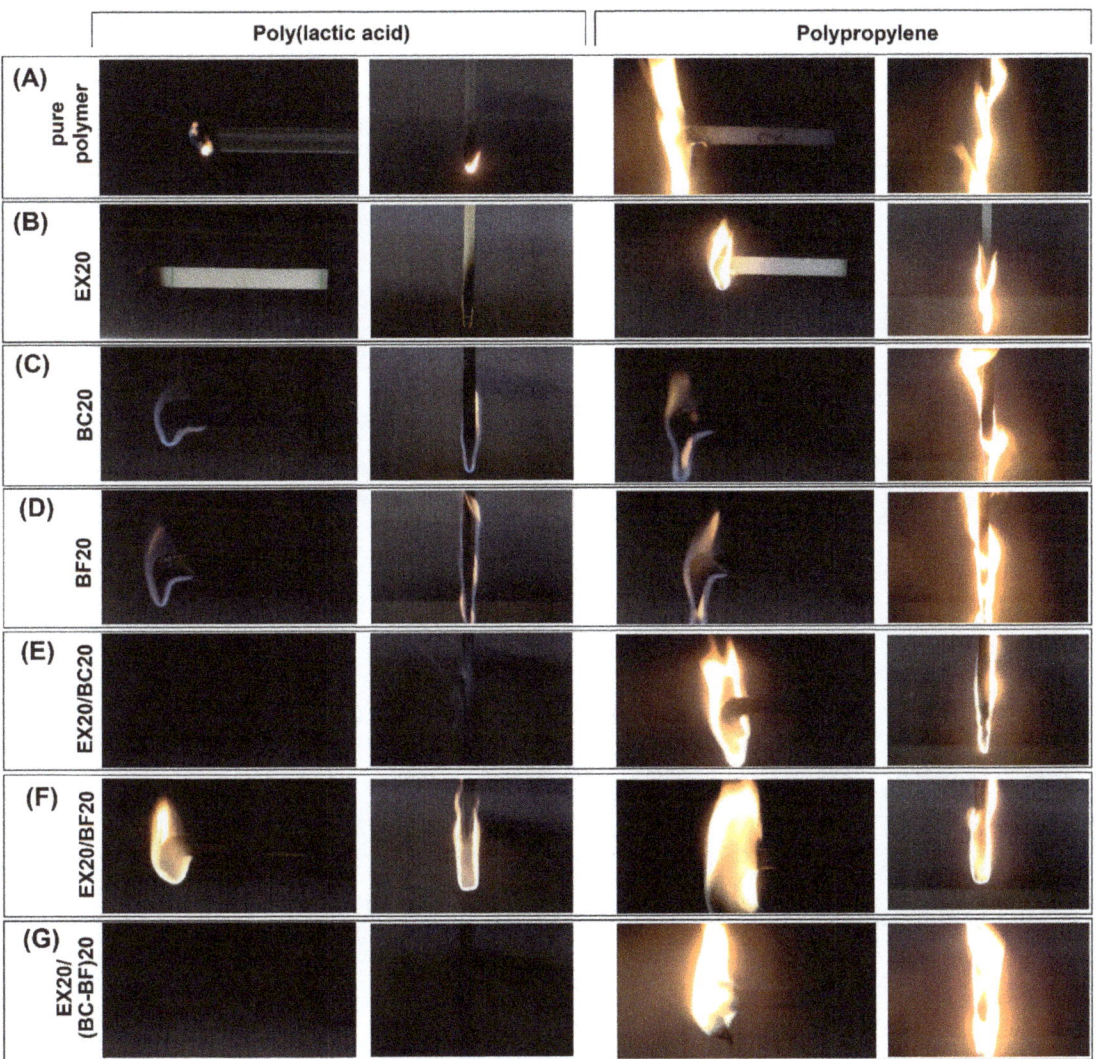

Figure 11. The sample appearance during the UL-94 tests (horizontal and vertical). The photos are grouped according to the type of fillers used.

Table 3. The results of UL-94 tests performed in horizontal and vertical positions.

Sample	Horizontal Measurement		Vertical Measurement			Rating
	Burning Speed (mm/min)	Dripping	T1 * (s)	T2 * (s)	Dripping	
PLA-based composites						
PLA pure	21.4	Yes	10	Total burn	Yes	HB
PLA/EX20	SE *	Yes	0	0	No	V0
PLA/BC20	32.2	Yes	Total burn	-	Yes	HB
PLA/BF20	32.6	Yes	Total burn	-	Yes	HB
PLA/EX20/BC20	SE	No	0	0	No	V0
PLA/EX20/BF20	25.9	Yes	Total burn	-	No	HB
PLA/EX20/(BC-BF)20	SE	No	0	3	No	V0
PP-based composites						
PP pure	34.8	Yes	Total burn	-	Yes	HB
PP/EX20	22.1	Yes	3	Total burn	Yes	HB
PP/BC20	25.9	Yes	Total burn	-	Yes	HB
PP/BF20	27.7	Yes	Total burn	-	Yes	HB
PP/EX20/BC20	18.8	Yes	5	Total burn	Yes	HB
PP/EX20/BF20	27.3	No	Total burn	-	No	HB
PP/EX20/(BC-BF)20	16.7	Yes	Total burn	-	Yes	HB

* T1/T2—burning time after 1st and 2nd ignition; * SE—self-extinguishing.

The results for the PLA and PP-based samples are extremely different. The modification of PLA was evidently more favorable for PLA composites since, for selected composites, it was possible to obtain a V-0 flammability rating. It was possible for all of the prepared specimens to withstand the HB (horizontal burning) requirements since the highest burning speed recorded for pure PP was around 35 mm/min. For other samples, the burning speed varied from 16.7 to 32.6; however, there were no visible trends related to the sample composition. In general, the flammability of hybrid samples was lower. For PLA/EX20, PLA/EX20/BC20, and PLA/EX20/(BC-BF)20, the flame was rapidly self-extinguishing, while for the rest of the PLA-based materials, the tested specimens were completely burned. This behavior confirmed the importance of the APP addition, while the presence of polymer fillers, even at high quantities, did not improve flame retardancy. For PP-based samples, the results are less favorable since, for all specimens, the material burned along the whole sample length. The only visible improvement was observed for the PP/EX20/BF20 composite, where the material did not drip from the burning specimen. For other materials, the molten polymer dripped from the sample and ignited the cotton cloth placed under the specimen.

The results revealed that the addition of phosphorus-based APP flame retardants is more efficient for PLA-based samples. Interestingly, the obtained hybrid samples are characterized by V-0 class. This must be considered a very good result since the APP weight content in all samples was only 20%. The less favorable observed properties of the PP-based samples were mainly associated with a lack of oxygen present in the PP chain structure as well as the lower density of PP, which leads to a lower volumetric proportion of APP in the structure of the material. The lower volume of the flame-retardant additionally reduces its efficiency.

The vertical test results reflect the trends observed in the horizontal position. A self-extinguishing flame was observed again for the same three samples containing pure APP filler (EX20), EX20/BC20, and EX20(BC-BF)20 fillers. The remaining samples burned completely. It is clear that the presence of the EX additive is necessary to decrease the flammability of PLA, however, the use of fibrous reinforcement (BF) inhibits the flame retardancy mechanism. The use of hybrid systems with the addition of EX, BC, and BF seems to be the most effective in this approach since they help to maintain a low flammability rate, simultaneously improving the mechanical characteristics of the composite. Unfortunately, the results for the flammability tests of the PP samples confirmed a lack of improvement.

All specimens were totally burned. Except for the PLA sample, all the PP samples melted intensely, initiating the cotton fabric ignition.

The conducted research shows that the use of biocarbon without the addition of APP does not limit the flammability of the composite. However, unlike the addition of the BF filler, it seems that the use of BC supports the mechanism of phosphorus-based additives. Previous research indicates that, in the case of BF reinforcement, the mechanism of the burning process of composite materials is subject to unfavorable changes. The research conducted by Tang et al. [91] suggests that, for PP/BF materials, the thermal conductivity of composites is increased, which leads to the heating and decomposition of the material in a large volume. The consequence of this behavior is the inability to form a permanent char layer that could prevent the propagation of the thermal decomposition of the material. Unfortunately, there are no similar measurements for PLA-based samples. However, taking into account the results of the cited works, it can be stated with some certainty that for the fine dispersion of small BC particles, the thermal propagation to the inside of the material must be significantly limited in relation to the relatively long BF fibers. Despite the presence of BF fibers in the hybrid composite, the formation of the charred layer is facilitated by the presence of BC particles.

Summarizing the obtained results, it is worth noting that it is not clear what the main mechanism limiting the flammability of composites contacting the EX/BC system is. Taking into account the results of previous work by Wang et al. [56], the physical mechanism that reduces the flammability of materials with the addition of BC and the active FR is associated with the increased thermal conductivity of the composite. The increased heat transfer leads to faster decomposition of FR, which increases the efficiency of the APP compound. In our opinion, this mechanism is not decisive since most of the previous studies, where carbon fillers with much better thermal conductivity were used, revealed that the addition of carbon black, graphite, or graphene leads to the formation of a protective char layer [92,93]. Nevertheless, the discussion on this topic has only a theoretical character. The planned further research will include a more detailed analysis of a selected group of materials to determine a presumed mechanism of operation of systems containing biocarbon.

4. Conclusions

The conducted research confirmed that it is possible to combine the sustainable nature of polymer compounds with their flame retardancy effects. The reinforcing system containing BC particles and BF fibers seems to be an effective additive improving the mechanical properties of materials with the addition of APP flame retardant. As the results show, the addition of BC filler plays an important role in increasing the efficiency of EX operation, while the system containing EX and BF is characterized by increased flammability. Unfortunately, the beneficial effects of using the new filler system are observed only for PLA, while the flammability of PP is not improved. The current research related to the use of APP indicates that the most likely cause of the observed flammability differences is the higher efficiency of APP additive in materials containing oxygen in the chain structure. The obtained research results are very promising; therefore, as part of further experiments, it is planned to use a similar system for technical polymers, such as nylons or styrene polymers.

Supplementary Materials: The following supporting information can be downloaded at: https://www.mdpi.com/article/10.3390/polym14194086/s1, Figure S1: The storage modulus/tan δ plots for (A) pure PLA and (B) pure PP sample; Table S1: The basic data collected during the thermogravimetric (TGA) measurements; Table S2: The results of DSC measurements from 1st heating and cooling stage; Table S3: The list of mechanical properties obtained during the static tensile/flexural measurements and Izod impact tests; Table S4: Results obtained during testing of PLA and PP-based composites using a PCFC microcalorimeter.

Author Contributions: Conceptualization, J.A.; methodology, J.A.; validation, J.A.; formal analysis, J.A.; investigation, J.A. and S.M.; resources, J.A.; data curation, J.A.; writing—original draft preparation, J.A. and S.M.; writing—review and editing, J.A. and S.M.; visualization, J.A.; supervision, J.A.;

project administration, J.A.; and funding acquisition, J.A. All authors have read and agreed to the published version of the manuscript.

Funding: This research was partly funded by the National Research and Development Centre of Poland, the project "Development of hybrid biodegradable composites' manufacturing technology for the automotive industry" (LIDER/25/0148/L-8/16/NCBR/2017), and by The National Agency for Academic Exchange NAWA as part of the Bekker programme, grant number PPN/BEK/2019/1/00161/DEC/1 "Injection molding of hybrid composites reinforced with a biobased filler system".

Data Availability Statement: Not applicable.

Acknowledgments: The authors would like to thank Ewa Mazurek and Dorota Grabowicz of HSH Chemie for kindly supplying the flame retardants. We would also like to thank Kacper Barwik for help with sample preparation.

Conflicts of Interest: The authors declare no conflict of interest.

References

1. Salasinska, K.; Celiński, M.; Mizera, K.; Kozikowski, P.; Leszczyński, M.K.; Gajek, A. Synergistic effect between histidine phosphate complex and hazelnut shell for flammability reduction of low-smoke emission epoxy resin. *Polym. Degrad. Stab.* **2020**, *181*, 109292. [CrossRef]
2. Kairyte, A.; Kremensas, A.; Vaitkus, S.; Członka, S.; Strakowska, A. Fire suppression and thermal behavior of biobased rigid polyurethane foam filled with biomass incinerationwaste ash. *Polymers* **2020**, *12*, 683. [CrossRef] [PubMed]
3. Hull, T.R.; Law, R.J.; Bergman, Å. *Environmental Drivers for Replacement of Halogenated Flame Retardants*; Elsevier: Amsterdam, The Netherlands, 2014; ISBN 9780444538093. [CrossRef]
4. Abbasi, G.; Li, L.; Breivik, K. Global Historical Stocks and Emissions of PBDEs. *Environ. Sci. Technol.* **2019**, *53*, 6330–6340. [CrossRef] [PubMed]
5. Lavandier, R.; Quinete, N.; Hauser-Davis, R.A.; Dias, P.S.; Taniguchi, S.; Montone, R.; Moreira, I. Polychlorinated biphenyls (PCBs) and Polybrominated Diphenyl ethers (PBDEs) in three fish species from an estuary in the southeastern coast of Brazil. *Chemosphere* **2013**, *90*, 2435–2443. [CrossRef]
6. Xiong, P.; Yan, X.; Zhu, Q.; Qu, G.; Shi, J.; Liao, C.; Jiang, G. A Review of Environmental Occurrence, Fate, and Toxicity of Novel Brominated Flame Retardants. *Environ. Sci. Technol.* **2019**, *53*, 13551–13569. [CrossRef]
7. Woo, Y.; Cho, D. Effect of aluminum trihydroxide on flame retardancy and dynamic mechanical and tensile properties of kenaf/poly(lactic acid) green composites. *Adv. Compos. Mater.* **2013**, *22*, 451–464. [CrossRef]
8. Yanagisawa, T.; Kiuchi, Y.; Iji, M. Enhanced flame retardancy of polylactic acid with aluminum tri-hydroxide and phenolic resins. *Kobunshi Ronbunshu* **2009**, *66*, 49–54. [CrossRef]
9. Nishida, H.; Fan, Y.; Mori, T.; Oyagi, N.; Shirai, Y.; Endo, T. Feedstock recycling of flame-resisting poly(lactic acid)/aluminum hydroxide composite to L,L-lactide. *Ind. Eng. Chem. Res.* **2005**, *44*, 1433–1437. [CrossRef]
10. Cao, X.; Chi, X.; Deng, X.; Sun, Q.; Gong, X.; Yu, B.; Yuen, A.C.Y.; Wu, W.; Li, R.K.Y. Facile synthesis of phosphorus and cobalt co-Doped graphitic carbon nitride for fire and smoke suppressions of polylactide composite. *Polymers* **2020**, *12*, 1106. [CrossRef]
11. Mngomezulu, M.E.; Luyt, A.S.; Chapple, S.A.; John, M.J. Poly(lactic acid)-starch/Expandable Graphite (PLA-starch/EG) Flame Retardant Composites. *J. Renew. Mater.* **2018**, *6*, 26–37. [CrossRef]
12. Fukushima, K.; Murariu, M.; Camino, G.; Dubois, P. Effect of expanded graphite/layered-silicate clay on thermal, mechanical and fire retardant properties of poly(lactic acid). *Polym. Degrad. Stab.* **2010**, *95*, 1063–1076. [CrossRef]
13. Wen, X.; Liu, Z.; Li, Z.; Zhang, J.; Wang, D.Y.; Szymańska, K.; Chen, X.; Mijowska, E.; Tang, T. Constructing multifunctional nanofiller with reactive interface in PLA/CB-g-DOPO composites for simultaneously improving flame retardancy, electrical conductivity and mechanical properties. *Compos. Sci. Technol.* **2020**, *188*, 107988. [CrossRef]
14. Yue, X.; Li, C.; Ni, Y.; Xu, Y.; Wang, J. Flame retardant nanocomposites based on 2D layered nanomaterials: A review. *J. Mater. Sci.* **2019**, *54*, 13070–13105. [CrossRef]
15. Vahidi, G.; Bajwa, D.S.; Shojaeiarani, J.; Stark, N.; Darabi, A. Advancements in traditional and nanosized flame retardants for polymers—A review. *J. Appl. Polym. Sci.* **2021**, *138*, 50050. [CrossRef]
16. Murariu, M.; Bonnaud, L.; Yoann, P.; Fontaine, G.; Bourbigot, S.; Dubois, P. New trends in polylactide (PLA)-based materials: "Green" PLA-Calcium sulfate (nano)composites tailored with flame retardant properties. *Polym. Degrad. Stab.* **2010**, *95*, 374–381. [CrossRef]
17. Sun, Y.; Sun, S.; Chen, L.; Liu, L.; Song, P.; Li, W.; Yu, Y.; Fengzhu, L.; Qian, J.; Wang, H. Flame retardant and mechanically tough poly(lactic acid) biocomposites via combining ammonia polyphosphate and polyethylene glycol. *Compos. Commun.* **2017**, *6*, 1–5. [CrossRef]
18. Yu, S.; Xiang, H.; Zhou, J.; Zhu, M. Enhanced flame-retardant performance of poly(lactic acid) (PLA) composite by using intrinsically phosphorus-containing PLA. *Prog. Nat. Sci. Mater. Int.* **2018**, *28*, 590–597. [CrossRef]

19. Jia, Y.W.; Zhao, X.; Fu, T.; Li, D.F.; Guo, Y.; Wang, X.L.; Wang, Y.Z. Synergy effect between quaternary phosphonium ionic liquid and ammonium polyphosphate toward flame retardant PLA with improved toughness. *Compos. Part B Eng.* **2020**, *197*, 108192. [CrossRef]
20. Mazur, K.; Singh, R.; Friedrich, R.P.; Genç, H.; Unterweger, H.; Sałasińska, K.; Bogucki, R.; Kuciel, S.; Cicha, I. The Effect of Antibacterial Particle Incorporation on the Mechanical Properties, Biodegradability, and Biocompatibility of PLA and PHBV Composites. *Macromol. Mater. Eng.* **2020**, *305*, 2000244. [CrossRef]
21. Jiang, S.C.; Yang, Y.F.; Ge, S.B.; Zhang, Z.F.; Peng, W.X. Preparation and properties of novel flame-retardant PBS wood-plastic composites. *Arab. J. Chem.* **2018**, *11*, 844–857. [CrossRef]
22. Zhang, Y.; Hu, Y.; Wang, J.; Tian, W.; Liew, K.M.; Zhang, Y.; Wang, B. Engineering carbon nanotubes wrapped ammonium polyphosphate for enhancing mechanical and flame retardant properties of poly(butylene succinate). *Compos. Part A Appl. Sci. Manuf.* **2018**, *115*, 215–227. [CrossRef]
23. Prabhakar, M.N.; ur Rehman Shah, A.; Song, J.-I. Improved flame-retardant and tensile properties of thermoplastic starch/flax fabric green composites. *Carbohydr. Polym.* **2017**, *168*, 201–211. [CrossRef]
24. Wu, K.; Hu, Y.; Song, H.L.L.; Wang, Z. Flame retardancy and thermal degradation of intumescent flame retardant starch-based biodegradable composites. *Ind. Eng. Chem. Res.* **2009**, *48*, 3150–3157. [CrossRef]
25. Qi, J.; Pan, Y.; Luo, Z.; Wang, B. Facile and scalable fabrication of bioderived flame retardant based on adenine for enhancing fire safety of fully biodegradable PLA/PBAT/TPS ternary blends. *J. Appl. Polym. Sci.* **2021**, *138*, 50877. [CrossRef]
26. Ma, M.; Wang, X.; Liu, K.; Chen, S.; Shi, Y.; He, H.; Wang, X. Achieving simultaneously toughening and flame-retardant modification of poly(lactic acid) by in-situ formed cross-linked polyurethane and reactive blending with ammonium polyphosphate. *J. Mater. Sci.* **2022**, *57*, 5645–5657. [CrossRef]
27. Schirp, A.; Su, S. Effectiveness of pre-treated wood particles and halogen-free flame retardants used in wood-plastic composites. *Polym. Degrad. Stab.* **2016**, *126*, 81–92. [CrossRef]
28. Seefeldt, H.; Braun, U. A new flame retardant for wood materials tested in wood-plastic composites. *Macromol. Mater. Eng.* **2012**, *297*, 814–820. [CrossRef]
29. Stark, N.M.; White, R.H.; Mueller, S.A.; Osswald, T.A. Evaluation of various fire retardants for use in wood flour-polyethylene composites. *Polym. Degrad. Stab.* **2010**, *95*, 1903–1910. [CrossRef]
30. Salasinska, K.; Mizera, K.; Barczewski, M.; Borucka, M.; Gloc, M.; Celiński, M.; Gajek, A. The influence of degree of fragmentation of Pinus sibirica on flammability, thermal and thermomechanical behavior of the epoxy-composites. *Polym. Test.* **2019**, *79*, 106036. [CrossRef]
31. Wang, S.; Zhong, J.; Gu, Y.; Li, G.; Cui, J. Mechanical properties, flame retardancy, and thermal stability of basalt fiber reinforced polypropylene composites. *Polym. Compos.* **2020**, *41*, 4181–4191. [CrossRef]
32. Ying, S.; Zhou, X. Chemical and thermal resistance of basalt fiber in inclement environments. *J. Wuhan Univ. Technol. Mater. Sci. Ed.* **2013**, *28*, 560–565. [CrossRef]
33. Li, Z.; Ma, J.; Ma, H.; Xu, X. Properties and Applications of Basalt Fiber and Its Composites. *IOP Conf. Ser. Earth Environ. Sci.* **2018**, *186*, 012052. [CrossRef]
34. Chen, N.; Pilla, S. A comprehensive review on transforming lignocellulosic materials into biocarbon and its utilization for composites applications. *Compos. Part C Open Access* **2022**, *7*, 100225. [CrossRef]
35. Chang, B.P.; Rodriguez-Uribe, A.; Mohanty, A.K.; Misra, M. A comprehensive review of renewable and sustainable biosourced carbon through pyrolysis in biocomposites uses: Current development and future opportunity. *Renew. Sustain. Energy Rev.* **2021**, *152*, 111666. [CrossRef]
36. Greg, L.; Martinez, M. Canadian Research Allows Ford to Use McDonald's Coffee Chaff in Headlights. Plastics News. Available online: https://www.plasticsnews.com/news/ford-using-mcdonalds-coffee-chaff-headlights (accessed on 4 December 2019).
37. Tadele, D.; Roy, P.; Defersha, F.; Misra, M.; Mohanty, A.K. A comparative life-cycle assessment of talc- and biochar-reinforced composites for lightweight automotive parts. *Clean Technol. Environ. Policy* **2020**, *22*, 639–649. [CrossRef]
38. Seredynski, P. Ford's Mielewski Envisions Broad Portfolio of Renewable Materials. SAE International. Available online: https://www.sae.org/news/2020/09/2020-spe-fords-mielewski-envisions-renewable-materials-portfolio (accessed on 18 September 2020).
39. Botta, L.; Teresi, R.; Titone, V.; Salvaggio, G.; La Mantia, F.P.; Lopresti, F. Use of biochar as filler for biocomposite blown films: Structure-processing-properties relationships. *Polymers* **2021**, *13*, 3953. [CrossRef]
40. Diaz, C.A.; Shah, R.K.; Evans, T.; Trabold, T.A.; Draper, K. Thermoformed containers based on starch and starch/coffee waste biochar composites. *Energies* **2020**, *13*, 6034. [CrossRef]
41. Behazin, E.; Misra, M.; Mohanty, A.K. Sustainable Biocomposites from Pyrolyzed Grass and Toughened Polypropylene: Structure-Property Relationships. *ACS Omega* **2017**, *2*, 2191–2199. [CrossRef]
42. Zhang, Q.; Zhang, D.; Xu, H.; Lu, W.; Ren, X.; Cai, H.; Lei, H.; Huo, E.; Zhao, Y.; Qian, M.; et al. Biochar filled high-density polyethylene composites with excellent properties: Towards maximizing the utilization of agricultural wastes. *Ind. Crops Prod.* **2020**, *146*, 112185. [CrossRef]
43. Andrzejewski, J.; Misra, M.; Mohanty, A.K. Polycarbonate biocomposites reinforced with a hybrid filler system of recycled carbon fiber and biocarbon: Preparation and thermomechanical characterization. *J. Appl. Polym. Sci.* **2018**, *135*, 46449. [CrossRef]

44. Watt, E.; Abdelwahab, M.A.; Snowdon, M.R.; Mohanty, A.K.; Khalil, H.; Misra, M. Hybrid biocomposites from polypropylene, sustainable biocarbon and graphene nanoplatelets. *Sci. Rep.* **2020**, *10*, 10714. [CrossRef] [PubMed]
45. Sałasińska, K.; Borucka, M.; Celiński, M.; Gajek, A.; Zatorski, W.; Mizera, K.; Leszczyńska, M.; Ryszkowska, J. Thermal stability, fire behavior, and fumes emission of polyethylene nanocomposites with halogen-free fire retardants. *Adv. Polym. Technol.* **2017**, *37*, 2394–2410. [CrossRef]
46. Ogunsona, E.O.; Misra, M.; Mohanty, A.K. Sustainable biocomposites from biobased polyamide 6,10 and biocarbon from pyrolyzed miscanthus fibers. *J. Appl. Polym. Sci.* **2017**, *134*, 1–11. [CrossRef]
47. Ogunsona, E.O.; Codou, A.; Misra, M.; Mohanty, A.K. Thermally Stable Pyrolytic Biocarbon as an Effective and Sustainable Reinforcing Filler for Polyamide Bio-composites Fabrication. *J. Polym. Environ.* **2018**, *26*, 3574–3589. [CrossRef]
48. Codou, A.; Misra, M.; Mohanty, A.K. Sustainable biocomposites from Nylon 6 and polypropylene blends and biocarbon—Studies on tailored morphologies and complex composite structures. *Compos. Part A Appl. Sci. Manuf.* **2020**, *129*, 105680. [CrossRef]
49. Andrzejewski, J.; Aniśko, J.; Szulc, J. A comparative study of biocarbon reinforced polyoxymethylene and polyamide: Materials performance and durability. *Compos. Part A Appl. Sci. Manuf.* **2022**, *152*, 106715. [CrossRef]
50. Mazur, K.; Kuciel, S.; Salasinska, K. Mechanical, fire, and smoke behaviour of hybrid composites based on polyamide 6 with basalt/carbon fibres. *J. Compos. Mater.* **2019**, *53*, 3979–3991. [CrossRef]
51. Sałasińska, K.; Cabulis, P.; Kirpluks, M.; Kovalovs, A.; Kozikowski, P.; Barczewski, M.; Celiński, M.; Mizera, K.; Gałecka, M.; Skukis, E.; et al. The Effect of Manufacture Process on Mechanical Properties and Burning Behavior of Epoxy-Based Hybrid Composites. *Materials* **2022**, *15*, 301. [CrossRef]
52. Prabhakar, M.N.; Naga Kumar, C.; Dong Woo, L.; Jung-IL, S. Hybrid approach to improve the flame-retardant and thermal properties of sustainable biocomposites used in outdoor engineering applications. *Compos. Part A Appl. Sci. Manuf.* **2022**, *152*, 106674. [CrossRef]
53. Izwan, S.M.; Sapuan, S.M.; Zuhri, M.Y.M.; Mohamed, A.R. Thermal stability and dynamic mechanical analysis of benzoylation treated sugar palm/kenaf fiber reinforced polypropylene hybrid composites. *Polymers* **2021**, *13*, 2961. [CrossRef]
54. Snowdon, M.R.; Wu, F.; Mohanty, A.K.; Misra, M. Comparative study of the extrinsic properties of poly(lactic acid)-based biocomposites filled with talc: Versus sustainable biocarbon. *RSC Adv.* **2019**, *9*, 6752–6761. [CrossRef]
55. Li, W.; Zhang, L.; Chai, W.; Yin, N.; Semple, K.; Li, L.; Zhang, W.; Dai, C. Enhancement of flame retardancy and mechanical properties of polylactic acid with a biodegradable fire-retardant filler system based on bamboo charcoal. *Polymers* **2021**, *13*, 2167. [CrossRef]
56. Wang, S.; Zhang, L.; Semple, K.; Zhang, M.; Zhang, W.; Dai, C. Development of biodegradable flame-retardant bamboo charcoal composites, part ii: Thermal degradation, gas phase, and elemental analyses. *Polymers* **2020**, *12*, 2238. [CrossRef]
57. Andrzejewski, J.; Gapiński, B.; Islam, A.; Szostak, M. The influence of the hybridization process on the mechanical and thermal properties of polyoxymethylene (POM) composites with the use of a novel sustainable reinforcing system based on biocarbon and basalt fiber (BC/BF). *Materials* **2020**, *13*, 3496. [CrossRef]
58. Kufel, A.; Para, S.; Kuciel, S. Basalt/glass fiber polypropylene hybrid composites: Mechanical properties at different temperatures and under cyclic loading and micromechanical modelling. *Materials* **2021**, *14*, 5574. [CrossRef]
59. Yan, X.; Shen, H.; Yu, L.; Hamada, H. Polypropylene–glass fiber/basalt fiber hybrid composites fabricated by direct fiber feeding injection molding process. *J. Appl. Polym. Sci.* **2017**, *134*, 45472. [CrossRef]
60. Rabe, S.; Sanchez-Olivares, G.; Pérez-Chávez, R.; Schartel, B. Natural keratin and coconut fibres from industrial wastes in flame retarded thermoplastic starch biocomposites. *Materials* **2019**, *12*, 344. [CrossRef]
61. Andrzejewski, J.; Mohanty, A.K.; Misra, M. Development of hybrid composites reinforced with biocarbon/carbon fiber system. The comparative study for PC, ABS and PC/ABS based materials. *Compos. Part B Eng.* **2020**, *200*, 108319. [CrossRef]
62. Chen, Y.; Li, L.; Qian, L. The pyrolysis behaviors of phosphorus-containing organosilicon compound modified ammonium polyphosphate with different phosphorus-containing groups, and their different flame-retardant mechanisms in polyurethane foam. *RSC Adv.* **2018**, *8*, 27470–27480. [CrossRef]
63. Schirp, A.; Schwarz, B. Influence of compounding conditions, treatment of wood particles with fire-retardants and artificial weathering on properties of wood-polymer composites for façade applications. *Eur. J. Wood Wood Prod.* **2021**, *79*, 821–840. [CrossRef]
64. Cayla, A.; Rault, F.; Giraud, S.; Salaün, F.; Sonnier, R.; Dumazert, L. Influence of ammonium polyphosphate/lignin ratio on thermal and fire behavior of biobased thermoplastic: The case of Polyamide 11. *Materials* **2019**, *12*, 1146. [CrossRef]
65. Codou, A.; Anstey, A.; Misra, M.; Mohanty, A.K. Novel compatibilized nylon-based ternary blends with polypropylene and poly(lactic acid): Morphology evolution and rheological behaviour. *RSC Adv.* **2018**, *8*, 15709–15724. [CrossRef]
66. Kandola, B.K.; Pornwannachai, W.; Ebdon, J.R. Flax/pp and flax/pla thermoplastic composites: Influence of fire retardants on the individual components. *Polymers* **2020**, *12*, 2452. [CrossRef]
67. Dobrzyńska-Mizera, M.; Dutkiewicz, M.; Sterzyński, T.; Di Lorenzo, M.L. Isotactic polypropylene modified with sorbitol-based derivative and siloxane-silsesquioxane resin. *Eur. Polym. J.* **2016**, *85*, 62–71. [CrossRef]
68. Barczewski, M.; Dobrzyńska-Mizera, M.; Dudziec, B.; Sterzyński, T. Influence of a sorbitol-based nucleating agent modified with silsesquioxanes on the non-isothermal crystallization of isotactic polypropylene. *J. Appl. Polym. Sci.* **2014**, *131*, 1–9. [CrossRef]

69. Barczewski, M.; Mysiukiewicz, O.; Andrzejewski, J.; Piasecki, A.; Strzemięcka, B.; Adamek, G. The inhibiting effect of basalt powder on crystallization behavior and the structure-property relationship of α-nucleated polypropylene composites. *Polym. Test.* **2021**, *103*, 107372. [CrossRef]
70. Wu, Q.; Cui, X.; Mu, C.; Sun, J.; Gu, X.; Li, H.; Zhang, S. Toward a new approach to synchronously improve the fire performance and toughness of polylactic acid by the incorporation of facilely synthesized ammonium polyphosphate derivatives. *Compos. Part A Appl. Sci. Manuf.* **2021**, *150*, 106595. [CrossRef]
71. Li, D.F.; Zhao, X.; Jia, Y.W.; He, L.; Wang, X.L.; Wang, Y.Z. Dual effect of dynamic vulcanization of biobased unsaturated polyester: Simultaneously enhance the toughness and fire safety of Poly(lactic acid). *Compos. Part B Eng.* **2019**, *175*, 107069. [CrossRef]
72. Li, H.; Huneault, M.A. Effect of nucleation and plasticization on the crystallization of poly(lactic acid). *Polymer* **2007**, *48*, 6855–6866. [CrossRef]
73. Kulinski, Z.; Piorkowska, E.; Gadzinowska, K.; Stasiak, M. Plasticization of poly(L-lactide) with poly(propylene glycol). *Biomacromolecules* **2006**, *7*, 2128–2135. [CrossRef]
74. Martin, O.; Averous, L.; Piorkowska, E.; Kulinski, Z.; Galeski, A.; Masirek, R.; Pluta, M.; Piorkowska, E.; Kulinski, Z.; Piorkowska, E.; et al. Citrate esters as plasticizers for poly(lactic acid). *Polymer* **2015**, *2*, 209–218. [CrossRef]
75. Phelps, J.H.; Abd El-Rahman, A.I.; Kunc, V.; Tucker, C.L. A model for fiber length attrition in injection-molded long-fiber composites. *Compos. Part A Appl. Sci. Manuf.* **2013**, *51*, 11–21. [CrossRef]
76. Lingesh, B.V.; Rudresh, B.M.; Ravi Kumar, B.N.; Reddappa, H.N. Effect of fiber loading on mechanical, physical behavior of thermoplastic blend composites. *Mater. Today Proc.* **2022**, *54*, 245–250. [CrossRef]
77. Evens, T.; Bex, G.J.; Yigit, M.; De Keyzer, J.; Desplentere, F.; Van Bael, A. The influence of mechanical recycling on properties in injection molding of fiber-reinforced polypropylene. *Int. Polym. Process.* **2019**, *34*, 398–407. [CrossRef]
78. Sang, L.; Han, S.; Li, Z.; Yang, X.; Hou, W. Development of short basalt fiber reinforced polylactide composites and their feasible evaluation for 3D printing applications. *Compos. Part B Eng.* **2019**, *164*, 629–639. [CrossRef]
79. Barczewski, M.; Mysiukiewicz, O.; Lewandowski, K.; Nowak, D.; Matykiewicz, D.; Andrzejewski, J.; Skórczewska, K.; Piasecki, A. Effect of Basalt Powder Surface Treatments on Mechanical and Processing Properties of Polylactide-Based Composites. *Materials* **2020**, *13*, 5436. [CrossRef]
80. Andrzejewski, J.; Barczewski, M.; Szostak, M. Injection Molding of Highly Filled Polypropylene-based Biocomposites. Buckwheat Husk and Wood Flour Filler: A Comparison of Agricultural and Wood Industry Waste Utilization. *Polymers* **2019**, *11*, 1881. [CrossRef]
81. Van Dommelen, J.A.W.; Brekelmans, W.A.M.; Baaijens, F.P.T. Multiscale modeling of particle-modified polyethylene. *J. Mater. Sci.* **2003**, *38*, 4393–4405. [CrossRef]
82. Ammar, O.; Bouaziz, Y.; Haddar, N.; Mnif, N. Talc as Reinforcing Filler in Polypropylene Compounds: Effect on Morphology and Mechanical Properties. *Polym. Sci.* **2017**, *3*, 1–7. [CrossRef]
83. Andrzejewski, J.; Nowakowski, M. Development of Toughened Flax Fiber Reinforced Composites. Modification of Poly(lactic acid)/Poly(butylene adipate-co-terephthalate) Blends by Reactive Extrusion Process. *Materials* **2021**, *14*, 1523. [CrossRef]
84. Pastorek, M.; Kovalcik, A. Effects of thermal annealing as polymer processing step on poly(lactic acid). *Mater. Manuf. Process.* **2018**, *33*, 1674–1680. [CrossRef]
85. Nagarajan, V.; Zhang, K.; Misra, M.; Mohanty, A.K. Overcoming the fundamental challenges in improving the impact strength and crystallinity of PLA biocomposites: Influence of nucleating agent and mold temperature. *ACS Appl. Mater. Interfaces* **2015**, *7*, 11203–11214. [CrossRef] [PubMed]
86. Nagarajan, V.; Mohanty, A.K.; Misra, M. Crystallization behavior and morphology of polylactic acid (PLA) with aromatic sulfonate derivative. *J. Appl. Polym. Sci.* **2016**, *133*, 1–11. [CrossRef]
87. Feng, Y.; Ma, P.; Xu, P.; Wang, R.; Dong, W.; Chen, M.; Joziasse, C. The crystallization behavior of poly(lactic acid) with different types of nucleating agents. *Int. J. Biol. Macromol.* **2018**, *106*, 955–962. [CrossRef] [PubMed]
88. Chen, Y.; Wang, W.; Liu, Z.; Yao, Y.; Qian, L. Synthesis of a novel flame retardant containing phosphazene and triazine groups and its enhanced charring effect in poly(lactic acid) resin. *J. Appl. Polym. Sci.* **2017**, *134*, 44660. [CrossRef]
89. Wu, N.; Yu, J.; Lang, W.; Ma, X.; Yang, Y. Flame retardancy and toughness of poly(lactic acid)/GNR/SiAHP composites. *Polymers* **2019**, *11*, 1129. [CrossRef]
90. Salasinska, K.; Mizera, K.; Celiński, M.; Kozikowski, P.; Borucka, M.; Gajek, A. Thermal properties and fire behavior of polyethylene with a mixture of copper phosphate and melamine phosphate as a novel flame retardant. *Fire Saf. J.* **2020**, *115*, 103137. [CrossRef]
91. Tang, C.; Xu, F.X.; Li, G. Combustion performance and thermal stability of basalt fiber-reinforced polypropylene composites. *Polymers* **2019**, *11*, 1826. [CrossRef]
92. Isitman, N.A.; Kaynak, C. Nanoclay and carbon nanotubes as potential synergists of an organophosphorus flame-retardant in poly(methyl methacrylate). *Polym. Degrad. Stab.* **2010**, *95*, 1523–1532. [CrossRef]
93. Dittrich, B.; Wartig, K.A.; Hofmann, D.; Mülhaupt, R.; Schartel, B. Flame retardancy through carbon nanomaterials: Carbon black, multiwall nanotubes, expanded graphite, multi-layer graphene and graphene in polypropylene. *Polym. Degrad. Stab.* **2013**, *98*, 1495–1505. [CrossRef]

Article

Characterization of Thermoplastic Starch Plasticized with Ternary Urea-Polyols Deep Eutectic Solvent with Two Selected Fillers: Microcrystalline Cellulose and Montmorillonite

Magdalena Zdanowicz [1,*] and Kamila Sałasińska [2,3]

[1] Center of Bioimmobilisation and Innovative Packaging Materials, Faculty of Food Sciences and Fisheries, West Pomeranian University of Technology, Szczecin, Janickiego 35, 71-270 Szczecin, Poland
[2] Faculty of Materials Science and Engineering, Warsaw University of Technology, Wołoska 141, 02-507 Warsaw, Poland
[3] Central Institute for Labour Protection—National Research Institute, Department of Chemical, Biological and Aerosol Hazards, Czerniakowska 16, 00-701 Warsaw, Poland
* Correspondence: mzdanowicz@zut.edu.pl

Abstract: The aim of the study was to prepare and characterize composite materials based on thermoplastic starch (TPS)/deep eutectic solvent (DES). Potato starch was plasticized with ternary DES: urea:glycerol:sorbitol and modified with the selected fillers: microcrystalline cellulose and sodium montmorillonite. Films were prepared via twin-screw extrusion and thermocompression of the extrudates. Then, the physicochemical properties of the TPS films were examined. The ternary DES effectively plasticized the polysaccharide leading to a highly amorphous structure of the TPS (confirmed via mechanical tests, DMTA and XRD analyses). An investigation of the behavior in water (swelling and dissolution degree) and water vapor transmission rate of the films was determined. The introduction of the two types of fillers resulted in higher tensile strength and better barrier properties of the composite TPS films. However, montmorillonite addition exhibited a higher impact than microcrystalline cellulose. Moreover, a cone calorimetry analysis of the TPS materials revealed that they showed better fire-retardant properties than TPS plasticized with a conventional plasticizer (glycerol).

Keywords: cone calorimetry; deep eutectic solvents; fillers; fire behavior; microcrystalline cellulose; montmorillonite; thermoplastic starch

1. Introduction

Starch plasticizers are low molecular polar compounds that enable the polysaccharide processing into more amorphous "plastic-like" materials called thermoplastic starch (TPS). The TPS materials can be obtained e.g., via gelatinization of starch in aqueous media casting and evaporation of water or via extrusion [1]. The modification via extrusion is solventless (water, more often in form as starch moisture, acts as a co-plasticizer), does not require water evaporation from casted materials and allows processing the material on a large scale in a short time, which is crucial issue from the industrial point of view. Due to native starch having a glass transition close to its degradation temperature, it needs a special additive to facilitate its modification. Starch plasticizers disrupt H-bonding between polysaccharide chains and form new ones with OH groups of anhydroglucosidic units of the polymer. The most common starch plasticizers are glycerol (G) or other higher polyalcohols (sorbitol, maltitol) [2–4], amides [5–7], sugars [8,9] or carboxylic acids (in a smaller amount, they can also act as crosslinkers) [10]. In the last two decades, interest in ionic liquids (ILs) as polysaccharide processing media rapidly increased [11–13]. Due to the high cost of ILs and often non-green character, deep eutectic solvents are used as more eco-friendly tailorable, alternative media not only for starch treatment (e.g., plasticizing, dissolution) [14–19] but also for polysaccharides processing such as their extraction [20,21], biomass pretreatment (e.g., delignification, fractionization, and separation) [22–24]. DES are mixtures that exhibited phase transition temperatures at much lower temperature that their components. In comparison with ILs, DES are easy-to-prepare media from "greener" (often from natural sources) and cheaper components obtained via mixing [15].

TPS materials are highly hydrophilic, thus some fillers are introduced into polymer matrix not only to increase barrier properties and decrease moisture sensitivity but also to facilitate mechanical properties or add some functionality (i.e., conductivity or antimicrobial properties [25]). However, there are only few works that describe studies of mutual interactions between starch, DES as a plasticizer and a filler [26–30].

In the previous work, we confirmed that DES based on urea, glycerol, and sorbitol (UGS) effectively plasticizes starch via thermocompression [31]. In this work, ternary DES UGS is used for starch processing via extrusion for the first time. Moreover, two selected fillers were added: mineral sodium montmorillonite (M) and organic microcrystalline cellulose (C) to investigate their influence on TPS/DES materials. Fillers were added in two concentrations: 5 and 10 pph on dry starch. Mechanical tests, sorption tests, water vapor transmission rate (WVTR) determination, and dynamical mechanical thermal analysis (DMTA), and X-ray diffractometry (XRD) for TPS and TPS composites were performed. In our previous work we studied the fire behavior of TPS plasticized with ternary DES based on choline chloride:resorcinol:urea with lignin addition [30]. However, there are only few works describing the fire behavior of TPS materials [32–34]. Due to M being known as a fire retardant, a cone calorimetry analysis for TPS with clay addition was performed. The results were compared to starch plasticized with a conventional plasticizer (glycerol).

2. Materials and Methods

2.1. Materials

Potato starch with a moisture content of 15.5 wt% (average molecular weight 4.37×10^7 g/mol; 29.7 wt% amylose content) was supplied by Nowamyl SA (Nowogard, Poland). Urea—U (≥98%) and glycerol—G (99%) were purchased from Chempur (Poland) and sorbitol—S (99%) from Alfa Aesar (Kandel, Germany). As fillers were used one selected organic filler—microcrystalline cellulose—C (particle size 20 μm purchased from Sigma–Aldrich), and sodium montmorillonite—M (Cloisite Na$^+$, Southern Clay Products, Gonzales, TX, USA) as an inorganic filler.

2.2. Eutectic Mixtures Preparation

Eutectic mixtures were prepared as follows: selected components: U, G and S at molar ratio 2:1:1, respectively, were placed in a glass reactor heated up to 95 °C and stirred until the homogenous pellucid liquid was obtained; then the mixture was poured into a glass vial and placed in a vacuum chamber (105 °C, 250 mbars, 1 h) to remove the remaining moisture and then kept in sealed vials. The melting point of the DES UGS according to DSC analysis is 59 °C [31].

2.3. Preparation of TPS Films via Extrusion

Preparation of TPS films was as follows: premixtures of starch, filler and DES were processed after 1 day of storage. The systems were extruded with a laboratory twin-screw co-rotational extruder with an L/D ratio of 40:1 and a screw diameter of 16 mm (PRISM Eurolab Digital, Thermo Electron Co., Waltham, MA, USA). The temperature profile from the feed throat to the nozzle was 80/100/110/120 × 7 °C, and the screw speed was maintained at 150 rpm. After processing, extrudates were pelletized with a granulator (Prism Varicut), obtaining ca. 3 mm pellets. After 14 days of storage at ambient conditions in sealed PE bags, pellets were thermocompressed at 120 °C under 12 tons (153 bars), cooled down under pressure to ca. 85 °C and stored in a climate chamber (25 °C, RH 50%) for 48 h before testing.

2.4. Mechanical Tests

Mechanical tests for TPS films were performed using Instron 5982 (Instron, Norwood, MA, USA, load cell 1 kN) according to ASTM D822-02 standard. The films (thickness 0.50–0.65 mm) were cut into 10 mm wide, 120 mm long stripes. The initial grip separation was 50 mm, and the crosshead speed was set to 10 mm/min. At least eight replicated samples for each system were tested and the mechanical parameters (EB—elongation at break, TS—tensile strength and YM—Young's modulus) were calculated with the Bluehill 3 software.

2.5. DMTA Measurements

The measurements were carried out with a film tension clamp at a frequency of 1 Hz, a heating rate of 3 °C/min, and a temperature range from −90 to 140 °C using Dynamic Mechanical Analyzer Q800 (TA Instruments, New Castle, DE, USA). The analysis was conducted twice for each material.

2.6. Microstructure Analysis

The microstructure of composites was examined using a scanning electron microscope TM3000 (Hitachi, Tokyo, Japan), equipped with a backscattered electron detector (BSE). The samples were fixed to carbon tape and coated with gold using POLARON SC7640 (Quorum Technologies Ltd., Newhaven, UK). Images were taken with an accelerating voltage of 5–15 kV. The magnification of 1000× was used.

2.7. X-ray Diffraction Analysis of TPS Films

The crystallinity of TPS films was analyzed using the XRD technique (X'pert Pro, PANalytical, Almelo, The Netherland, operated at the CuK(alfa) wavelength 1.54 Å). The d-spacing was calculated from the Braggs formula (2 d sinθ = n λ, where λ = 0.154 nm) being the order of reflection, θ the angle of refraction).

2.8. Moisture Sorption, Water Sorption and Swelling Degrees

Samples (three samples for each test) with dimensions of 20 mm × 20 mm were prepared for moisture sorption and swelling tests. The test was performed as in previous work [35]. Samples were placed in a vacuum dryer (250 mbars) at 65 °C/24 h. Mass of dried samples, as well as stored in a climate chamber and kept at 25 °C and 75% RH (for moisture sorption evaluation) or immersed in distilled water for 24 h (for swelling behavior investigation), were determined.

2.9. Forced Flaming Fire Behavior

Burning behavior was evaluated by cone calorimeter examinations conducted on Fire Testing Technology apparatus (East Grinstead, UK,) following the ISO 5660-1 and ISO 5660-2 procedures. The horizontally oriented samples were irradiated at a heat flux of 35 kW/m^2, and spark ignition was employed to ignite the pyrolysis products. An optical system with a silicon photodiode and a helium-neon laser provided a continuous survey of smoke. The burning process during tests was photographed using a digital camera, EOS 400 D, from Canon Inc. (Tokyo, Japan).

2.10. Statistical Analysis

Statistical analysis of the mechanical test data for extruded samples was subjected to one-way analysis of variance, and the significant difference was determined by the significance difference test (*t*-Student's test).

3. Results

3.1. Mechanical Test Results

Table 1 lists mechanical tensile test results for extruded TPS films. It can be seen from the collected data that nonionic ternary DES based on urea, glycerol, and sorbitol led to better mechanical properties than choline chloride/U DES [27] or starch plasticized only with glycerol (TPS-G) [36]. The introduction of S into plasticizing mixtures can increase the tensile strength of the TPS films due to a higher amount of H-bonds formation. TS value is similar to TPS/DES obtained via only thermocompression (5.3 MPa). However, EB is about 50% higher [31]. It can be caused by better distribution of plasticizer in the polymer matrix and more untangled polymer chains formation during extrusion. Comparing TPS film with the composites, both additives affected mechanical properties leading to increased TS and YM and decreased EB. However, adding 5 pph of the filler did not affect the TS significantly ($p < 0.05$). The higher content of the filler caused a higher TS value. In turn, 10 pph of M addition led to slightly higher TS than 10 pph of C.

Table 1. Mechanical test results.

Sample	Tensile Strength [MPa]	Elongation at Break [%]	Young's Modulus [MPa]	Thickness [mm]
TPS	4.6 (±0.26) [d]	148 (±11.6) [a]	112 (±14.3) [d]	0.61 (±0.04) [a]
TPS/5M	4.7 (±0.23) [d]	122 (±7.2) [b]	114 (±12.4) [d]	0.55 (±0.00) [b]
TPS/10M	6.2 (±0.30) [a]	99 (±14.0) [d]	232 (±15.2) [a]	0.52 (±0.04) [d]
TPS/5C	5.5 (±0.33) [c]	124 (±10.0) [b]	192 (±9.9) [c]	0.57 (±0.04) [c]
TPS/10C	5.9 (±0.32) [b]	116 (±6.8) [c]	211 (±16.4) [b]	0.61 (±0.04) [a]

a–d—averages marked with the same letters do not differ significantly from each other for $p < 0.05$.

Based on the literature results for TPS composites, the increase in tensile strength depends on the plasticizer as well as the type of filler. For example, for TPS composites plasticized with glycerol where unmodified M was added, an increase in TS was from 6.0 to 7.6 MPa for 10% of M addition [37] or from 6.9 to 7.3 MPa for 8% addition of M [38] but when M is activated with plasticizer the increase in TS is more pronounced, due to intercalation of M platelets with the modifier [7,39,40]. Interestingly, adding the filler to the TPS matrix at high content (5 and 10 pph) did not affect it. A more discernible influence of the additive on TPS/DES was registered for our previous work, where TPS/ternary DES (choline chloride:resorcinol:urea) was prepared with dissolved lignin presence [30]. It could be caused by better distribution of lignin that was in dissolved form.

3.2. Dynamical Mechanical Thermal Analysis (DMTA) Results

In this work, DMTA is represented with tan delta and storage modulus curves presented in Figure 1. The TPS films revealed two main relaxation regions. The first with maxima at a range from −18 to −12 °C is related to secondary β-relaxation (T_β) of the domain rich in plasticizer's molecules, and the second α-relaxation (T_α) with higher intensity (maxima at range 36–63 °C) and is assigned to a polysaccharide-rich domain. Similar results were also observed for TPS/UGS obtained via thermocompression [31]. The β-relaxation for starch plasticized with ternary DES occurred at a much higher temperature than glycerol (−52 °C) and even quite higher than DES based on sorbitol and choline chloride (−21 °C) [41], indicating that sorbitol with urea and glycerol formed stronger H-bonding than with choline chloride. It can result from U presence due to the amine group can form stronger hydrogen bonds than choline salt. Films with the fillers exhibited much higher temperatures of α-relaxations. It is caused by a more rigid structure with restricted mobility of polymer chains with the fillers [2].

In the case of E' curves, there is a rapid drop of the parameter at ca. 20 °C for TPS without additive and at a range of 20–28 °C depending on the filler's type and content. This E' drop is assigned to a region of glass transition of the TPS. In the case of composite films E' curve at higher temperatures is under TPS without filler. This behavior can be caused by clay's platelets that slip between polymer/plasticizer matrix [27]. Moreover, applied plasticizer as a eutectic mixture may facilitate this behavior. Low E' values of composite films at higher temperatures indicated that filler presence does not affect the processing of

TPS/DES. The impact of the solid fillers presented in this work is more noticeable than for TPS/DES with lignin addition where the additive was introduced into the polymer matrix in dissolved form in the DES [30].

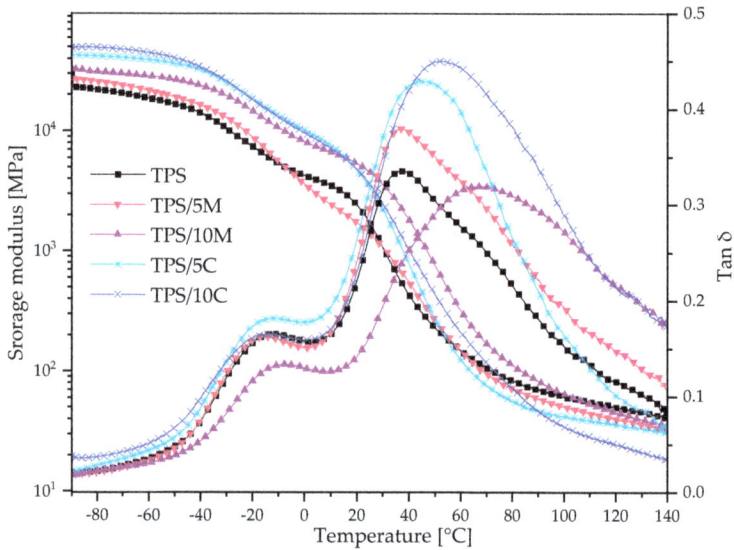

Figure 1. Storage modulus and Tan δ curves of TPS films.

3.3. Behavior of TPS Films in Moisture, Water, and Determination WVTR

The swelling degrees for extruded TPS films are in the range of 71.1–87.4, and these values are lower than TPS/UGS obtained via thermocompression (221%) (Table 2) [31]. Introducing the fillers into the polysaccharide matrix lowered swelling and dissolution in water, moisture sorption as well as WVTR values. The higher the filler content, the lower values of the parameters. Comparing the type of fillers, the aluminosilicate addition led to better barrier properties than microcrystalline cellulose.

Table 2. Swelling and dissolution degrees of TPS films distilled water, moisture adsorption degree and WVTR at 75% for 24 h.

Sample	Swelling Degree [%]	Dissolution Degree [%]	Moisture Sorption Degree at RH 75% [%]	WVTR$_{RH75\%}$ [g/m$^2 \cdot$24 h]
TPS	87.4 (±0.9)	22.5 (±0.07)	24.0 (±0.07)	381 (±21.0)
TPS/M5	84.1 (±0.8)	21.6 (±1.02)	24.0 (±0.04)	352 (±8.2)
TPS/M10	71.1 (±0.0)	19.9 (±1.21)	23.6 (±0.40)	345 (±0.3)
TPS/C5	79.2 (±1.1)	20.9 (±1.15)	23.5 (±0.14)	369 (±3.2)
TPS/C10	77.7 (±9.9)	20.4 (±0.07)	23.0 (±0.08)	358 (±4.2)

3.4. SEM Results

The morphology of the composites with the highest fillers amount is demonstrated in Figure 2. As can be seen, the material appearance is relatively smooth and homogenous, indicating a highly amorphous structure without swollen starch granules. Observation showed also relatively uniformly distributed particles; however, agglomerates, especially in the case of inorganic filler, were observed. Moreover, C and M particles significantly differed in size and shape. It can be noticed, especially for C, that the filler is well embedded in the matrix. In turn, a lack of pore at the interface between fillers and polysaccharide matrix, indicating good adhesion. This may be related to the improvement in the mechanical properties of composite materials.

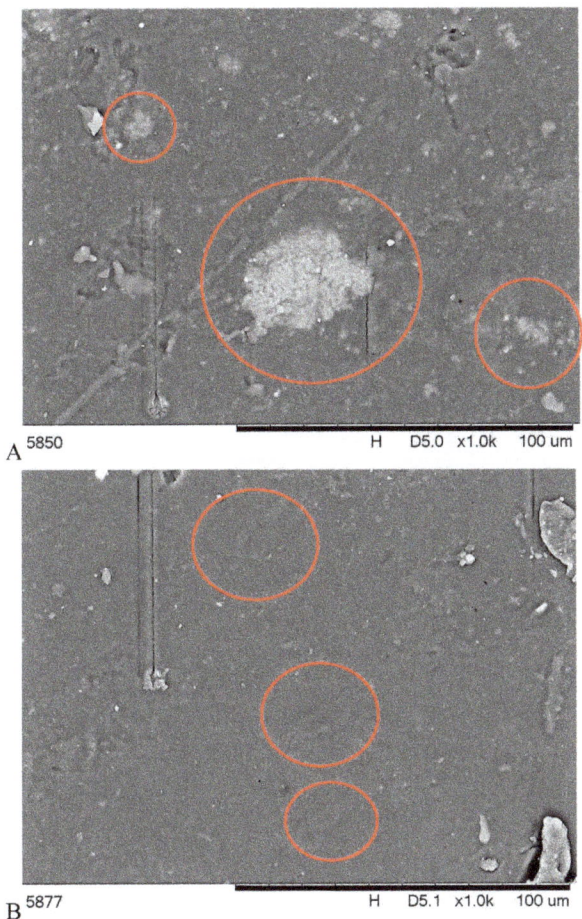

Figure 2. SEM micrographs for: (**A**)—TPS/10M and (**B**)—TPS/10C), filler particles are marked in red circles.

3.5. X-ray Diffractometry Results

Figure 3 shows diffractograms of native granular starch, The TPS film without filler, the pristine clay and the composite TPS films. The diffractogram of native potato starch granules reveals peaks at 5.4, 17.1, 19.5, 22.5, 24.0, and 26.0° that are characteristic of B-type semi-crystallinity of the polysaccharide. After extrusion with UGS, starch underwent plasticization effectively, which is confirmed with a flat diffractogram indicating the highly amorphous structure of TPS. For pristine M peak with high intensity at 7.0° is assigned to the interlayer spacing of clay platelets (d_{001} 1.26 nm). It can be seen that for the composite materials, there is a shift towards a lower contact value at 4.9° (1.80 nm) which indicates the intercalation of clay platelets in the polymeric matrix. This can result from better mechanical properties for TPS with M than with C (Table 1). The increase of the clay platelets' distance can be caused by shearing forces during extrusion, facilitating the intercalation of processing starch with the presence of plasticizing agent causing swelling and partial untangling of polysaccharide chains between the filler platelets. For composite TPS, peaks appeared at 9.8, 14.9, 19.7, 24.8, and 30.1°, assigned to V-type crystallinity. The filler presence can induce the formation of post-processing crystallinity in small helical regions of starch.

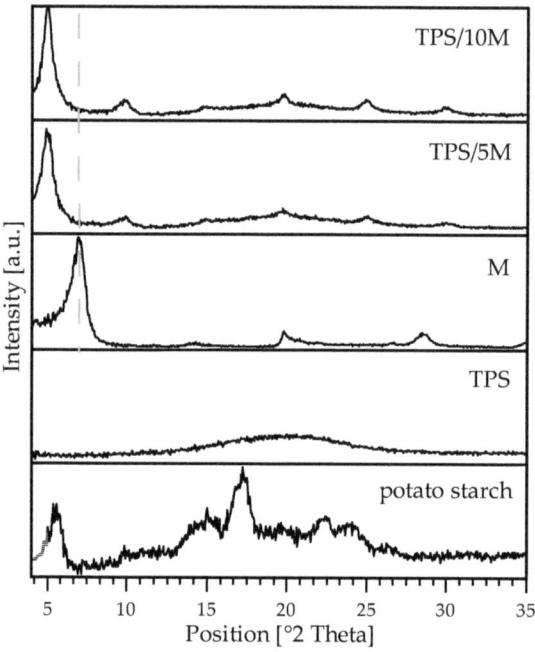

Figure 3. XRD patterns of native granular potato starch, TPS, sodium montmorillonite (M), and composite TPS films with 5 pph (TPS/5M) and 10 pph of M (TPS/10M).

3.6. Cone Calorimetry Results

Cone calorimetry is widely employed to assess the fire hazards of polymeric materials due to its capability to simulate real fire hazards [41]. According to the literature aluminosilicates are known as fillers with good fire retardant properties; therefore, the sample with montmorillonite addition (TPS/10M) was chosen for the investigation. Thermoplastic starch modified with a conventional plasticizer (TPS-G) was included in the test for comparison. The present study continues our work on the investigation of the fire retardant of thermoplastic starch with novel plasticizers based on DESs and its composites [30]. Figure 4 shows the heat release rate (HRR) and total smoke release (TSR) curves of TPS plasticized with ternary DES: UGS with 10 pph of sodium montmorillonite and TPS-G, while the detailed data are listed in Table 3.

Table 3. The cone calorimeter results of TPS/10M and TPS-G.

Sample	TTI [s]	PHRR [kW/m^2]	MARHE [kW/m^2]	PHRR/ t PHRR, [kW/m^{2}s^1]	THR [MJ/m^2]	EHC [MJ/kg]	MLR [g/(s·m^2)]
TPS/10M	79	132	71	1.6	44	12	5.3
TPS-G	54	317	167	4.9	60	15	9.2

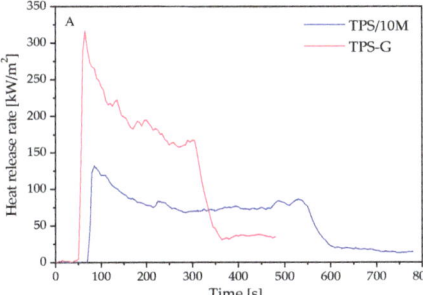 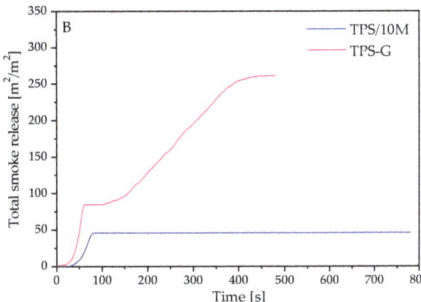

Figure 4. Representative curves of: (**A**)—heat release rate, (**B**)—total smoke release of investigated TPS materials.

During the cone calorimeter test, by oxygen consumption, the heat release rate (HRR) is specified as a crucial parameter to evaluate the intensity of fires [42]. As shown in Figure 4A, TPS-G exhibited a time to ignition (TTI) of 54 s, and a peak heat release rate (PHRR) of 317 kW/m^2. The HRR curve presents a peak at the beginning of the burning and then gradually drops, which is characteristic of thick and able-to-char samples. The char-forming ability of thermoplastic starch, as one of the flame retardant mechanisms, was described in our previous work [30]. In turn, the curve of TPS/10M consists of two peaks at the beginning and end of burning, from which the first one yielded the PHRR. This curve course is typical for thick charring samples with additional peaks at the end of burning, and the second peak may be induced by cracking char or a growth in the effective pyrolysis [42]. Importantly, the PHRR of TPS/10M is much smaller compared to TPS-G and equal to 132 kW/m^2, while TTI is as much as 79 s. Since the maximum average rate of heat emission (MARHE) is extrapolated from the maximum HRR, the value obtained for TPS/10M was also lower (reduction by 57%). The MARHE parameter is one of the essential indicators enabling flame spread evaluation. Another parameter used to assess the influence of modification on the fire growth rate is PHRR/t $_{PHRR}$ [43], and the value obtained for the TPS/10M was three times lower compared to TPS-G. The delayed TTI, as well as reduced PHRR, MARHE and PHRR/t $_{PHRR}$, demonstrate the improvement in the flame retardancy performance of TPS/10M.

Total heat release (THR) corresponds to the total heat output up to the defined point, which in this case, suits about 120 s after flaming burning has been completed. THR demonstrated a similar trend to PHRR, reaching 44 MJ/m^2 and 60 MJ/m^2 for TPS/10M and TPS-G, respectively. The lower THR observed for TPS/10M may follow from incomplete combustion caused by the formation of more char or reduced combustion efficiency [44]. Since the mass loss rate (MLR) and the effective heat of combustion (EHC) were also lower, both activities are assumable. Some charring polymers, which consist of carbon and heteroatoms, may reduce the EHC via accompanying fuel dilution [45]. The polymer containing a significant amount of carbon was a resource for the formation of char, while the sugar units degraded to non-flammable volatiles (H_2O, CO_2), favoring the development of its cellular structure. This effect was also supported by the presence of nitrogen from urea. The nitrogen compounds can promote the generation of a layer of gaseous products, which protects the material from the heat of the flaming zone, and volatile products can also act as radical interceptors [46]. Furthermore, nitrogen compounds form acids, which can catalyze the dehydration reaction of organic material, leading to char formation. The beneficial effect of montmorillonite on developing a durable and effective barrier preventing the exchange of mass and energy has been described in the literature [47,48]. Moreover, the influence of the lowered polymer share, as one of the main combustible components, in favor of an inorganic filler with higher thermal stability could not be excluded. Both carbonaceous char forming during the burning and inert residue from inorganic fillers reduce fuel release and, consequently, THR [42]. Compared to calcium montmorillonite, sodium montmorillonite used in the study is characterized by a greater ability to absorb water, which may be released during burning [49].

Excluding above mentioned key parameters, materials are assessed by the smoke generated, which can dramatically reduce the visibility during a fire, thus making it more difficult to escape. Smoke mainly consists of unstable carbon particles and cyclic compounds [41].

Total smoke release (TSR) is a crucial parameter to evaluate the smoke emission during the cone calorimetry test. Figure 4B demonstrates that the replacement of part of the glycerol by urea and sorbitol caused a decrease in TSR. In the case of ternary plasticizer UGS, the maximum value of 46 m^2/m^2 was reached after several dozen seconds from the start of the test and remained at this level until the end of the measurement, while for G, the TSR was increasing, reaching a maximum of 261 m^2/m^2 in 470 s. Such reduction is mainly attributed to the capability of bonding unstable carbon particles in residue, as evidenced by the increased char yield, preventing them from escaping to the flame zone, as well as diluting by non-combustible gases. The photographic report made for TPS/10M during the cone calorimetry test, presented in Figure 5, demonstrates the creation of char on the surface of the polymer. A horizontally oriented sample subjected to the stream of heat flux underwent thermal decomposition and ignition. The employed modification resulted in forming a char layer, which was swelled and congealed, while growth char caused a reduction in the burning process intensity. At the end of the test, an increase in intensity, caused by a growth in the pyrolysis products emission due to a slight displacement of the sample, was observed. The photographs of the residue present a few centimetres high swollen structure and are consistent with the conducted analysis.

Figure 5. *Cont.*

Figure 5. Burning behavior of TPS/10M during cone calorimetry tests.

4. Conclusions

In the presented work mutual interactions between starch matrix, ternary deep eutectic solvent made with urea, glycerol and sorbitol (at molar ratio 2:1:1) used as plasticizer and two types of fillers: microcrystalline cellulose (as organic filler) and sodium montmorillonite (as inorganic filler) were investigated. DES effectively plasticized the polysaccharide leading to its highly amorphous structure. The introduction of the fillers improved mechanical properties, and decreased sorption degrees as well as WVTR values. As DMTA results indicated, the addition of the fillers did not affect the processability of the composites. XRD analysis revealed that the d-spacing of MMT-Na platelets was slightly enlarged in the polymer matrix after extrusion. SEM analysis confirmed that fillers were well embedded in TPS matrix. The cone calorimetry analysis of the TPS materials revealed that composite material plasticized with DES with the clay presence showed better fire retardant properties than TPS plasticized with a conventional plasticizer (glycerol).

Author Contributions: Conceptualization, formal analysis, investigation, writing—original draft preparation, methodology, resources, M.Z., writing, formal analysis (cone calorimetry), K.S. All authors have read and agreed to the published version of the manuscript.

Funding: This research was funded by The National Centre for Research and Development (Warsaw, Poland), [grant number: TANGO-V-A/0047/2021-00]. Project's title: "Development of multifunctional biodegradable polysaccharides-based materials intended for plants cultivation" and partially financed by National Science Centre, Poland [SONATA 9 grant number: 2015/17/D/ST8/01290].

Institutional Review Board Statement: Not applicable.

Informed Consent Statement: Not applicable.

Data Availability Statement: The data presented in this study are available on request from the corresponding author.

Acknowledgments: The authors would like to thank to Marta Rokosa for helping in statistical analysis and Gohar Khachatryan (University of Agriculture in Krakow) for determination of starch average molecular weight.

Conflicts of Interest: The authors declare no conflict of interest.

References

1. Nafchi, A.M.; Moradpour, M.; Saeidi, M.; Alias, A.K. Thermoplastic starches: Properties, challenges and prospects. *Starch* **2013**, *65*, 61–72. [CrossRef]
2. Rico, M.; Rodriguez-Llamazares, S.; Barral, L.; Bouza, R.; Montero, B. Processing and characterization of polyols plasticized-starch reinforced with microcrystalline cellulose. *Carbohydr. Polym.* **2016**, *149*, 83–93. [CrossRef] [PubMed]
3. Chandhury, A.L.; Miler, M.; Torley, P.J.; Sopade, P.A.; Halley, P.J. Amylose content and chemical modification effects on the extrusion of thermoplastic starch from maize. *Carbohydr. Polym.* **2008**, *74*, 907–913.
4. Mikus, P.-Y.; Alix, S.; Soulestin, J.; Lacrampe, M.; Krawczak, P.; Coqueret, X.; Dole, P. Deformation mechanisms of plasticized starch materials. *Carbohydr. Polym.* **2014**, *114*, 450–457. [CrossRef] [PubMed]
5. Ma, X.; Yu, J. The plasticizers containing amide groups for thermoplastic starch. *Carbohydr. Polym.* **2004**, *57*, 197–203. [CrossRef]
6. Huang, M.; Yu, J.; Ma, X. High mechanical performance MMT-urea and formamide-plasticized thermoplastic cornstarch thermoplastic cornstarch biodegradable nanocomposites. *Carbohydr. Polym.* **2006**, *63*, 393–399. [CrossRef]

7. Rychter, P.; Kot, M.; Bajer, K.; Rogacz, D.; Šišková, A.; Kapuśniak, J. Utilization of starch films plasticized with urea as fertilizer for improvement of plant growth. *Carbohydr. Polym.* **2016**, *137*, 127–138. [CrossRef]
8. Zdanowicz, M.; Staciwa, P.; Spychaj, T. Low transition temperature mixtures (LTTM) containing sugars as potato starch plasticizers. *Starch* **2019**, *71*, 1900004. [CrossRef]
9. Gao W Liu, P.; Li, X.; Qiu, L.; Hou, H.; Cui, B. The co-plasticization effects of glycerol and small molecular sugars on starch-based nanocomposite films prepared by extrusion blowing. *Int. J. Biol. Macromol.* **2019**, *133*, 1175–1181. [CrossRef]
10. Jiugao, J.; Ning, W.; Ma, X. The effects of citric acid on the properties of thermoplastic starch plasticized by glycerol. *Starch* **2005**, *57*, 494–504. [CrossRef]
11. Ptak, S.; Zarski, A.; Kapuśniak, J. The Importance of Ionic Liquids in the Modification of Starch and Processing of Starch-Based Materials. *Materials* **2020**, *13*, 4479. [CrossRef] [PubMed]
12. Zdanowicz, M.; Spychaj, T. Ionic liquids as starch plasticizers or solvents. *Polimery* **2011**, *56*, 861–864. [CrossRef]
13. Xie, F.; Flanagan, B.M.; Li, M.; Sangwan, P.; Truss, R.W.; Halley, P.J.; Strounina, E.; Whittaker, A.K.; Gidley, M.; Dean, K.M.; et al. Characteristics of starch-based films plasticized by glicerol and by the ionic liquid 1-ethyl-3-methylimidazolium acetate: A comparative study. *Carbohydr. Polym.* **2014**, *111*, 841–848. [CrossRef] [PubMed]
14. Montilla-Buitrago, C.E.; Gómez-López, R.A.; Solanilla-Duque, J.F.; Héctor, S.; Serna-Cock, L.; Villada-Castillo, H.S. Effect of plasticizers on properties, retrogradation, and processing of extrusion-obtained thermoplastic starch: A review. *Starch* **2021**, *79*, 210060. [CrossRef]
15. Zdanowicz, M.; Wilpiszewska, K.; Spychaj, T. Deep eutectic solvents for polysaccharides processing: A review. *Carbohydr. Polym.* **2018**, *200*, 361–380. [CrossRef]
16. Yu, J.; Liu, X.; Xu, S.; Shao, P.; Li, J.; Chen, Z.; Wang, X.; Lin, Y.; Renard, C.M.G.C. Advances in Green Solvents for Production of Polysaccharide-based Packaging Films: Insights of Ionic Liquids and Deep Eutectic Solvents. *Comp. Rev. Food Sci. Food Saf.* **2022**. [CrossRef]
17. Chen, Y.-L.; Zhang, X.; You, T.-T.; Xu, F. Deep Eutectic Solvents (DESs) for Cellulose Dissolution: A Mini-Review. *Cellulose* **2019**, *26*, 205–213. [CrossRef]
18. Lončarić, M.; Jakobek, L.; Molnar, M. Deep Eutectic Solvents in the Production of Biopolymer-Based Materials. *Croat. Chem. Acta* **2021**, *94*, P1–P8. [CrossRef]
19. Zdanowicz, M. Influence of Urea Content in Deep Eutectic Solvents on Thermoplastic Starch Films' Properties. *Appl. Sci.* **2023**, *13*, 1383. [CrossRef]
20. Morais, E.S.; Lopes, A.M.d.C.; Freire, M.G.; Freire, C.S.R.; Coutinho, J.A.P.; Silvestre, A.J.D. Use of Ionic Liquids and Deep Eutectic Solvents in Polysaccharides Dissolution and Extraction Processes towards Sustainable Biomass Valorization. *Molecules* **2020**, *25*, 3652. [CrossRef]
21. Saini, A.; Kumar, A.; Panesar, P.S.; Thakur, A. Potential of Deep Eutectic Solvents in the Extraction of Value-added Compounds from Agro-industrial By-products. *Appl. Food Res.* **2022**, *2*, 100211. [CrossRef]
22. Bjelić, A.; Hočevar, B.; Grilc, M.; Novak, U.; Likozar, B. A Review of Sustainable Lignocellulose Biorefining Applying (Natural) Deep Eutectic Solvents (DESs) for Separations, Catalysis and Enzymatic Biotransformation Processes. *Rev. Chem. Eng.* **2022**, *38*, 243–272. [CrossRef]
23. Mamilla, J.L.K.; Novak, U.; Grilc, M.; Likozar, B. Natural Deep Eutectic Solvents (DES) for Fractionation of Waste Lignocellulosic Biomass and Its Cascade Conversion to Value-Added Bio-Based Chemicals. *Biomass Bioener.* **2019**, *120*, 417–425. [CrossRef]
24. Olugbemide, A.D.; Oberlintner, A.; Novak, U.; Likozar, B. Lignocellulosic Corn Stover Biomass Pre-Treatment by Deep Eutectic Solvents (DES) for Biomethane Production Process by Bioresource Anaerobic Digestion. *Sustainability* **2021**, *13*, 10504. [CrossRef]
25. Sousa Martinez de Freitas, A.; Bernardo de Silva, A.P.; Stieven Montagna, L.; Araujo Nogueira, I.; Carvahlo, N.K.; Siqueira de Feria, V.; Lemes, A.P. Thermoplastic starch nanocomposites: Sources, production and applications—A review. *J. Biomat. Sci. Polym. Ed.* **2022**, *33*, 900–945. [CrossRef] [PubMed]
26. Zdanowicz, M.; Johansson, C. Impact of additives on mechanical and barrier properties of starch-based films plasticized with deep eutectic solvents. *Starch Stärke* **2017**, *69*, 1700030. [CrossRef]
27. Adamus, J.; Spychaj, T.; Zdanowicz, M.; Jędrzejewski, R. Thermoplastic starch with deep eutectic solvents and montmorillonite as a base for composite materials. *Ind. Crop. Prod.* **2018**, *123*, 278–284. [CrossRef]
28. Skowrońska, D.; Wilpiszewska, K. The Effect of Montmorillonites on the Physicochemical Properties of Potato Starch Films Plasticized with Deep Eutectic Solvent. *Polymers* **2022**, *23*, 16008. [CrossRef]
29. Grylewicz, A.; Spychaj, T.; Zdanowicz, M. Thermoplastic starch/wood biocomposites processed with deep eutectic solvents. *Comp. Part A* **2019**, *121*, 517–524. [CrossRef]
30. Zdanowicz, M.; Sałasińska, K.; Lewandowski, K.; Skórczewska, K. Thermoplastic Starch/Ternary Deep Eutectic Solvent/Lignin Materials: Study of Physicochemical Properties and Fire Behavior. *ACS Sus. Chem. Eng.* **2022**, *10*, 4579–4587. [CrossRef]
31. Zdanowicz, M. Deep eutectic solvents based on urea, polyols and sugars for starch treatment. *Int. J. Biol. Macromol.* **2021**, *179*, 387–939. [CrossRef] [PubMed]
32. Wang, D.; Wang, Y.; Li, T.; Zhang, S.; Ma, P.; Shi, D.; Chen, M.; Dong, W. A bio-based flame-retardant starch based on phytic acid. *ACS Sus. Chem. Eng.* **2020**, *8*, 10265–10274. [CrossRef]
33. Prabhakar, M.N.; Shah, A.R.; Song, J.-I. Improved Flame-Retardant and Tensile Properties of Thermoplastic Starch/Flax Fabric Green Composites. *Carbohydr. Polym.* **2017**, *168*, 201–211. [CrossRef]

34. Bocz, K.; Szolnoki, B.; Władyka-Przybylak, M.; Bujnowicz, K.; Harakály, G.; Bodzay, B.; Zimonyi, E.; Toldy, A.; Marosi, G. Flame Retardancy of Biocomposites Based on Thermoplastic Starch. *Polimery* **2013**, *58*, 385–394. [CrossRef]
35. Zdanowicz, M.; Jędrzejewski, R.; Pilawka, R. Deep eutectic solvents as simultaneous plasticizing and crosslinking agents for starch. *Int. J. Biol. Macromol.* **2019**, *129*, 1040–1046. [CrossRef] [PubMed]
36. Zdanowicz, M.; Staciwa, P.; Jędrzejewski, R.; Spychaj, T. Sugar Alcohol-Based Deep Eutectic Solvents as Potato Starch Plasticizers. *Polymers* **2019**, *11*, 1385. [CrossRef] [PubMed]
37. Zhang, Y.; Liu, Q.; Hrymak, A.; Han, J.H. Characterization of extruded thermoplastic starch reinforced by montmorillonite nanoclay. *J. Polym. Environ.* **2013**, *21*, 122–131. [CrossRef]
38. Ma, X.; Yu, J.; Wang, N. Production of thermoplastic starch/ MMT-sorbitol nanocomposites by dual-melt extrusion processing. *Macromol. Mat. Eng.* **2007**, *292*, 723–728. [CrossRef]
39. Wang, X.; Zhang, X.; Liu, H.; Wang, N. Impact of pre-processing of montmorillonite on the properties of melt-extruded thermoplastic starch/montmorillonite nanocomposites. *Starch* **2009**, *61*, 489–494. [CrossRef]
40. Huang, M.; Yu, J. Structure and Properties of Thermoplastic Corn Starch/Montmorillonite Biodegradable Composites. *J. Appl. Polym. Sci.* **2006**, *99*, 170–176. [CrossRef]
41. Liu, L.; Qian, M.; Song, P.; Huang, G.; Yu, Y.; Fu, S. Fabrication of Green Lignin-Based Flame Retardants for Enhancing the Thermal and Fire Retardancy Properties of Polypropylene/Wood Composites. *ACS Sus. Chem. Eng.* **2016**, *4*, 2422–2431. [CrossRef]
42. Schartel, B.; Hull, T.R. Development of fire-retarded materials—Interpretation of cone calorimeter data. *Fire Mater.* **2007**, *311*, 327–354. [CrossRef]
43. Schartel, B.; Wilkie, C.A.; Camino, G. Recommendations on the scientific approach to polymer flame retardancy: Part 2—Concepts. *J. Fire Sci.* **2016**, *35*, 3–20. [CrossRef]
44. Günther, M.; Levchik, S.V.; Schartel, B. Bubbles and collapses: Fire phenomena of flame-retarded flexible polyurethane foams. *Polym. Adv. Technol.* **2020**, *31*, 2185–2198. [CrossRef]
45. Schartel, B. Phosphorus-based Flame Retardancy Mechanisms—Old Hat or a Starting Point for Future Development? *Materials* **2010**, *3*, 4710–4745. [CrossRef]
46. Troitzsch, J. *International Plastics Flammability Handbook*; Hanser Publisher: New York, NY, USA, 1990; Chapter 5.
47. Yang, F.; Nelson, G.L. Combination effect of nanoparticles with flame retardants on the flammability of nanocomposites. *Polym. Degr. Stab.* **2011**, *96*, 270–276. [CrossRef]
48. Bourbigot, S.; Vanderhart, D.L.; Gilman JWBellayer, S.; Stretz, H.; Paul, R. Solid state NMR characterization and flammability of styrene–acrylonitrile copolymer montmorillonite nanocomposite. *Polymer* **2004**, *45*, 7627–7638. [CrossRef]
49. Önal, M. Physicochemical properties of bentonite: An overview. *Commun. Fac. Sci. Univ. Ank. Ser. B* **2006**, *52*, 7–21. [CrossRef]

Disclaimer/Publisher's Note: The statements, opinions and data contained in all publications are solely those of the individual author(s) and contributor(s) and not of MDPI and/or the editor(s). MDPI and/or the editor(s) disclaim responsibility for any injury to people or property resulting from any ideas, methods, instructions or products referred to in the content.

Article

Comprehensive Analysis of the Influence of Expanded Vermiculite on the Foaming Process and Selected Properties of Composite Rigid Polyurethane Foams

Mateusz Barczewski [1], Maria Kurańska [2,*], Kamila Sałasińska [3,4], Joanna Aniśko [1], Joanna Szulc [5], Izabela Szafraniak-Wiza [6], Aleksander Prociak [2], Krzysztof Polaczek [2], Katarzyna Uram [2], Karolina Surmacz [2] and Adam Piasecki [6]

[1] Institute of Materials Technology, Poznan University of Technology, Piotrowo 3, 61-138 Poznan, Poland
[2] Department of Chemistry and Technology of Polymers, Cracow University of Technology, Warszawska 24, 31-155 Cracow, Poland
[3] Faculty of Materials Science and Engineering, Warsaw University of Technology, Wołoska 141, 02-507 Warsaw, Poland
[4] Central Institute for Labour Protection—National Research Institute, Department of Chemical, Biological and Aerosol Hazards, 00-701 Warsaw, Poland
[5] Faculty of Chemical Technology and Engineering, Bydgoszcz University of Technology, Seminaryjna 3, 85-326 Bydgoszcz, Poland
[6] Institute of Materials Engineering, Faculty of Materials Engineering and Technical Physics, Poznan University of Technology, Piotrowo 3, 61-138 Poznan, Poland
* Correspondence: maria.kuranska@pk.edu.pl

Abstract: This article presents the results of research on obtaining new polyurethane (PUR) foams modified with thermally expanded vermiculite. The filler was added in amount of 3 wt.% up to 15 wt.%. The additionally applied procedure of immersion the non-organic filler in H_2O_2 was performed to increase the exfoliation effect of thermally treated mineral and additional oxidation the surfaces. The effect of fillers on foaming process, cell structure, thermal insulation, apparent density, compressive strength, thermal properties, and flammability are assessed. The foaming process of PUR foams modified with vermiculite was comparable for all systems, regardless of the content of the filler. A slight increase in reactivity was observed, confirmed by a faster decrease in dielectric polarization for the system with modified vermiculite by H_2O_2. The modification of the reference system with the vermiculite increased the content of closed cells from 76% to 91% for the foams with the highest vermiculite content. Coefficient of thermal conductivity of reference foam and foams modified with vermiculite was in the range 24–26 mW/mK. The use of vermiculite up to 15 wt.% did not influence significantly on mechanical properties and flammability, which from an economic point of view is important because it is possible to reduce the cost of materials by introducing a cheap filler without deteriorating their properties.

Keywords: polyurethane foams; PUR; vermiculite; rigid foams; thermal insulation

1. Introduction

Rigid polyurethane (PUR) foams are mainly used as high-performance thermal insulation of buildings, refrigerators and transmission pipes [1]. Rigid PUR foams with apparent densities of about 30 to 200 kg/m^3 withstand temperatures between −196 °C and 130 °C [2]. The thermal conductivity value of closed cells PUR foams ranges from 0.02 W/m·K to 0.03 W/m·K (influenced by the gas filling the foam cells [3]), which is a lower value compared to other commonly used thermal insulation materials, such as mineral wool (0.037–0.055 W/m·K), cellulose (0.040–0.065 W/m·K), expanded polystyrene (0.03–0.04 W/m·K), and extruded polystyrene (0.034–0.044 W/m·K) [1]. Currently, one of the main trends in research on rigid foams concerns the use of filler to modify foam properties, such as increasing mechanical strength or reducing thermal conductivity, flammability, or apparent density [1]. A promising filler that could find application in polyurethanes, including rigid and elastomeric solid materials and foams, is vermiculite (VMT) [4–7].

VMT is a commonly used layered silicate characterized by a single-layer structure in a 2:1 system, which consists of two layers of silicon oxygen tetrahedron sandwiched by layers of magnesium oxygen octahedron. The single layer is approximately 1 nm thick, while the interlayer spacing is usually around 1.4 nm [8]. As a result of the partial replacement of the silicon-oxygen tetrahedron sheet by aluminum, vermiculite has a negative charge and cations are found in the structure of the mineral, e.g., Ca^{2+}, K^+, or Mg^{2+}, which maintain electrical balance in the interlayer [8,9]. This mineral is found at various latitudes and is mainly mined in South Africa [10], China [11], and Brazil [12]. Owing to its structure, this material offers a possibility of intensive volume growth after high-temperature heating, resulting in a product in the form of expanded vermiculite. Thermal expansion occurs perpendicularly to VMT sheets, and the product obtained after a thermal modification has a concertina-like, highly porous structure [8,13]. Thanks to the separation of sheets of VMT, including expanded VMT, this material is widely used in construction, as acoustic and thermal insulation, agriculture, and as a filler for polymeric materials [7,8,13–15]. The use of plate-shaped fillers allows for an increase in the barrier properties of polymers modified with them; this effect can be used both to increase the effectiveness of flame-retardant systems [5,16] and to reduce oxygen diffusion into closed cells of PUR foams, slowing down insulation aging [4]. To obtain the monosheets of the filler from vermiculite, it is necessary to break its packet and complex structure. This process may occur in thermal treatment, leading to the formation of expanded vermiculite, or exfoliation by organofunctionalization [8,14]. Most studies relate to the implementation of VMT organofunctionalization, which results from the possibility of giving new functional features and obtaining a controlled structure of the filler. Expanded vermiculite after mechanical processing (grinding and sieving) is characterized by a much lower price and the possibility of easy process implementation in industrial conditions. It should be mentioned that compared with expanded VMT, VMT nanosheets with larger specific surface areas and more reactive sites provide opportunities for exfoliated VMT to serve also as a nanocomposite material for nanofluidic channels and intelligent responses [8].

Among the published works on the modification of PUR with different varieties of vermiculite, most of the research has been aimed at improving the dispersion of the filler, increasing the mechanical and thermal properties of the final materials, as well as improving the barrier effects. In the studies of Zhang et al. [7], it was shown that a deliberate modification of VMT by cation exchange with octadecyl trimethyl ammonium bromide allowed increasing the ability of OVMT to disperse better in polyurethane, as well as to create of additional physical cross-links in the polycarbonate polyol structure that are constituent elements of PU soft-segments. As a result, materials with significantly increased (by more than 50%) tensile strength were obtained. In turn, Park et al. [4] modified the filler based on cation exchange with long-chain quaternary ammonium, allowing for increasing the filler dispersion in methylene diphenyl isocyanate (MDI). In the case of porous materials, the final properties of foams depend not only on the type of filler, dispersion, and polymer-filler interface interactions but also on the modification of the cell structure. Umasankar Patro and co-workers studied the effect of a nanometric exfoliated vermiculite addition on the properties of MDI-based rigid foams [6]. Investigations have shown that with an addition of 8 pphp of VMT, the target cell area was reduced by approximately 50%, with a simultaneous increase in the share of closed cells. This translated into a reduction in the thermal conductivity of composite foams compared to unmodified PUR foams and improved mechanical properties.

The present study assesses the effect of a micrometric filler in the form of expanded vermiculite, as a low-cost filler, on the thermal and mechanical properties and the flammability of rigid PUR foams. Moreover, research was undertaken to verify the validity of the application of an additional treatment consisting of the immersion of the filler in a concentrated hydrogen peroxide solution [17]. This process aimed to increase the exfoliation of the inorganic filler's structure and the filler surface's reactivity with the isocyanate component. The possibility of implementing a simple modification procedure and its impact on the thermoset matrix of PUR foams were analyzed.

2. Experimental

2.1. Materials

Polyether polyol based on sorbitol Rokopol RF-551, having a hydroxyl value of 400–440 mgKOH/g, a water content of 0.10 wt.%, a viscosity of 3000–5000 mPa·s, and a functionality of 4.5, was supplied by PCC Rokita S.A. (Brzeg Dolny, Poland). Polycat 9 produced by Evonik Industries AG (Essen, Germany), was used as a catalyst. Niax silicone L-6633 supplied by Momentive Performance Materials Inc. (Waterford, NY, USA) was used as a stabilizer of the foam structure.

Polymeric 4,4'-diphenylmethane diisocyanate (PMDI) with a free isocyanate groups content of 31 wt.% was supplied by Minova Ekochem S.A. (Siemianowice Śląskie, Poland). Water was used as a chemical blowing agent, which in reaction with isocyanate generates carbon dioxide. LANXESS (Cologne, Germany) supplied a flame retardant, triethyl phosphate (TEP).

Thermally expanded vermiculite with a particle size of up to 1.6 mm was provided by Perlit Polska (Puńców, Poland). According to the producer's data, the thermal treatment was carried out at a temperature of 1260 °C, and the chemical composition of the inorganic filler was as follows: 38.0–49.0% SiO_2, 20–23.5% MgO, 12–17.5% Al_2O_3, 0.3–5.4% Fe_2O_3, 5.2–7.9% K_2O, 0–1.2% FeO, 0.7–1.5% CaO, 0–0.8% Na_2O, 0–1.5% TiO_2, 0–0.5 Cr_2O_3, 0.1–0.3% MnO, 0–0.6% Cl, 0–0.6% CO_2, 0–0.2% S.

2.2. Filler Preparation

The use of thermal expansion usually allows for the process of water release and exfoliation of the vermiculite structure [8,13]. The additionally applied procedure of immersion of the non-organic filler in H_2O_2 was to boost the exfoliation effect of thermally treated mineral and additionally oxidize the surface, increasing the amount of isocyanate-reactive hydroxyl groups. The process was carried out within 24 h, which, according to the literature data, allows for obtaining an effective exfoliation effect of unmodified vermiculite [17]. Then, VMT was dried at the temperature of 80 °C for 48 h, and the remaining water was evaporated. The untreated thermally expanded vermiculite filler is marked with W, while the filler treated with hydrogen peroxide is designated as WO in the further part of the present study.

2.3. Preparation of Rigid Polyurethane Foams

A reference rigid PUR foam and the products modified with W and WO were prepared by a single-step method. The polyol premix consisting of a polyol, catalyst, surfactant, blowing agent, and vermiculite was mixed for 30 s. Next, the polyol premix and isocyanate were mixed for 6 s and poured into an open mold (250 mm × 250 mm), where they expanded freely in the vertical direction. The mass of vermiculite was 3%, 6%, 9%, 12%, and 15% of the polyol mass. The isocyanate index was 1.1. The materials were conditioned for 24 h at room temperature before being cut and tested. The formulation used for the preparation of PUR foams is shown in Table 1.

Table 1. The formulation of the reference polyurethane foam.

Component	Mass, g
Rokopol RF-551	100
PMDI	181
L6633	1.5
Polycat 9	2
Water	4
TEP	20
Vermiculite	3, 6, 9, 12, 15% of the mass of the polyol

2.4. Methods

The particle size of the W and WO fillers was assessed by a laser particle sizer Fritsch ANALYSETTE 22 apparatus (Idar-Oberstein, Germany) operated in the range of 0.08–2000 µm. The cumulative size distribution Q3(x) and adequate histogram dQ3(x) were considered during the analysis.

The parameters of the porous structure of the inorganic fillers subjected to different treatments, such as nitrogen adsorption isotherms at -196.15 °C and surface area, were determined using an accelerated surface area and porosimetry apparatus Micromeritics ASAP® 2420 (Norcross, GA, USA) by Brunauer–Emmett–Teller (BET) method. All samples were degassed at 120 °C for 12 h in a vacuum chamber prior to measurements. The specific surface area was determined by the multipoint BET method using adsorption data under relative pressure (p/p0).

The crystallographic structure of the materials was analyzed by the X-ray diffraction (XRD) with Cu Kα radiation (l = 1.54 Å) Panalytical, Empyrean model (Almelo, The Netherlands). The conditions of the XRD measurements were as follows: voltage 45 kV, anode current 40 mA, 2 Theta range from 5° to 40°, time per step 60.214 s, step size 0.0165°.

The viscosities of the polyol premixes filled with various amounts of vermiculite were found using a rotational rheometer MCR 301 from Anton Paar (Graz, Austria) operated with a 25 mm parallel plates measuring system with a gap of 0.3 mm. All specimens were pre-sheared before testing for 1 min with a shear rate of $1\ \text{s}^{-1}$ and a subsequent relaxation time of 2 min. The measurements were realized in the constant shear mode using 0.1, 1, and $10\ \text{s}^{-1}$ shear rates at 30 °C. The presented dynamic viscosity results are mean values from the 300 s experiment.

The foaming process was analyzed using the foam qualification system FOAMAT from Format Messtechnik GmbH (Karlsruhe, Germany), which allows determining changes in characteristic parameters of PUR reaction mixture, such as the temperature and dielectric polarization, during the foaming process.

The apparent density was measured as the ratio of the mass and volume of the samples according to ISO 845. The content of closed cells in the samples was measured in accordance with ISO 4590. The cell structure was examined with the use of a scanning electron microscope Hitachi S-4700 (Tokyo, Japan). The anisotropy index was calculated as the ratio of the cell heights and widths.

The compressive strength at 10% deformation was analyzed in accordance with ISO 826. The compressive strength of the foams was measured using an AllroundLine model Z005 TH from Zwick Roell (Austria) instrument in two directions, parallel and perpendicular to the rise direction of the foams. The compressive force was applied at a speed of 2 mm/s, axially in a normal direction to a square surface. The heat conduction coefficient tests were carried out using foam samples with dimensions of $200 \times 200 \times 50$ mm and a FOX 200 apparatus produced in accordance with the ISO 8301 standard at an average temperature of 10 °C. The temperature of the cold plate was 0 °C, and the temperature of the warm plate was 20 °C.

The additional analysis of filler dispersion in the polyurethane matrix was conducted using the scanning electron microscope MIRA3 from Tescan (Brno, Czech Republic). The measurements were performed with an accelerated voltage of 12 kV in backscattered electrons (BSE) and the secondary electron (SE) mode. The thin carbon coating (~20 nm) was deposited on samples using JEE 4B vacuum evaporator from Jeol (Tokio, Japonia).

The color of the PUR samples was evaluated according to the International Commission on Illumination (CIE) through L*a*b* coordinates [18]. In this system, L* is the color lightness (L* = 0 for black and L* = 100 for white), a* is the green(−)/red(+) axis, and b* is the blue(−)/yellow(+) axis. The color was determined by optical spectroscopy using a MiniScan MS/S-4000S spectrophotometer from HunterLab (Reston, VA, USA) and placed in a specially designed light trap chamber. The total color difference parameter ΔE* was calculated according to the following formulation [19]:

$$\Delta E^* = [(\Delta L^*)^2 + (\Delta a^*)^2 + (\Delta b^*)^2]^{0.5} \quad (1)$$

The thermal properties of PUR were examined by thermogravimetric analysis (TGA) with the temperature set between 25 °C and 900 °C at a heating rate of 10 °C·min^{-1} under nitrogen atmospheres using a TG 209 F1 apparatus from Netzsch (Germany). Samples having masses of 10 mg ± 0.1 mg were placed in Al_2O_3 pans. The initial decomposition temperature Ti was determined as the temperature at which the mass loss was 5%. Additionally, temperatures at 10, 25, and 50% mass loss were found.

The limiting oxygen index (LOI) was determined according to ISO 4589-2:2017. Burning behavior was evaluated with the use of a cone calorimeter from Fire Testing Technology (East Grinstead, UK). The samples (100 × 100 × 25 mm) were placed in aluminum foil and tested horizontally at an applied heat flux of 35 kW/m^2, in conformity with the ISO 5660 standard. Spark ignition was used to ignite the pyrolysis products. Next, the residues were photographed using a digital camera EOS 400 D from Canon Inc. (Tokyo, Japan).

3. Results

3.1. Filler Characterization

The cumulative size distribution Q3(x) and adequate histograms dQ3(x) made for the inorganic fillers used in this study are presented in Figure 1. An analysis of the graphs shows that the ground and sieved expanded W is characterized by larger particle sizes compared to WO. Additional treatment using hydrogen peroxide allows increasing the content of particles with smaller sizes, which is probably due to the destruction of the concertina-shaped vermiculite structure and an additional exfoliation effect, confirmed by XRD evaluation. Both vermiculite grades (W and WO) exhibit two particle size distribution modes due to the fraction of finely divided filler sheets formed during the grinding process.

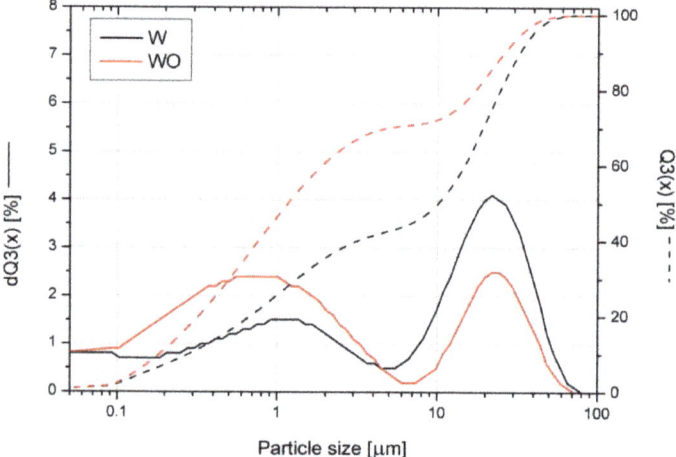

Figure 1. Particle size distributions for inorganic fillers.

In Figure 2, X-ray diffraction patterns of expanded vermiculite and vermiculite are also treated with hydrogen peroxide. The peaks at 2θ = 9°, 21.0°, 26.8°, and 34.3° in both vermiculites correspond to d-spacing of 9.8 Å, 4.2 Å, 3.3 Å, and 2.6 Å, respectively. The peak at 9° was shifted slightly to 8.96° after the treatment, which caused a d-spacing shift from 9.82° to 9.86°. This may be understood as an additional exfoliating effect of a chemical treatment [6]. The phenomenon may improve the dispersion of the filler in a polyol composition and the efficiency of modification of the final PUR foam.

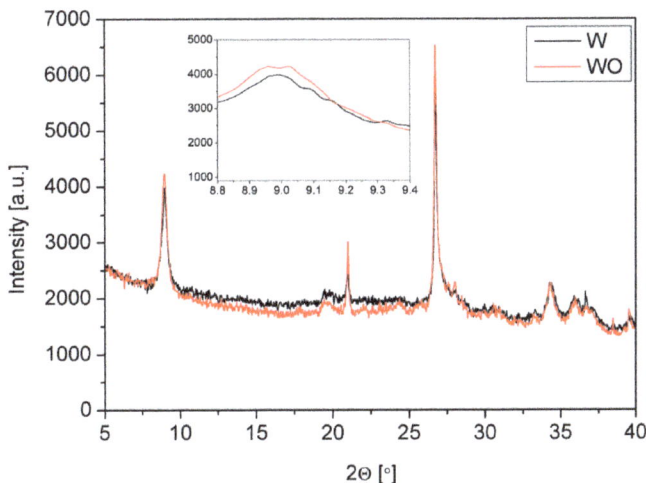

Figure 2. XRD patterns of thermally expanded vermiculite (W) and vermiculite additionally treated grade with H_2O_2 (WO).

BET surface area (S_{BET}), t-Plot external surface area (S_{EXT}), t-Plot micropore area (S_{MIC}), desorption average pore width (4 V/A), single point desorption total pore volume of pores less than 12.1 nm diameter at p/p_0 = 0.98325121 (V_P), t-Plot micropore volume (V_{MIC}) values are presented in Table 2. Figure 3a shows N_2 adsorption-desorption isotherms of the filler before and after the H_2O_2 treatment. The course of the a(p/p$_0$) curve is typical of expanded hydrous phyllosilicates [20,21]. The physicochemical properties found in our experiment are in good agreement with the literature [20]. S_{BET} increased after the two-step treatment (Table 2). Both fillers correspond to similar courses of the curves without other inflections and hysteresis loops, which may be related to significant changes in the modified material structure. Figure 3b presents a pore volume vs. pore diameter plot. It can be concluded that for both fillers, the pore diameter is below 40 nm. Therefore, the primary mechanism of adsorption results from mesopores adsorption. When p/p_0 increases above 0.8, the adsorption increases significantly, suggesting the presence of micropores [22]. Both vermiculites can be described as type II in the Brauner classification. The measured adsorption in the whole considered range is higher for WO. The additional chemical treatment improved the physicochemical properties of the filler, including the specific surface area, which should cause an improved reactivity toward the chemically hardened polymer.

Table 2. Characteristic of vermiculite grades based on BET analysis.

Material	S_{BET}, m²/g	S_{EXT}, m²/g	S_{MIC}, m²/g	4V/A, nm	V_P, cm³/g	V_{MIC}, cm³/g
W	12.58	12.77	0.81	14.29	0.048556	0.000257
WO	17.09	16.56	0.53	12.64	0.054028	0.000072

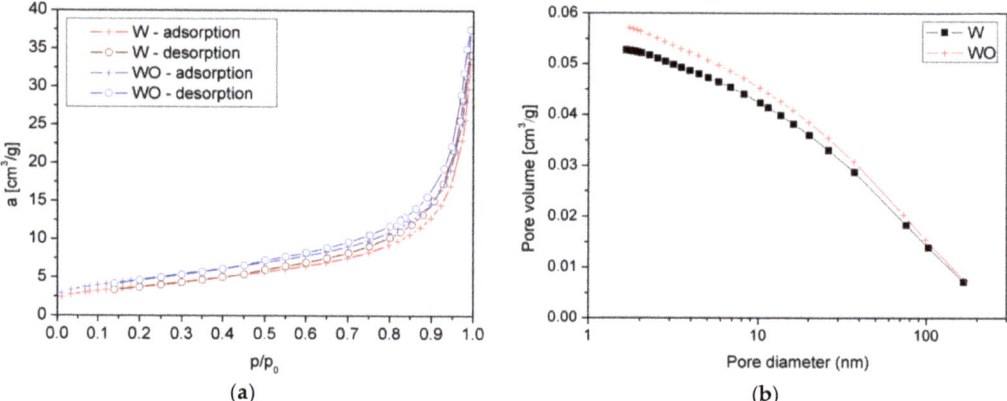

Figure 3. Nitrogen adsorption-desorption isotherms of thermally expanded vermiculite (W) and vermiculite additionally treated with H_2O_2 (WO).

3.2. Rheological Properties

Figure 4 shows the results of viscosity measurements of polyol premixes containing various contents of expanded vermiculite without treatment (W) and after an additional peroxide treatment (WO).

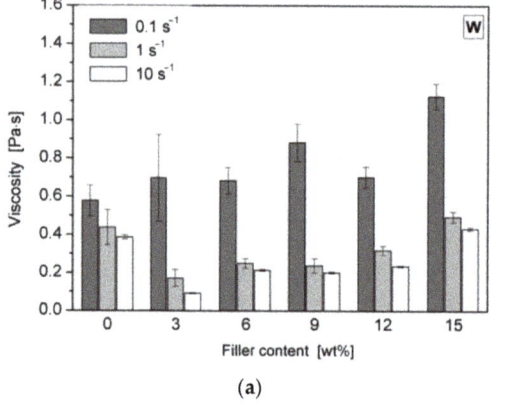
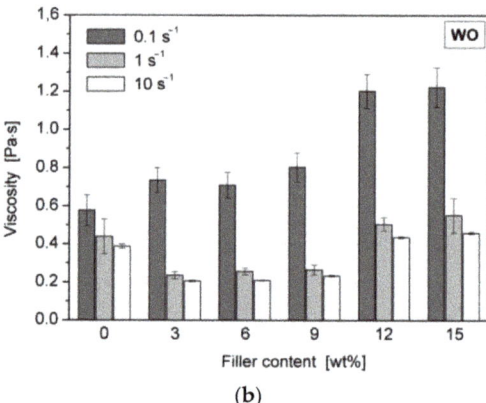

Figure 4. Rheological properties of polyol-W (**a**) and -WO (**b**) premixes with various filler contents measured with 0.1, 1, and 10 s^{-1} shear rates.

It can be seen that an increasing content of the filler in polyol premix increases the dynamic viscosity of the composition, while at higher shear rates, the unfilled polyol shows lower viscosities. Usually, introducing a powder filler increases the viscosity [5,23]. It should be emphasized that the employed filler has an expanded form and was additionally fragmented with the use of a high-speed grinder and sieved. Therefore, in the considered case, the plate-shaped geometry of the filler and the measurement procedure, including the pre-shearing of the polyol, caused the observed changes in the average viscosity values. Consequently, at the lowest shear rate, the filler probably did not align itself with the flow direction, while the higher shear rates made the vermiculite sheets orient themselves, reducing the viscosity with respect to the unfilled composition. The compositions containing W exhibited greater spreads of the recorded mean viscosity values, which is understandable because of larger filler particles and a broader particle size distribution. It should be noted that the addition of vermiculite significantly increased the viscosity of the compositions only for a low shear rate. Therefore, the systems studied here may find spray-forming applications without affecting the processing conditions.

3.3. Foaming Process of PUR Systems Modified with Vermiculite

A modification of a PUR system with fillers can have an influence on the reactivity of the system. Changes in the reactivity of the PUR system were analyzed using the FOAMAT device. The reactivity of the PUR system is illustrated by the changes in the dielectric polarization curve. The systems with higher reactivity are characterized by a faster dielectric polarization reduction. The changes in dielectric polarization, as well as the temperature of the reaction mixture during the foaming process, are shown in Figure 5.

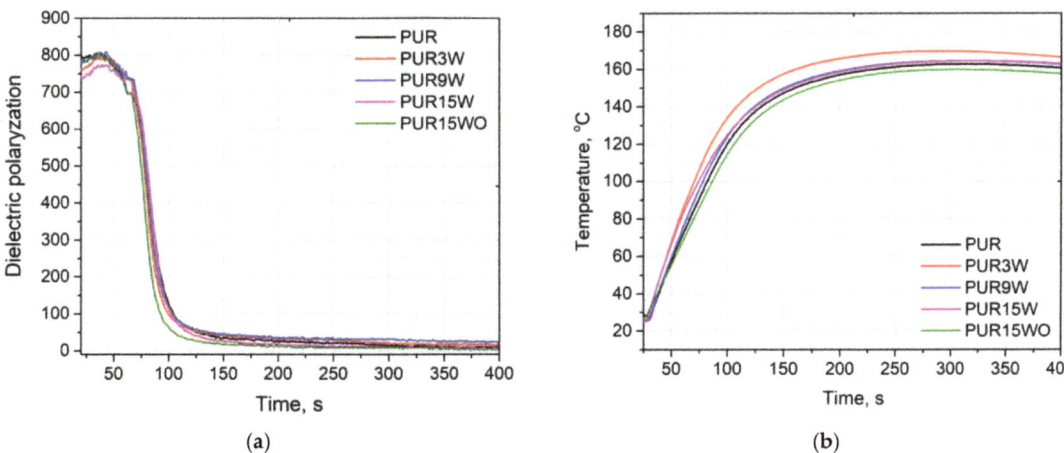

Figure 5. Influence of vermiculite on dielectric polarization (**a**) and temperature (**b**).

The results indicate that a modification of the reference system with vermiculite, regardless of its content, did not significantly affect the reactivity of the PUR system, which was confirmed by observations of the dielectric polarization and temperature changes. A similar effect was observed in our earlier work, where the PUR foams were modified with thermoset polyester composite waste [24]. A slight influence on the foaming process is extremely important as the foaming process determines the cell structure of the foam and its subsequent functional properties. In the literature, an effect of decreased reactivity of the PUR foam modified with a waste filler has been described. Formela et al. [25] applied the brewer's spent grain and ground tire rubber in rigid PUR foams, causing modification, which resulted in a decrease in the reaction rate. The rise time and tack-free time of the rigid PUR foams modified with 20 wt.% of the brewer's spent grain were two and almost three times longer than those of the unmodified system, respectively.

3.4. Properties of Rigid PUR Foams Modified with Vermiculite

The cellular structure of porous materials has a significant effect on their properties and depends on many factors such as: premix viscosity, their modification by fillers, method of foaming etc. [26].

All the tested foams exhibited well-developed hexagonal cell structures (Figure 6). The modification of the reference system (PUR) with the vermiculite and modified vermiculite improved the morphology of the PUR composites, generally reducing cells' average diameters (Table 3). However, the changes are insignificant and within the measurement error. The contents of closed cells were higher for the foams modified with vermiculite (Figure 7). The content of closed cells is important from the thermal insulation properties point of view.

Closed-cell foams are characterized by lower values of the thermal conductivity coefficient compared to open-cell foams.

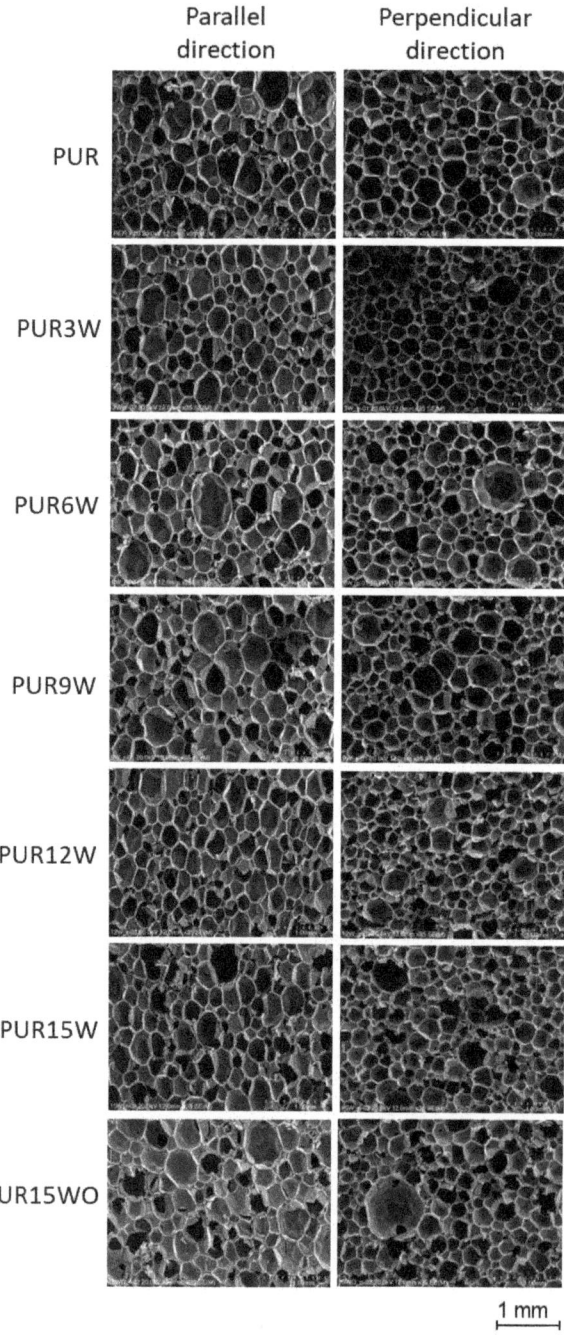

Figure 6. The cellular structure of PUR foams and foams modified with vermiculite.

Table 3. Influence of vermiculite on the diameter of cells in PUR foams.

Symbol	Parallel Direction, μm	Perpendicular Direction, μm	AI *
PUR	309.4 ± 105.6	281.3 ± 89.2	1.10
PUR3W	287.4 ± 112.4	248.1 ± 83.9	1.16
PUR6W	303.8 ± 122.4	254.2 ± 94.5	1.19
PUR9W	317.7 ± 119.3	273.7 ± 95.9	1.16
PUR12W	285.7 ± 105.6	260.9 ± 92.4	1.10
PUR15W	299.2 ± 107.1	264.6 ± 91.3	1.13
PUR15WO	291.1 ± 104.4	277.5 ± 103.6	1.05

AI *—anisotropy index calculated as a ratio of cell diameters measured in cross-sections parallel and perpendicular to the foam rise directions.

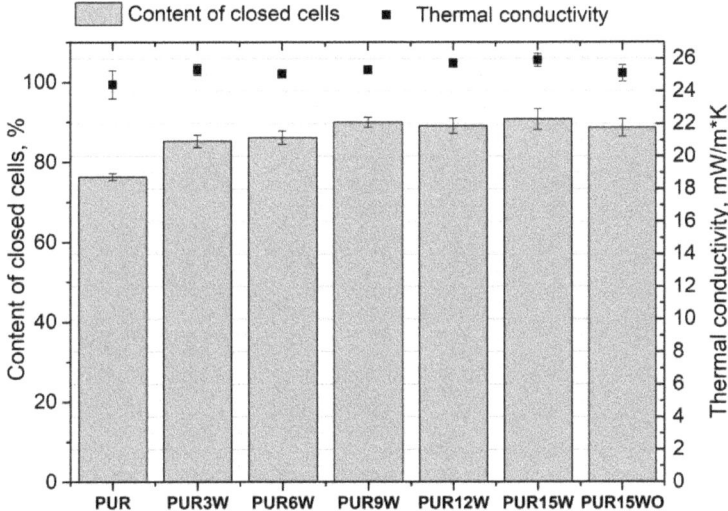

Figure 7. Content of closed cells and thermal conductivity of foams modified with vermiculite.

Based on the results presented in Figure 7, it can be observed that the content of closed cells increased from 76% for the reference material (PUR) to 91% for the foams with the highest vermiculite content (PUR15W). However, the value of the thermal conductivity coefficient is characterized by the highest value for a given material (PUR15W) despite the highest closed cell content. Such an effect may be related to the highest apparent density and relatively high AI of PUR15W material. It was observed that the material into which the WO was introduced has a higher apparent density, while the thermal conductivity of this material is lower than that of the material containing the same amount of W. This can be explained by the lowest AI, which means this PUR15WO foam has a less anisotropic structure. This structure limits heat transport through the foamed material. Depending on the type of filler, the effect on the variation of the foam's apparent density may be different. Natural fillers, such as flax and hemp fibers, can decrease the apparent density of foams as a result of the moisture present in them (carbon dioxide is generated in the reaction of water and isocyanate). In the case of fillers characterized by high density, e.g., carbon fibers, montmorillonite, or other inorganic fillers, the apparent density of PUR foams is increased [27].

The PUR foams modified with vermiculite were characterized by an apparent density (Figure 8) in the range of 35–39 kg/m^3. The foams with modified vermiculite (PUR15WO) had the highest apparent density. However, differences among tested foams are not significant, taking into account the standard deviation. The compressive strength (Figure 8) measured in the direction parallel to the direction of the foam growth is characterized by greater values than when measured in the perpendicular direction. These differences are due to the anisotropic nature of the cellular structure of the PUR foams obtained. The compressive strength results are comparable with those obtained for the reference material. There is a slight increase in the mechanical strength of the foams with the highest vermiculite content. This effect can be related to a slight increase in the apparent density of the materials with the highest filler content.

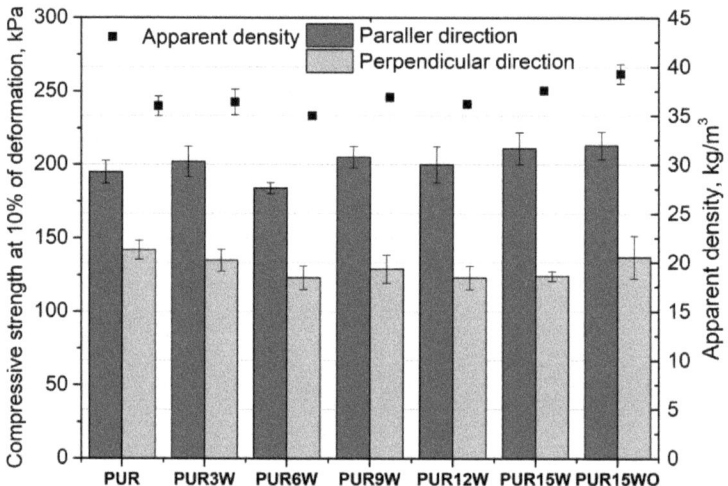

Figure 8. Apparent density and compressive strength of foams modified with vermiculite.

Figure 9 shows the collectively presented SEM images made for the reference sample (PUR), fillers (W, WO), and composites with the highest concentration of fillers (PUR15W, PUR15WO). The analysis was performed with SE and BSE modes to increase the visibility of the filler particles in the matrix. It can be concluded that the filler is well distributed in the polymer matrix. There are no torn-out inorganic fractions that could result from improper adhesion. Moreover, it should be emphasized that there are no agglomerated structures. Filler particles smaller than 1 μm are evenly distributed, while larger particles are embedded in the walls of the foam cells, especially in the nodes. Therefore, their localization does not weaken the cell structure, and no voids were noted in the interphase area, which could suggest a lack of adhesion between a polymer and a filler.

The aesthetics of the final products often plays an important role in selecting materials by design teams. The analysis of color, which is one of the primary criteria for the qualitative assessment of products, is essential from the point of view of the potential of selected product groups [28,29]. Table 4 summarizes the L*, a*, and b* chromatic parameters, describing the color in the CIELab space of the produced foams with different vermiculite contents. Additionally, the results of the total color change were calculated according to Equation (1). Even the smallest addition of the filler caused significant color changes, taking into account the criteria described in the ISO 2813 standard and the literature [29]. Based on the low values of standard deviations, all the foams were characterized by lower luminescence and had a brown shade with a uniform color. This also confirms the good compatibility and miscibility of the PUR-W/WO. It should be emphasized that in the case of the foams with the highest filler concentration, no significant changes in ΔE between the batches made with expanded and H_2O_2 treated with vermiculites were noted.

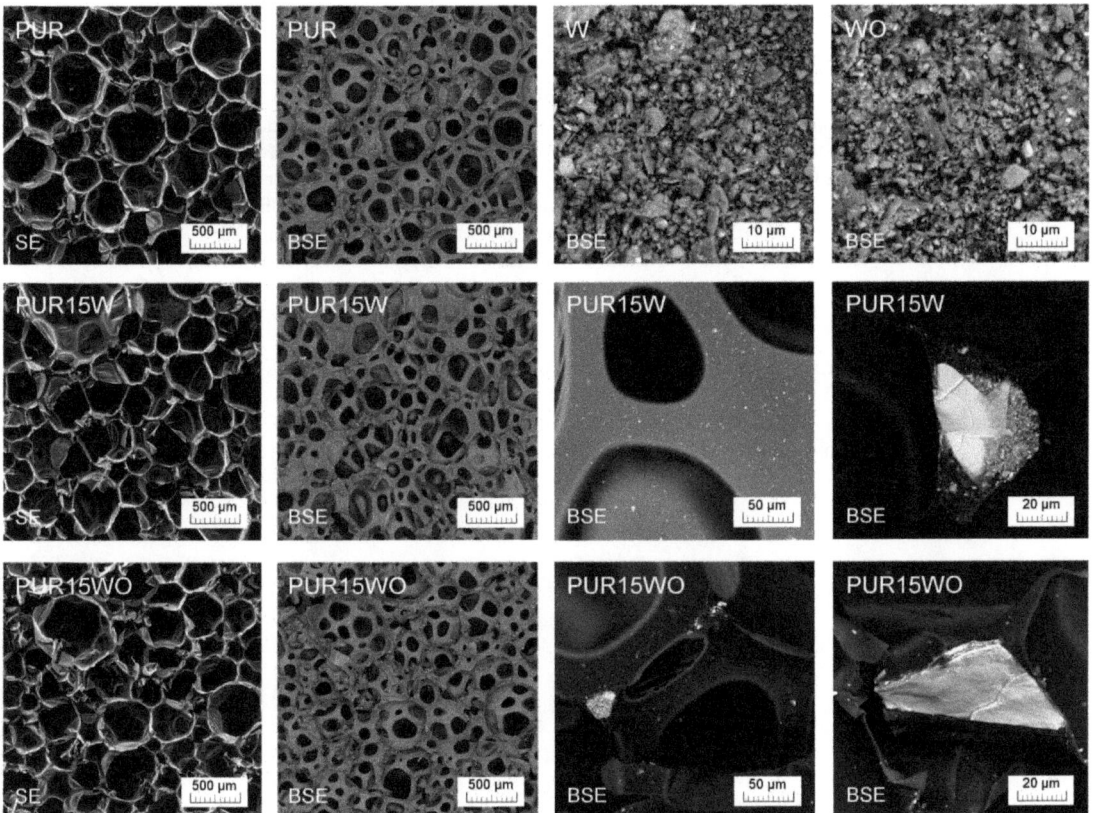

Figure 9. SEM images made for reference PUR and PUR 15W and PUR 15WO in SE and BSE mode showing the dispersion of the filler in a foamed polymeric matrix.

Table 4. CIELab color parameters and ΔE of PUR and PUR-W/WO composites.

Sample	L*	a*	b*	ΔE
PUR	85.62 ± 0.93	−3.05 ± 0.39	13.94 ± 0.85	-
PUR3W	79.08 ± 1.13	1.24 ± 0.16	14.44 ± 0.64	7.86
PUR6W	74.92 ± 1.74	3.29 ± 0.26	14.30 ± 0.64	12.46
PUR9W	70.97 ± 1.56	4.48 ± 0.34	14.89 ± 0.59	16.61
PUR12W	69.89 ± 1.06	5.42 ± 0.32	14.27 ± 0.50	17.89
PUR15W	65.61 ± 1.51	6.66 ± 0.40	15.03 ± 0.59	22.29
PUR15WO	65.46 ± 1.25	7.05 ± 0.36	16.29 ± 0.61	22.69

The results of the thermogravimetric analysis are presented in the form of TG, and DTG graphs in Figure 10. Table 5 collectively shows thermal parameters, such as temperature at 5%, 10%, 25%, and 50% mass loss, residual mass at 900 °C, and data describing peaks observed at the first derivative of the group. The courses of the TG and DTG curves (Figure 10) indicate the three-step course of the thermal degradation process of the rigid PUR foams. The first stage of degradation in the temperature range from 110 to 220 °C, with the maximum between 170 and 190 °C, is related to the evaporation of residual water and low molecular weight products in the PUR foam [30]. The observed dominant degradation stage with the maximum process intensity observed around 320 °C is associated with hard-segment decompositions. As demonstrated by Jiao et al. [30], in a narrow range between

320 °C and 350 °C, isocyanate monomers almost disappear. First, N-H bonds are degraded, resulting in the degradation of hard segments, then C-H bonds from methyl and methylene groups. The last step of decomposition observed in the temperature range of 370–420 °C corresponds to the degradation of ester bonds in polyols [31,32]. It should be emphasized that the introduction of the filler caused shifts in decomposition stages; however, it did not affect its mechanism, which proves that there were no significant changes in the chemical structure of the PUR composition. Based on the conducted research, it can be clearly stated that adding unmodified and hydrogen peroxide-modified vermiculite improved the thermal properties of the composites as compared to the unmodified PUR foam (enhanced $T_{5\%}$ and yield of residue). In the case of introducing expanded vermiculite into the PUR matrix, it is difficult to find a clear trend related to the amount of the filler and only an apparent effect of increasing thermal stability, especially distinct as evaluated at a residue (Table 5). On the other hand, using a two-step treatment based on thermal development and subsequent immersion in concentrated peroxide significantly improved thermal stability of PUR-based composite. This may be related to the improvement in the filler dispersion resulting from its structure modification described in the earlier paragraph, with the formation of sheets of reduced size.

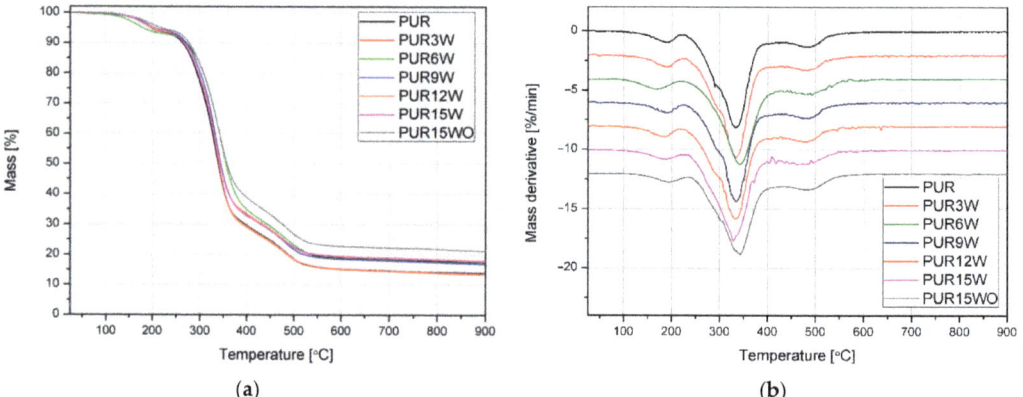

Figure 10. TG (**a**) and DTG (**b**) curves of polyurethane samples with various contents of vermiculite.

Table 5. Thermal parameters obtained by TGA for PUR and PUR-based composites modified with various vermiculite contents.

Sample	T5%, °C	T10%, °C	T25%, °C	T50%, °C	Residue at 900 °C, %	DTG 1st Peak, %/min; °C	DTG 2nd Peak, %/min; °C	DTG 3rd Peak, %/min; °C
PUR	198.7	263.8	304.2	339.7	13.76	−0.98; 190.5	−8.16; 332.2	−1.33; 481.6
PUR3W	199.8	266.7	307.6	341.0	13.49	−1.03; 190.0	−8.73; 333.5	−1.27; 472.7
PUR6W	181.7	270.9	319.0	356.7	17.38	−0.88; 170.2	−7.23; 343.0	−1.40; 493.0
PUR9W	205.1	271.4	312.4	345.5	16.91	−0.94; 190.2	−8.39; 335.1	−1.38; 486.8
PUR12W	199.7	266.8	308.6	344.1	17.86	−0.94; 188.2	−7.83; 335	−1.31; 481.9
PUR15W	204.8	266.6	307.1	343.8	17.61	−0.81; 187.4	−7.70; 329.2	−1.24; 460;8
PUR15WO	217.8	277.1	318.5	358.4	21.06	−0.75; 194.6	−6.88; 343.0	−1.37; 481.6

The cone calorimeter test is a small-scale test employed to observe a comprehensive set of fire features in a well-defined fire scenario [33]. The measurement provides the value of parameters, such as time to ignition (TTI), heat release rate (HRR), the maximum average rate of heat emission (MARHE), total heat release (THR), effective heat of combustion (EHC), specific extinction area (SEA), and total smoke release (TSR). The HRR curves for the materials investigated in this study are illustrated in Figure 11, while detailed data are summarized in Table 6.

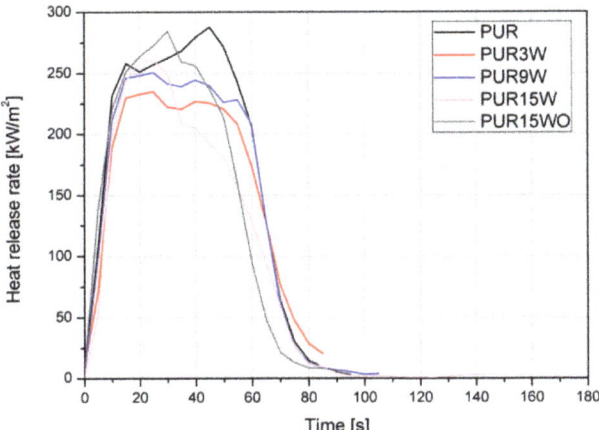

Figure 11. Representative heat release rate curves of PU foams modified with vermiculite.

Table 6. Cone calorimeter test results of the tested foams.

Sample	TTI, s	pHRR, kW/m²	MARHE, kW/m²	THR, MJ/m²	EHC, MJ/kg	SEA, m²/kg	TSR, m²/m²	Residue, %	LOI, %
PUR	4 ± 1	288 ± 19	233 ± 11	15 ± 1	16 ± 0	811 ± 22	768 ± 61	15.2 ± 2	21.5 ± 0.1
PUR3W	6 ± 3	237 ± 42	188 ± 34	13 ± 1	16 ± 0	787 ± 55	666 ± 78	13.3 ± 1	21.5 ± 0.1
PUR9W	4 ± 1	284 ± 19	222 ± 5	14 ± 0	16 ± 0	822 ± 25	707 ± 32	16.5 ± 3	21.5 ± 0.1
PUR15W	5 ± 1	266 ± 6	207 ± 11	13 ± 1	16 ± 0	824 ± 14	668 ± 39	21.4 ± 2	21.5 ± 0.1
PUR15WO	4 ± 1	274 ± 27	222 ± 24	14 ± 0	16 ± 0	810 ± 30	699 ± 10	19.6 ± 0	21.5 ± 0.1

The HRR curves suggest that all PUR foams ignited at a comparable time, which was confirmed by the time to ignition values presented in Table 6. Their cellular structure and low thermal conductivity strongly influence the burning behavior, and TTI reached only 5 ± 1 s. The heat release rate is an essential parameter to estimate fire development, intensity, and spreading. The trend of the HRR curve demonstrates the burning behavior of the materials as a function of time. The curve of the PUR exhibits quite a broad peak with a maximum average value of 288 kW/m². It can be observed that vermiculite led to a change in the curves trend from characteristic for thick non-charring to thick charring ones [33]. The lowest pHRR of 237 kW/m² (reduction of 18%) was obtained for the composites PUR3W, so the values were independent of the filler content and its modification. MARHE, as an indicator determined from HRR, is used to estimate the hazard of developing fires. Consequently, lower MARHE was obtained for samples with lower pHRR. Vermiculite is a filler known for its flame-retardant effects [34–36]; however, no change in LOI values as a result of W or WO addition was observed.

The integral of HRR over time expresses the total heat output, i.e., the THR [33]. The vermiculite addition caused a non-linear decrease in THR, suggesting incomplete combustion affected by char formation or reduced combustion efficiency [37]. Since there was no change in EHC, as well as according to the increased yield of residue (Table 6, Figure 10a), action probably occurred in the condensed phase. Moreover, the content of triethyl phosphate, which is a phosphorus flame retardant active mainly in the gas phase, was the same for all materials. The analysis of the photographs confirms that the presence of vermiculite facilitated the formation of a more compact structure, and the number of holes decreased with an increase in the amount of vermiculite (Figure 12). Similar to the carbonaceous char, inert residue from inorganic fillers works as a barrier and additionally replaces polymer, reducing the fuel release [30]. Probably, the residues of the investigated materials were the origin of both effects.

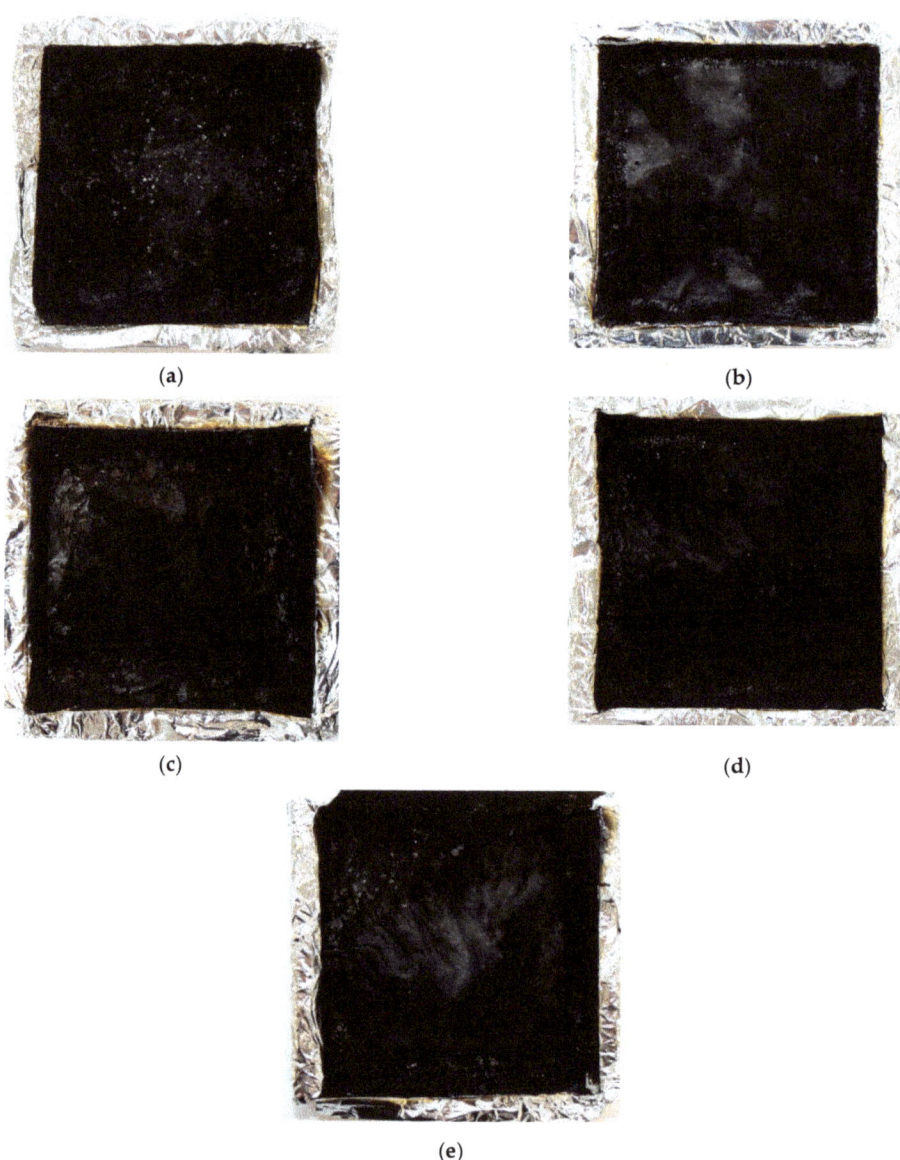

Figure 12. Photographs of samples after cone calorimetry tests (**a**) PUR, (**b**) PUR3W, (**c**) PUR9W, (**d**) PUR15W, (**e**) PUR15WO.

During a fire, smoke is of great importance as it reduces visibility and makes an escape more challenging [38–40]. Considering the SEA values together with the standard deviation, it can be concluded that the use of vermiculite did not change this parameter. The lowest SEA, which corresponds to the surface light-absorbing particles of smoke [39], was recorded for composites PUR3W and amounted to 787 m^2/kg. In turn, the TSR of all composites was reduced compared to the unmodified foam, and the highest decrease reached approximately 13% (PUR15W). Presumably, this is due to the increased amount of the material remaining in the condensed phase.

4. Conclusions

Rigid polyurethane foams modified with the thermally expanded vermiculite were successfully obtained. The apparent density of foams was comparable in the range of 35–39 kg/m^3. The impact of different filler contents on the foaming process, cellular structures, physical-mechanical properties, thermal stability, and flammability of porous composites was determined. It was found that a modification of the reference system with thermally expanded vermiculite did not significantly affect the reactivity of the polyurethane system, which was confirmed by similar trends observed for dielectric polarization changes. A slight increase in reactivity was observed, confirmed by a faster decrease in dielectric polarization for the polyurethane system with modified vermiculite by soaking in H_2O_2. The modification of the reference system with vermiculite and modified vermiculite improved the morphology of the porous polyurethane composites and increased the content of closed cells.

The coefficient of thermal conductivity of reference foam and foams modified with vermiculite was in the range 24–26 mW/mK, which makes such materials interesting for heat-insulating applications. The use of vermiculite up to 15 wt.% did not significantly influence other tested properties (thermal, mechanical, fire) of the foams but could make such modified materials cheaper and useful for industrial application.

Author Contributions: Conceptualization, M.B. and M.K.; methodology, M.B., M.K., K.S. (Kamila Sałasińska) and A.P. (Adam Piasecki); formal analysis, M.B., M.K. and K.S. (Kamila Sałasińska); investigation, M.B., K.S. (Kamila Sałasińska), K.S. (Karolina Surmacz), I.S.-W., J.A., K.U., K.P., A.P. (Adam Piasecki) and J.S.; data curation, M.B., M.K. and K.S. (Karolina Surmacz); writing—original draft preparation, M.B., M.K. and K.S.; writing—review and editing, M.B., M.K., K.S. (Kamila Sałasińska) and A.P. (Aleksander Prociak); visualization, M.K. and K.S. (Kamila Sałasińska); supervision, M.B., M.K., K.S. (Kamila Sałasińska) and A.P. (Aleksander Prociak). All authors have read and agreed to the published version of the manuscript.

Funding: The results presented in this paper were partially funded with grants for education allocated in Poznan University of Technology by the Ministry of Science and Higher Education in Poland executed under the subject of No 0513/SBAD/4774.

Institutional Review Board Statement: Not applicable.

Informed Consent Statement: Not applicable.

Data Availability Statement: The data presented in this study are available on request from the corresponding author.

Conflicts of Interest: The authors declare no conflict of interest.

References

1. Gama, N.V.; Ferreira, A.; Barros-Timmons, A. Polyurethane Foams: Past, Present, and Future. *Materials* **2018**, *11*, 1841. [CrossRef] [PubMed]
2. Demharter, A. Polyurethane Rigid Foam, a Proven Thermal Insulating Material for Applications between +130 °C and −196 °C. *Cryogenics* **1998**, *38*, 113–117. [CrossRef]
3. Kurańska, M.; Prociak, A.; Michałowski, S.; Zawadzińska, K. The Influence of Blowing Agents Type on Foaming Process and Properties of Rigid Polyurethane Foams. *Polimery* **2018**, *63*, 672–678. [CrossRef]
4. Park, Y.T.; Qian, Y.; Lindsay, C.I.; Nijs, C.; Camargo, R.E.; Stein, A.; Macosko, C.W. Polyol-Assisted Vermiculite Dispersion in Polyurethane Nanocomposites. *ACS Appl. Mater. Interfaces* **2013**, *5*, 3054–3062. [CrossRef] [PubMed]
5. Miedzińska, K.; Członka, S.; Strąkowska, A.; Strzelec, K. Vermiculite Filler Modified with Casein, Chitosan, and Potato Protein as a Flame Retardant for Polyurethane Foams. *Int. J. Mol. Sci.* **2021**, *22*, 10825. [CrossRef]
6. Patro, T.U.; Harikrishnan, G.; Misra, A.; Khakhar, D.V. Formation and Characterization of Polyurethane-Vermiculite Clay Nanocomposite Foams. *Polym. Eng. Sci.* **2008**, *48*, 1778–1784. [CrossRef]

7. Zhang, T.; Zhang, F.; Dai, S.; Li, Z.; Wang, B.; Quan, H.; Huang, Z. Polyurethane/Organic Vermiculite Composites with Enhanced Mechanical Properties. *J. Appl. Polym. Sci.* **2016**, *133*, 43219. [CrossRef]
8. Li, M.; Zhao, Y.; Ai, Z.; Bai, H.; Zhang, T.; Song, S. Preparation and Application of Expanded and Exfoliated Vermiculite: A Critical Review. *Chem. Phys.* **2021**, *550*, 111313. [CrossRef]
9. Muiambo, H.F.; Focke, W.W. Ion Exchanged Vermiculites with Lower Expansion Onset Temperatures. *Mol. Cryst. Liq. Cryst.* **2012**, *555*, 65–75. [CrossRef]
10. Muiambo, H.F.; Focke, W.W.; Atanasova, M.; van der Westhuizen, I.; Tiedt, L.R. Thermal Properties of Sodium-Exchanged Palabora Vermiculite. *Appl. Clay Sci.* **2010**, *50*, 51–57. [CrossRef]
11. Wang, M.; Liao, L.; Zhang, X.; Li, Z.; Xia, Z.; Cao, W. Adsorption of Low-Concentration Ammonium onto Vermiculite from Hebei Province, China. *Clays Clay Miner.* **2011**, *59*, 459–465. [CrossRef]
12. Bortoluzzi, E.C.; Velde, B.; Pernes, M.; Dur, J.C.; Tessier, D. Vermiculite, with Hydroxy-Aluminium Interlayer, and Kaolinite Formation in a Subtropical Sandy Soil from South Brazil. *Clay Miner.* **2008**, *43*, 185–193. [CrossRef]
13. Hillier, S.; Marwa, E.M.M.; Rice, C.M. On the Mechanism of Exfoliation of 'Vermiculite'. *Clay Miner.* **2013**, *48*, 563–582. [CrossRef]
14. Feng, J.; Liu, M.; Fu, L.; Zhang, K.; Xie, Z.; Shi, D.; Ma, X. Enhancement and Mechanism of Vermiculite Thermal Expansion Modified by Sodium Ions. *RSC Adv.* **2020**, *10*, 7635–7642. [CrossRef]
15. Barczewski, M.; Mysiukiewicz, O.; Hejna, A.; Biskup, R.; Szulc, J.; Michałowski, S.; Piasecki, A.; Kloziński, A. The Effect of Surface Treatment with Isocyanate and Aromatic Carbodiimide of Thermally Expanded Vermiculite Used as a Functional Filler for Polylactide-Based Composites. *Polymers* **2021**, *13*, 890. [CrossRef]
16. Barczewski, M.; Hejna, A.; Sałasińska, K.; Aniśko, J.; Piasecki, A.; Skórczewska, K.; Andrzejewski, J. Thermomechanical and Fire Properties of Polyethylene-Composite-Filled Ammonium Polyphosphate and Inorganic Fillers: An Evaluation of Their Modification Efficiency. *Polymers* **2022**, *14*, 2501. [CrossRef]
17. Obut, A.; Girgin, İ. Hydrogen Peroxide Exfoliation of Vermiculite and Phlogopite. *Miner. Eng.* **2002**, *15*, 683–687. [CrossRef]
18. International Commission on Illumination. *Recommendations on Uniform Color Spaces, Color-Difference Equations, Psychometric Color Terms*; C.I.E.: Vienna, Austria, 1978.
19. Grząbka-Zasadzińska, A.; Klapiszewski, Ł.; Borysiak, S.; Jesionowski, T. Thermal and Mechanical Properties of Silica—Lignin/Polylactide Composites Subjected to Biodegradation. *Materials* **2018**, *11*, 2257. [CrossRef]
20. Marcos, C. Effect of Water Immersion on Raw and Expanded Ugandan Vermiculite. *Minerals* **2021**, *12*, 23. [CrossRef]
21. Moraes, D.; Miranda, L.; Angélica, R.; Rocha Filho, G.; Zamian, J. Functionalization of Bentonite and Vermiculite after the Creation of Structural Defects through an Acid Leaching Process. *J. Braz. Chem. Soc.* **2018**, *29*, 320–327. [CrossRef]
22. Tang, Q.; Wang, F.; Tang, M.; Liang, J.; Ren, C. Study on Pore Distribution and Formation Rule of Sepiolite Mineral Nanomaterials. *J. Nanomater.* **2012**, *2012*, 382603. [CrossRef]
23. Kuranska, M.; Prociak, A.; Michalowski, S.; Cabulis, U.; Kirpluks, M. Microcellulose as a Natural Filler in Polyurethane Foams Based on the Biopolyol from Rapeseed Oil. *Polimery* **2016**, *61*, 625–632. [CrossRef]
24. Barczewski, M.; Kurańska, M.; Sałasińska, K.; Michałowski, S.; Prociak, A.; Uram, K.; Lewandowski, K. Rigid Polyurethane Foams Modified with Thermoset Polyester-Glass Fiber Composite Waste. *Polym. Test.* **2020**, *81*, 106190. [CrossRef]
25. Formela, K.; Hejna, A.; Zedler, Ł.; Przybysz, M.; Ryl, J.; Reza, M.; Piszczyk, Ł. Industrial Crops & Products Structural, Thermal and Physico-Mechanical Properties of Polyurethane/Brewers' Spent Grain Composite Foams Modi Fi Ed with Ground Tire Rubber. *Ind. Crops Prod.* **2017**, *108*, 844–852. [CrossRef]
26. Paciorek-Sadowska, J.; Borowicz, M.; Isbrandt, M. Effect of Evening Primrose (Oenothera Biennis) Oil Cake on the Properties of Polyurethane/Polyisocyanurate Bio-composites. *Int. J. Mol. Sci.* **2021**, *22*, 8950. [CrossRef]
27. Cabulis, U.; Kirpluks, M. Rapeseed Oil as Main Component in Synthesis of Bio-Polyurethane-Polyisocyanurate Porous Materials Modi Fi Ed with Carbon Fi Bers. *Polym. Test.* **2017**, *59*, 478–486. [CrossRef]
28. Członka, S.; Sienkiewicz, N.; Kairytè, A.; Vaitkus, S. Colored Polyurethane Foams with Enhanced Mechanical and Thermal Properties. *Polym. Test.* **2019**, *78*, 105986. [CrossRef]
29. Bociaga, E.; Trzaskalska, M. Influence of Polymer Processing Parameters and Coloring Agents on Gloss and Color of Acrylonitrile-Butadiene-Styrene Terpolymer Moldings. *Polimery* **2016**, *61*, 544–550. [CrossRef]
30. Jiao, L.; Xiao, H.; Wang, Q.; Sun, J. Thermal Degradation Characteristics of Rigid Polyurethane Foam and the Volatile Products Analysis with TG-FTIR-MS. *Polym. Degrad. Stab.* **2013**, *98*, 2687–2696. [CrossRef]
31. Gu, R.; Sain, M.M. Effects of Wood Fiber and Microclay on the Performance of Soy Based Polyurethane Foams. *J. Polym. Environ.* **2013**, *21*, 30–38. [CrossRef]
32. Cifarelli, A.; Boggioni, L.; Vignali, A.; Tritto, I.; Bertini, F.; Losio, S. Flexible Polyurethane Foams from Epoxidized Vegetable Oils and a Bio-Based Diisocyanate. *Polymers* **2021**, *13*, 612. [CrossRef] [PubMed]
33. Schartel, B.; Hull, T.R. Development of Fire-Retarded Materials—Interpretation of Cone Calorimeter Data. *Fire Mater.* **2007**, *31*, 327–354. [CrossRef]
34. Ren, Q.; Zhang, Y.; Li, J.; Li, J.C. Synergistic Effect of Vermiculite on the Intumescent Flame Retardance of Polypropylene. *J. Appl. Polym. Sci.* **2011**, *120*, 1225–1233. [CrossRef]
35. Wang, F.; Gao, Z.; Zheng, M.; Sun, J. Thermal Degradation and Fire Performance of Plywood Treated with Expanded Vermiculite. *Fire Mater.* **2016**, *40*, 427–433. [CrossRef]

36. Sałasińska, K.; Kirpluks, M.; Cabulis, P.; Kovalovs, A.; Skukis, E.; Kozikowski, P.; Celiński, M.; Mizera, K.; Gałecka, M.; Kalnins, K.; et al. Experimental Investigation of the Mechanical Properties and Fire Behavior of Epoxy Composites Reinforced by Fabrics and Powder Fillers. *Processes* **2021**, *9*, 738. [CrossRef]
37. Scharte, B. Phosphorus-Based Flame Retardancy Mechanisms-Old Hat or a Starting Point for Future Development? *Materials* **2010**, *3*, 4710–4745. [CrossRef] [PubMed]
38. Nie, S.; Zhang, M.; Yuan, S.; Dai, G.; Hong, N.; Song, L.; Hu, Y.; Liu, X. Thermal and Flame Retardant Properties of Novel Intumescent Flame Retardant Low-Density Polyethylene (LDPE) Composites. *J. Therm. Anal. Calorim.* **2012**, *109*, 999–1004. [CrossRef]
39. Konecki, M.; Półka, M. Analiza Zasięgu Widzialności w Dymie Powstałym w Czasie Spalania Materiałów Poliestrowych. *Polimery* **2006**, *51*, 293–300. [CrossRef]
40. Liu, L.; Qian, M.; Song, P.; Huang, G.; Yu, Y.; Fu, S. Fabrication of Green Lignin-Based Flame Retardants for Enhancing the Thermal and Fire Retardancy Properties of Polypropylene/Wood Composites. *ACS Sustain. Chem. Eng.* **2016**, *4*, 2422–2431. [CrossRef]

Article

Insights into Stoichiometry Adjustments Governing the Performance of Flexible Foamed Polyurethane/Ground Tire Rubber Composites

Adam Olszewski [1], Paulina Kosmela [1], Wiktoria Żukowska [1], Paweł Wojtasz [1], Mariusz Szczepański [1], Mateusz Barczewski [2], Łukasz Zedler [3], Krzysztof Formela [1] and Aleksander Hejna [1,*]

[1] Department of Polymer Technology, Gdańsk University of Technology, Narutowicza 11/12, 80-233 Gdańsk, Poland
[2] Institute of Materials Technology, Poznan University of Technology, Piotrowo 3, 61-138 Poznan, Poland
[3] Department of Molecular Biotechnology and Microbiology, Gdańsk University of Technology, Narutowicza 11/12, 80-233 Gdańsk, Poland
* Correspondence: ohejna12@gmail.com

Abstract: Polyurethanes (PU) are widely applied in the industry due to their tunable performance adjusted by changes in the isocyanate index—stoichiometric balance between isocyanate and hydroxyl groups. This balance is affected by the incorporation of modifiers of fillers into the PU matrix and is especially crucial for PU foams due to the additional role of isocyanates—foaming of the material. Despite the awareness of the issue underlined in research works, the contribution of additives into formulations is often omitted, adversely impacting foams' performance. Herein, flexible foamed PU/ground tire rubber (GTR) composites containing 12 different types of modified GTR particles differing by hydroxyl value (L_{OH}) (from 45.05 to 88.49 mg KOH/g) were prepared. The impact of GTR functionalities on the mechanical, thermomechanical, and thermal performance of composites prepared with and without considering the L_{OH} of fillers was assessed. Formulation adjustments induced changes in tensile strength (92–218% of the initial value), elongation at break (78–100%), tensile toughness (100–185%), compressive strength (156–343%), and compressive toughness (166–310%) proportional to the shift of glass transition temperatures (3.4–12.3 °C) caused by the additional isocyanates' reactions yielding structure stiffening. On the other hand, formulation adjustments reduced composites' thermal degradation onset due to the inferior thermal stability of hard segments compared to soft segments. Generally, changes in the composites' performance resulting from formulation adjustments were proportional to the hydroxyl values of GTR, justifying the applied approach.

Keywords: flexible polyurethane foams; ground tire rubber; waste; filler modification; interfacial interactions

Citation: Olszewski, A.; Kosmela, P.; Żukowska, W.; Wojtasz, P.; Szczepański, M.; Barczewski, M.; Zedler, Ł.; Formela, K.; Hejna, A. Insights into Stoichiometry Adjustments Governing the Performance of Flexible Foamed Polyurethane/Ground Tire Rubber Composites. *Polymers* 2022, 14, 3838. https://doi.org/10.3390/polym14183838

Academic Editor: Sándor Kéki

Received: 18 August 2022
Accepted: 9 September 2022
Published: 14 September 2022

Publisher's Note: MDPI stays neutral with regard to jurisdictional claims in published maps and institutional affiliations.

Copyright: © 2022 by the authors. Licensee MDPI, Basel, Switzerland. This article is an open access article distributed under the terms and conditions of the Creative Commons Attribution (CC BY) license (https://creativecommons.org/licenses/by/4.0/).

1. Introduction

Polyurethanes (PU) are commonly applied in multiple branches of the industry due to their broad range of easily adjustable properties [1,2]. They are present on the market in the form of foams, elastomers, coatings, adhesives, and others, which enables applications, e.g., in construction, building, automotive, electronic, furniture, or household industries, as seats, cushions, thermal and acoustic insulation materials, sealants, adhesives, coatings, membranes, textiles, varnishes, wheels, and others [3–5]. Irrespectively of their type, their performance is driven by their formulations, which are often quite complex and consist of two main components—polyols and isocyanates, but also surfactants, blowing agents, catalysts, chain extenders, and crosslinking agents [6,7]. The detailed type and content of particular compounds significantly impact the structure and performance of polyurethanes [8]. Nevertheless, irrespective of the additional compounds, the most critical is the stoichiometry of the polyaddition reaction between isocyanates and polyols, which generates urethane groups [9]. The ratio of these components, particularly their functional groups, called the isocyanate index, is the main parameter quantitatively describing polyaddition stoichiometry [10]. Its value is most critical for PU foams since the isocyanates, except for the generation of urethanes, are also responsible for the foaming of material in the case of chemical foaming with carbon dioxide generated in the reaction between isocyanates and water [10]. It is a popular approach because it eliminates using physical blowing agents, often harmful to the environment [11].

The isocyanate index can be affected by the presence of electron-donating groups, mostly hydroxyl and amine, due to their high reactivity [12]. Therefore, the presence of moisture in additional compounds containing such functionalities impacts the balance between isocyanate and hydroxyl groups and should be considered when developing PU formulation. The moisture contribution is more accessible to consider due to the high reactivity of isocyanates with water molecules [13]. However, research works often do not consider the impact of other compounds, mainly fillers applied in the manufacturing of PU-based composites. Multiple materials investigated as fillers for PU composites contain nucleophile groups, which can react with isocyanates [14]. As a result, the stoichiometric balance of polyaddition yielding urethane groups is affected along with the structure and performance of resulting composites. The filler-induced disturbance of the isocyanate index affecting the foaming kinetics and cellular structure of PU foams has been reported by Członka et al. [15] and Bryśkiewicz et al. [16]. In order to prepare foamed PU composites deliberately, the contribution of fillers to the isocyanate index should be considered.

Nevertheless, the impact of filler functionalities is not as straightforward as in the case of moisture. The irregularities are primarily attributed to the complex structure of fillers, either plant-based like cellulose or wood flour or others like leather waste or recycled plastics and rubber [17–20]. Contrary to the water or polyols particles, the accessibility of hydroxyl groups present in the structure of fillers is significantly lower, primarily due to the steric hindrance [13].

Keeping in mind these issues, it is essential to consider the impact of fillers' chemical structure during the development of foamed PU-based composites to maximize their efficiency and repeatability. In order to adjust PU formulations properly, the contribution of fillers' functionalities to the overall isocyanate index should be assessed. In our previous work [21], we proposed the method for the determination of hydroxyl value (L_{OH}) of ground tire rubber (GTR) developed by the modification of the method for the determination of free isocyanate group content by titration with dibutylamine, according to ASTM D-2572. The proposed method involves a chemical reaction of GTR with toluene diisocyanate (TDI), so it mirrored the conditions during the preparation of PU/GTR composites. Subsequently, one type of GTR was incorporated into foamed flexible PU matrix, and its contribution to the isocyanate index was considered or not during formulation development [22]. The properties of unfilled PU foams and composites containing deactivated GTR (with blocked hydroxyl groups) were compared. Static tensile and compression tests were performed to evaluate the actual impact of GTR functional groups on the composites' performance. Tensile-based dependencies indicated that around 23–33% of GTR hydroxyls reacted with isocyanates, while compression tests suggested higher values in the 48–57% range. Therefore, obtained results confirmed the above-mentioned incomplete reactivity of fillers' functional groups during PU preparation.

The presented research work is an extension of the previous study. It deals with the mechanical, thermomechanical, and thermal performance of flexible foamed PU/GTR composites containing twelve different types of GTR previously subjected to thermomechanical modification in a twin-screw extruder. As presented in previous works [23–26], such a treatment induces noticeable changes in the chemical structure of GTR, including oxidation reactions leading to the formation of electron-donating groups, which can react with isocyanates. Therefore, the impact of treatment temperature (from 50 to 200 °C) and screw speed (from 50 to 300 rpm) on the GTR hydroxyl value and the performance of resulting

PU/GTR composites were assessed. To evaluate the actual impact of GTR functionalities on the prepared composites, two approaches to PU preparation were applied, with and without considering the L_{OH} of GTR fillers in calculations of the isocyanate index.

2. Materials and Methods

2.1. Materials

During the preparation of flexible foamed PU/GTR composites, Rokopol®F3000 and Rokopol®V700 produced by the PCC Group (Brzeg Dolny, Poland), as well as glycerol obtained from Sigma Aldrich (Saint Louis, MI, USA), were applied as polyol components. The first two compounds are homogenous, clear, liquid polyether polyols and glycerol-based polyoxyalkylene triols. Rokopol®F3000 and Rokopol®V700 are characterized by L_{OH} in the range of 53–59 and 225–250 mg KOH/g, respectively. Their molecular weights are 3000 and 700 g/mol, respectively, while dynamic viscosity at 25 °C ranges from 460 to 520 and 220 to 270 mPa·s, respectively. These compounds are characterized by acid values below 0.1 mg KOH/g, water content below 0.1%, and color below 50 Hazen units. The purity of applied glycerol was min. 99.5%, its hydroxyl value equaled 1800 mg KOH/g, while density was 1.26 g/cm^3. As the isocyanate component was applied SPECFLEX NF 434 purchased from M. B. Market Ltd. (Baniocha, Poland), methylene diphenyl diisocyanate-based component characterized by the brown color, free isocyanate content of 29.5%, dynamic viscosity at 25 °C of 66 mPa·s, and specific gravity at 25 °C of 1.21. Applied PU composition also included three catalysts. The first one was PC CAT® TKA30 from Performance Chemicals (Belvedere, UK)—potassium acetate dissolved in monoethylene glycol applied as a crosslinking catalyst and characterized by the specific gravity at 25 °C of 0.964. The second catalyst was Dabco 33LV from Air Products (Allentown, USA)—33 wt. % solution of 1,4-Diazabicyclo[2.2.2]octane in dipropylene glycol used as a gelling catalyst. It was characterized by the dynamic viscosity at 25 °C of 125 mPa·s, specific gravity at 25 °C of 1.03, and L_{OH} of 560 mg KOH/g. The last catalyst was dibutyltin dilaurate acquired from Sigma Aldrich (Saint Louis, MO, USA), characterized by specific gravity at 25 °C of 1.066 and tin content of 18.2–18.9%. Distilled water was applied as a chemical blowing agent for prepared PU foams. Ground tire rubber received from Recykl S.A. (Śrem, Poland) was applied as filler for PU-based composites. It was obtained in the process of ambient grinding of post-consumer tires (a mix of passenger cars and truck tires). Applied GTR was characterized by the L_{OH} of 61.7 mg KOH/g.

2.2. Modification of Ground Tire Rubber

Thermomechanical treatment of GTR was performed with an EHP 2 × 20 Sline co-rotating twin-screw extruder from Zamak Mercator (Skawina, Poland) as described in our previous work [21]. Briefly, GTR was extruded at the constant feeding rate of 2 kg/h, the barrel temperature was set at 50, 100, 150, or 200 °C, and the screw speed was 50, 100, or 300 rpm. Table 1 presents the impact of thermomechanical treatment on the hydroxyl value (L_{OH}) of GTR, determined using the modified test method for isocyanate groups, as described in our other work [21]. The highest hydroxyl values were noted for GTR modified at 100 and 150 °C, pointing to the optimal process temperature, which can accelerate the oxidation occurring during thermomechanical treatment, but on the other hand, may induce partial decomposition of material and reduce the content of functional groups contributing to the hydroxyl value. This effect of excessive GTR oxidation led to the generation of carboxyl groups, which show lower reactivity with isocyanates, hence lower hydroxyl value was noted in our previous work [27]. Considering the impact of screw speed, it shows contradictory effects because increasing this parameter leads to higher shear forces acting on the material, but for a shorter time. Therefore, the overall impact of extrusion parameters on the materials' properties is not straightforward, which was also reported by other researchers working on extrusion thermomechanical treatment of various materials [28–32].

Table 1. Hydroxyl values of thermomechanically modified GTR particles depending on the applied processing conditions.

GTR Treatment Parameters		L_{OH}, mg KOH/g	GTR Treatment Parameters		L_{OH}, mg KOH/g
Temperature, °C	Screw Speed, rpm		Temperature, °C	Screw Speed, rpm	
50	50	70.65 ± 2.37	150	50	71.13 ± 1.76
	100	69.78 ± 2.43		100	68.00 ± 1.60
	300	59.26 ± 2.81		300	76.55 ± 0.85
100	50	88.49 ± 2.97	200	50	45.05 ± 2.01
	100	75.71 ± 2.92		100	47.08 ± 1.50
	300	71.85 ± 2.01		300	55.21 ± 2.85

2.3. Preparation of Flexible Polyurethane/Ground Tire Rubber Composite Foams

Composite foams were prepared on a laboratory scale by a single-step method with the isocyanate index of 1:1. A predetermined amount of selected GTR filler (20 parts by weight to the mass of foam) was mixed with the polyols at 1000 rpm for 60 s to guarantee its proper distribution. Afterward, all components were mixed for 10 s at 1800 rpm and poured into a closed aluminum mold with dimensions of 20 × 10 × 4 cm. After demolding, the samples were conditioned at room temperature for 24 h. The amount of reaction mixture poured into the mold was adjusted to obtain foams with a similar level of apparent density, which noticeably affects cellular materials' performance. As a result, all foams were characterized by an apparent density of 187.5 ± 2.5 kg/m^3. Table 2 contains the details of foam formulations. Samples were coded according to the type of applied GTR and formulation variant. For example, the sample containing GTR modified at 100 °C, and 50 rpm including the hydroxyl value of GTR in the formulation, was coded as 100/50/OH.

Table 2. Formulations of PU/GTR composite foams investigated in the presented study.

Component	Type of Applied GTR (Temperature of Treatment/Screw Speed)											
	50/50		50/100		50/300		100/50		100/100		100/300	
	-	OH	-	OH	-	OH	-	OH	-	OH	-	OH
Rokopol F3000	26.10	26.38	26.10	26.37	26.10	26.27	26.10	26.03	26.10	26.33	26.10	26.13
Rokopol V700	26.10	26.38	26.10	26.37	26.10	26.27	26.10	26.03	26.10	26.33	26.10	26.13
Glycerol	0.60	0.63	0.60	0.63	0.60	0.63	0.60	0.63	0.60	0.63	0.60	0.63
DBTDL	0.50	0.48	0.50	0.48	0.50	0.48	0.50	0.47	0.50	0.48	0.50	0.47
33LV	0.30	0.32	0.30	0.32	0.30	0.32	0.30	0.31	0.30	0.32	0.30	0.32
TKA30	0.30	0.32	0.30	0.32	0.30	0.32	0.30	0.31	0.30	0.32	0.30	0.32
Water	0.30	0.27	0.30	0.27	0.30	0.27	0.30	0.27	0.30	0.27	0.30	0.27
GTR	20.00	16.18	20.00	16.18	20.00	16.90	20.00	16.90	20.00	16.15	20.00	16.90
pMDI	25.80	29.05	25.80	29.06	25.80	28.54	25.80	29.05	25.80	29.18	25.80	28.84
Component	Type of Applied GTR (Temperature of Treatment/Screw Speed)											
	150/50		150/100		150/300		200/50		200/100		200/300	
	-	OH	-	OH	-	OH	-	OH	-	OH	-	OH
Rokopol F3000	26.10	26.08	26.10	26.41	26.10	26.04	26.10	26.72	26.10	26.71	26.10	26.59
Rokopol V700	26.10	26.08	26.10	26.41	26.10	26.04	26.10	26.72	26.10	26.71	26.10	26.59
Glycerol	0.60	0.63	0.60	0.64	0.60	0.63	0.60	0.64	0.60	0.64	0.60	0.64
DBTDL	0.50	0.47	0.50	0.48	0.50	0.47	0.50	0.48	0.50	0.48	0.50	0.48
33LV	0.30	0.32	0.30	0.32	0.30	0.31	0.30	0.32	0.30	0.32	0.30	0.32
TKA30	0.30	0.32	0.30	0.32	0.30	0.31	0.30	0.32	0.30	0.32	0.30	0.32
Water	0.30	0.27	0.30	0.27	0.30	0.27	0.30	0.27	0.30	0.27	0.30	0.27
GTR	20.00	16.90	20.00	16.20	20.00	16.90	20.00	16.39	20.00	16.39	20.00	16.32
pMDI	25.80	28.94	25.80	28.97	25.80	29.04	25.80	28.12	25.80	28.14	25.80	28.47

2.4. Measurements

After conditioning, foamed polyurethane composites were cut into samples whose properties were later determined following the standard procedures.

The compressive strength of the studied samples was estimated following ISO 604. The cylindric samples with dimensions of 20 × 20 mm (height and diameter) were measured with a slide caliper with an accuracy of 0.1 mm. The compression test was performed on a Zwick/Roell Z020 tensile tester (Ulm, Germany) at a constant speed of 15%/min until reaching 60% deformation.

The tensile strength of foams was estimated following ISO 1798. The beam-shaped samples with 10 × 10 × 100 mm^3 dimensions were measured with a slide caliper with an accuracy of 0.1 mm. The tensile test was performed on a Zwick/Roell Z020 tensile tester (Ulm, Germany) at a constant speed of 500 mm/min.

Dynamical mechanical analysis (DMA) was performed using a Q800 DMA instrument from TA Instruments (New Castle, DE, USA) at a heating rate of 4 °C/min and the temperature range from −100 to 150 °C. Samples were cylindrical-shaped, with dimensions of 10 × 12 mm.

The thermogravimetric (TGA) analysis of GTR and composites was performed using the TG 209 F3 apparatus from Netzsch (Selb, Germany). Samples of foams weighing approx. 10 mg were placed in a ceramic dish. The study was conducted in an inert gas atmosphere—nitrogen in the range from 30 to 800 °C with a temperature increase rate of 10 °C/min.

3. Results and Discussion

Figure 1 presents the stress-strain curves for prepared composite foams and provides the results of performed tensile tests. It can be seen that all foams obtained without taking into account the hydroxyl values of GTR were characterized by tensile strength in the range of 134–211 kPa. When the L$_{OH}$ of ground tire rubber was considered during the preparation of foams' formulations, tensile strength was in the range of 188–339 kPa. Such an effect is associated with the reactions between hydroxyl groups present on the surface of rubber particles and isocyanates. Depending on their extent, hence the hydroxyl value of GTR, these reactions reduce the number of isocyanate groups contributing to polymerization reactions causing weakening of PU structure [27]. When GTR and its hydroxyl values were considered during foams' preparation, the cumulative amount of isocyanates in the system was higher, resulting in the enhancement of PU phase strength [33]. As presented in Table 1, the lowest L$_{OH}$ values were obtained for GTR particles modified at 200 °C, whose incorporation into PU foams yielded their highest tensile strengths in the range of 204–211 kPa. In the case of these samples, the GTR was less competitive for isocyanates, which did not deteriorate the foams' performance.

No direct influence of GTR treatment parameters on their mechanical performance was noted, as presented in Figure 2. The observed pattern was that the tensile strength and tensile toughness, which takes strength into account, showed lower values for the screw speed of 100 rpm compared to 50 and 300 rpm. Such an effect can be explained by the contradictory action of screw speed related to the materials' residence time in the extruder barrel and the magnitude of shear forces acting on the material [29]. For screw speed of 50 rpm, the material was subjected to a temperature of the extruder barrel and weaker shearing for a longer time, while for 300 rpm, stronger shearing acted on the material for a shorter time. Contradictory effects of screw speed were noted in other works dealing with extrusion-based particle modifications [30,34,35].

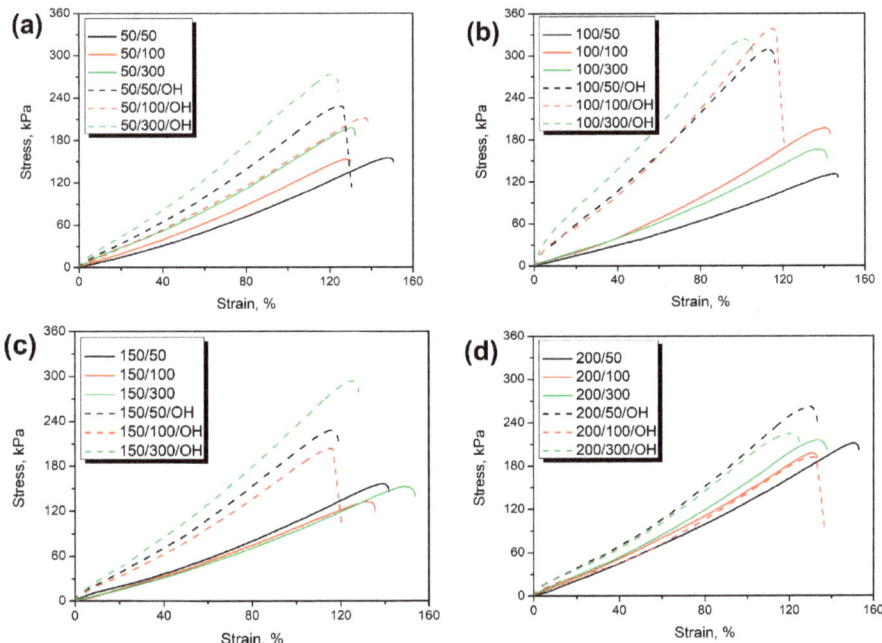

Figure 1. Stress-strain curves of composite foams containing GTR modified at (**a**) 50 °C, (**b**) 100 °C, (**c**) 150 °C, and (**d**) 200 °C.

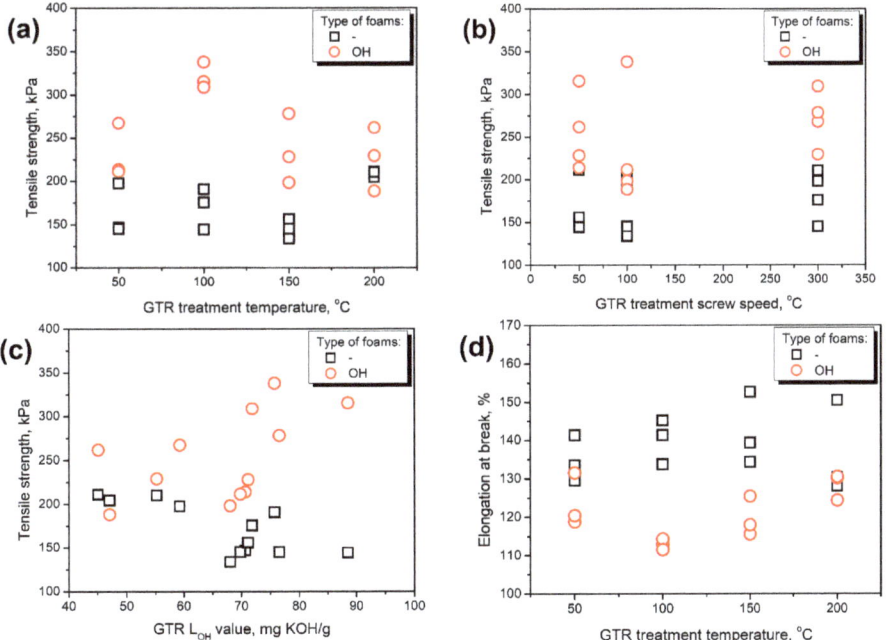

Figure 2. *Cont.*

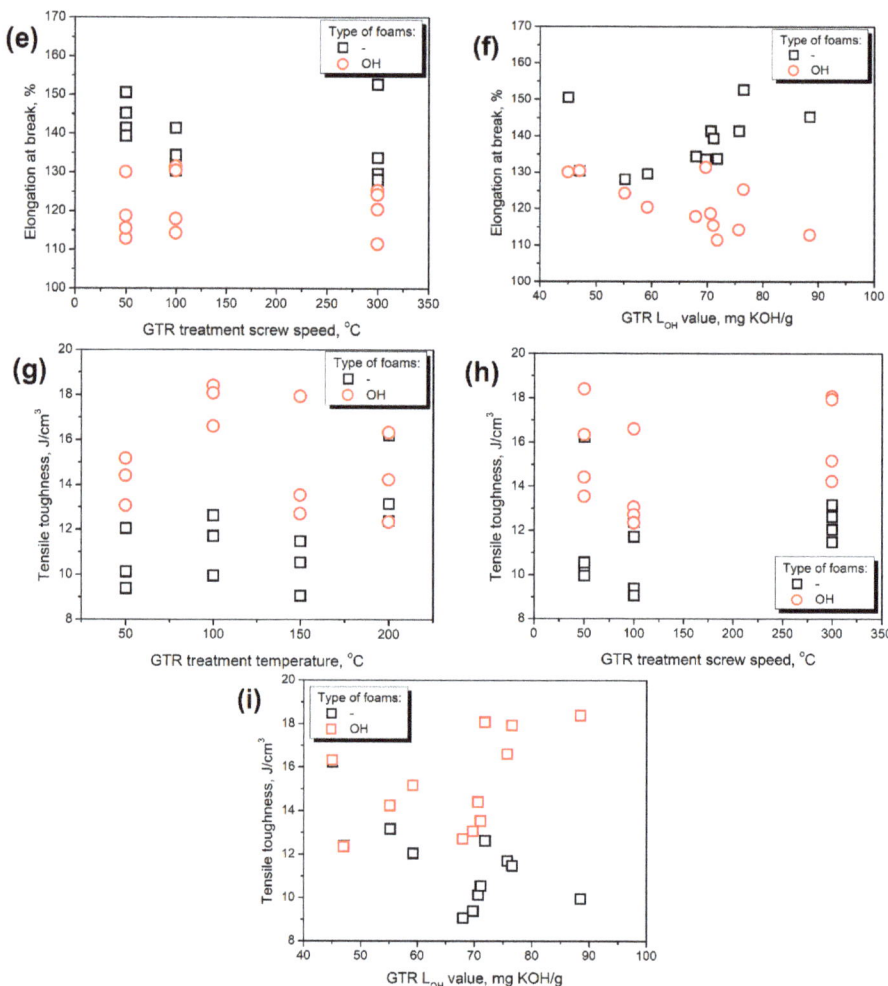

Figure 2. The impact of (**a,d,g**) GTR treatment temperature, (**b,e,h**) GTR treatment screw speed, and (**c,f,i**) GTR LOH value on the (**a,b,c**) tensile strength, (**d,e,f**) elongation at break, and (**g,h,i**) tensile toughness of prepared PU/GTR composite foams.

The strength of composites, their toughness, and elongation at break more significantly depended on the GTR L_{OH} values resulting from thermomechanical treatment, which is presented in more detail in Figure 2c,f,i. When the hydroxyl value of ground tire rubber was not considered, its increase yielded deterioration of mechanical strength and toughness, which can be attributed to the reduced amount of isocyanate groups taking part in polymerization reactions generating urethane groups. Due to the reduced amount of urethane groups, which act as crosslinks, foams could withstand more extensive deformations, slightly increasing elongation at break [36]. On the other hand, when hydroxyl values of ground tire rubber were considered, more urethane groups were generated. Hence, the PU phase was strengthened, which led to an increase in tensile strength and a drop in elongation at break.

Figure 2 also shows the evident differences in tensile performance between foams obtained with and without considering the hydroxyl values of GTR. Such an effect is related to the amount of additional isocyanate introduced into the system to compensate for the impact of hydroxyl groups on the surface of GTR particles. For more detailed analysis, Figure 3 shows the relationship between the enhancement of tensile strength and toughness, as well as the drop of elongation at break after formulation adjustments and the L_{OH} values of applied GTR. It can be seen that these relationships are predominantly proportional, which is in line with our previous work dealing with PU/GTR composites prepared with varying isocyanate indexes [22]. Clearly, the presented results confirm literature data suggesting additional crosslinking resulting from the excess of isocyanate groups in the system, which increases tensile strength, enhances toughness, and limits the mobility of polymer macromolecules reducing elongation at break [22,37].

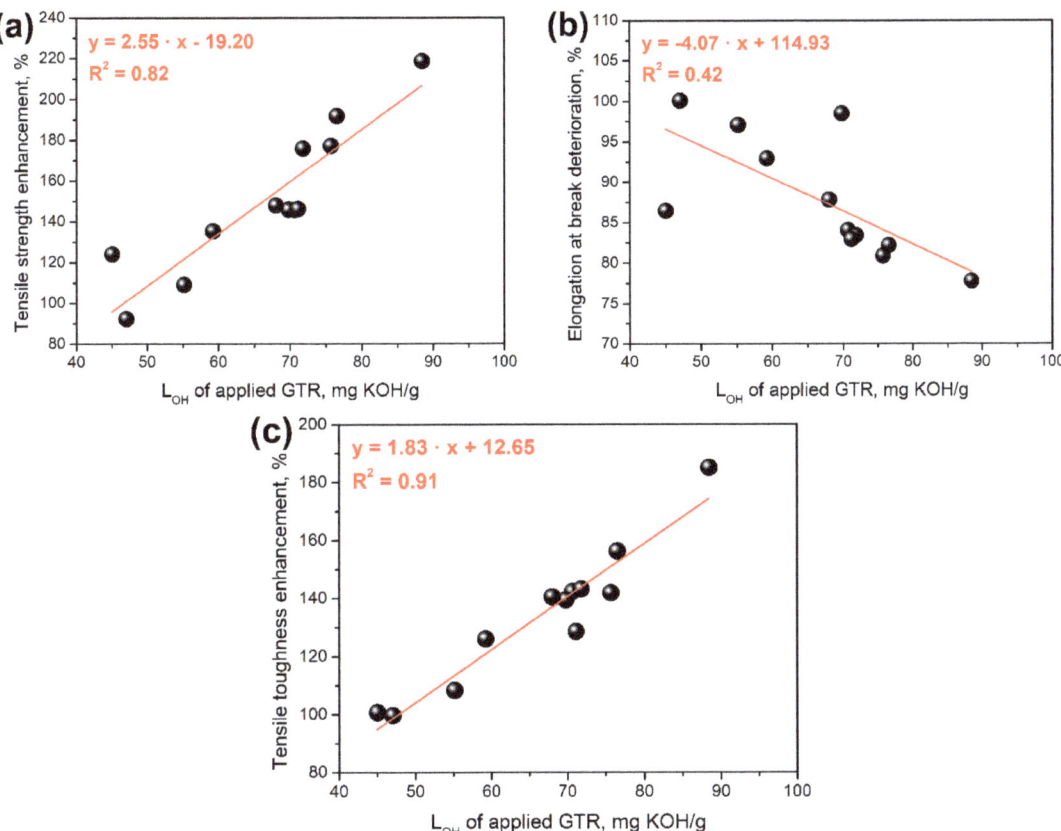

Figure 3. The impact of L_{OH} of applied GTR on the (**a**) tensile strength enhancement, (**b**) elongation at break deterioration, and (**c**) tensile toughness enhancement of PU/GTR composite foams after formulation adjustments.

Figure 4 shows the static compressive performance of prepared PU/GTR composites and their dependence on the type of introduced GTR particles and applied formulation adjustments. The shape of compressive stress-deformation curves is typical for porous materials, with the gradual strength increase at higher deformation resulting from buckling cell walls and approaching the bulky structure by foams [38]. The significant enhancement of materials' strength was observed after taking into L_{OH} values of applied rubber particles, which can be presented in Figure 5. It can be seen that the enhancement of composites' compressive performance was increasing with the GTR hydroxyl value despite some variations from proportionality. Compressive strength values at 50% deformation prior to the significant foams' densification were increased by 68–224% after formulation adjustments. The strengthening of composites could be attributed to the higher extent of crosslinking resulting from the additional isocyanate reactions [39]. Additional covalent bonds acted as network nodes and reduced the mobility of macromolecular chains inside composites [40]. Therefore, composites require higher external force to reach certain deformation, which the increase in toughness can quantitatively express.

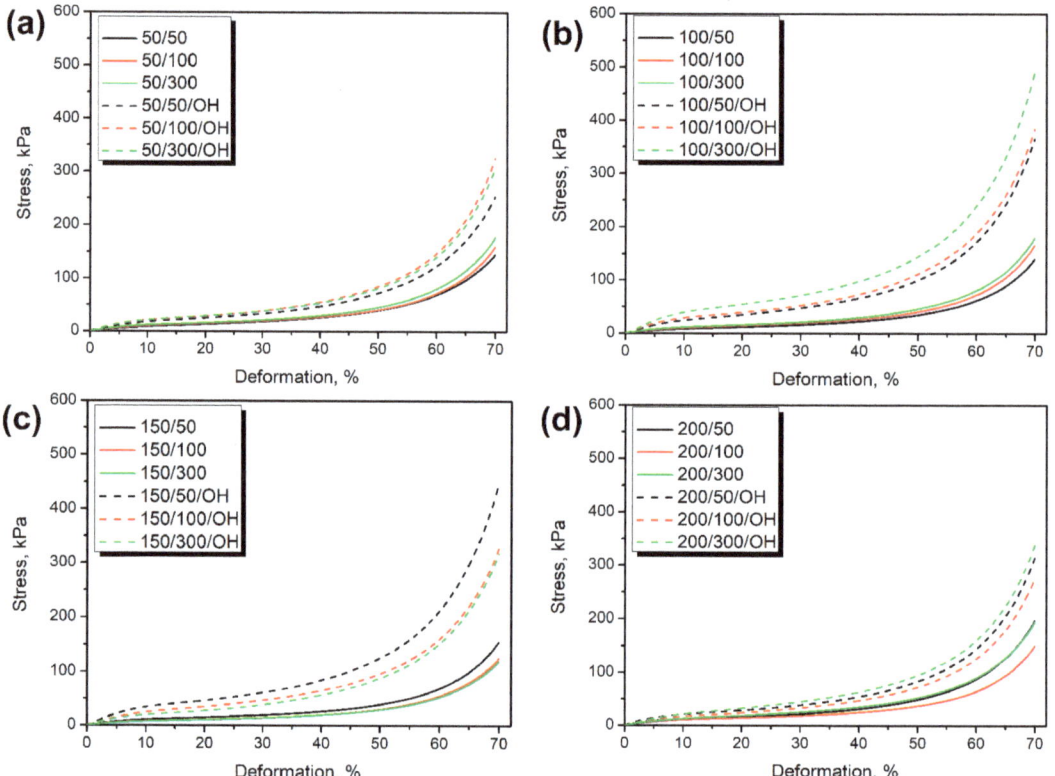

Figure 4. Stress-deformation curves of composite foams containing GTR modified at (**a**) 50 °C, (**b**) 100 °C, (**c**) 150 °C, and (**d**) 200 °C.

Figure 6 presents the impact of particular parameters of GTR thermomechanical treatment on the value of composites' compressive toughness, which increased after formulation adjustments along with the compressive strength. Similar to the static tensile tests, the most notable increase in compressive performance, both strength and toughness were noted for composites containing GTR modified at 100 and 150 °C, which can be associated with the highest hydroxyl values (see Table 1). No direct impact of screw speed can be noted. Considering the impact of hydroxyl groups on the surface of GTR particles, the most significant dependence can be observed. Without considering GTR L_{OH} values in formulations, the compressive toughness was lower for composites containing "hydroxyl-richer" GTR particles. Such an effect could be associated with the attraction of isocyanate groups by additional hydroxyls and the weakening of the PU structure [41]. Nevertheless, the performance deterioration was not very significant due to the only partial reactivity of GTR hydroxyl groups with isocyanates, which can be attributed to their lower reactivity compared to polyols, e.g., due to steric hindrance [13]. Compressive toughness increased with GTR hydroxyl values for OH type of foams due to the above-mentioned crosslinking enhancement. What is essential for all discussions related to compressive performance is that all composites were characterized by a similar apparent density, which plays a crucial role in cellular materials [42].

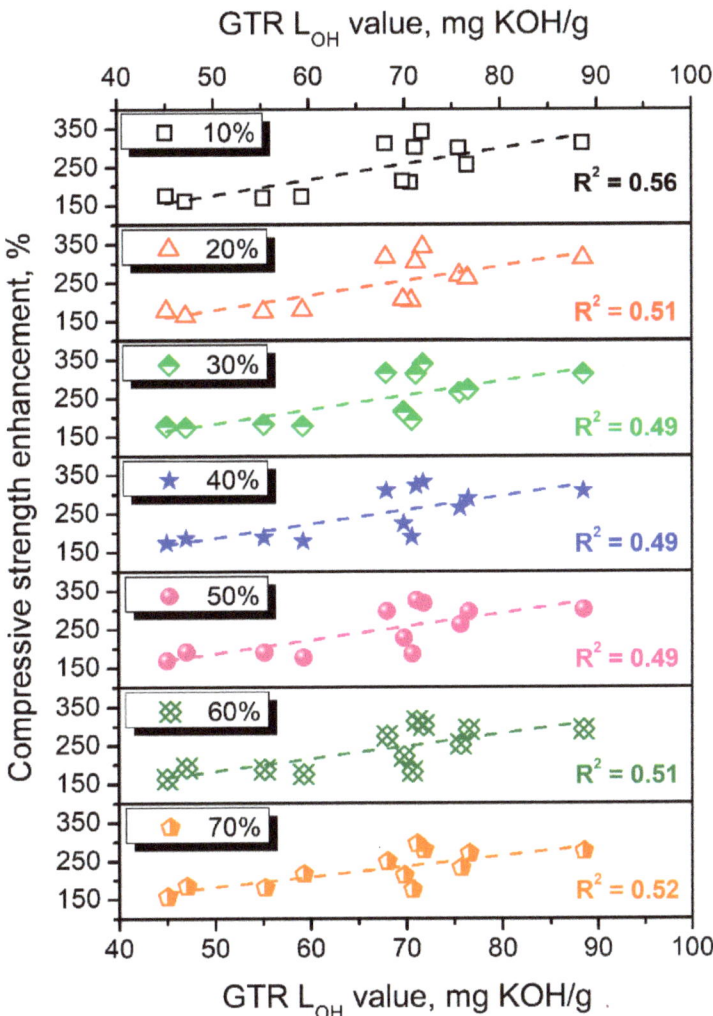

Figure 5. The enhancement of compressive strength at a different level of composites' deformation resulting from applied formulation adjustments.

Figure 7 presents the temperature plots of PU/GTR composites' storage modulus (E') obtained during dynamic mechanical analysis. All the analyzed materials show a typical drop of modulus attributed to the glass transition of polymer matrix [43]. The drop of E' was around two orders of magnitude, as noted in previous works [41,44]. Formulation adjustments resulting from additional hydroxyl groups present on the surface of GTR particles and increased isocyanate content caused stiffening of composites expressed by the E' rise, which can be seen in the presented graphs. For composites prepared without taking into account the impact of the GTR hydroxyl value, the storage modulus at 25 °C was in the range of 194–1073 kPa, while formulation adjustments shifted the E′ range to 823–4106 kPa. Similar to the enhancement of tensile strength, the stiffening of composites was attributed to the additional crosslinking reactions induced by a higher content of isocyanate groups in the system.

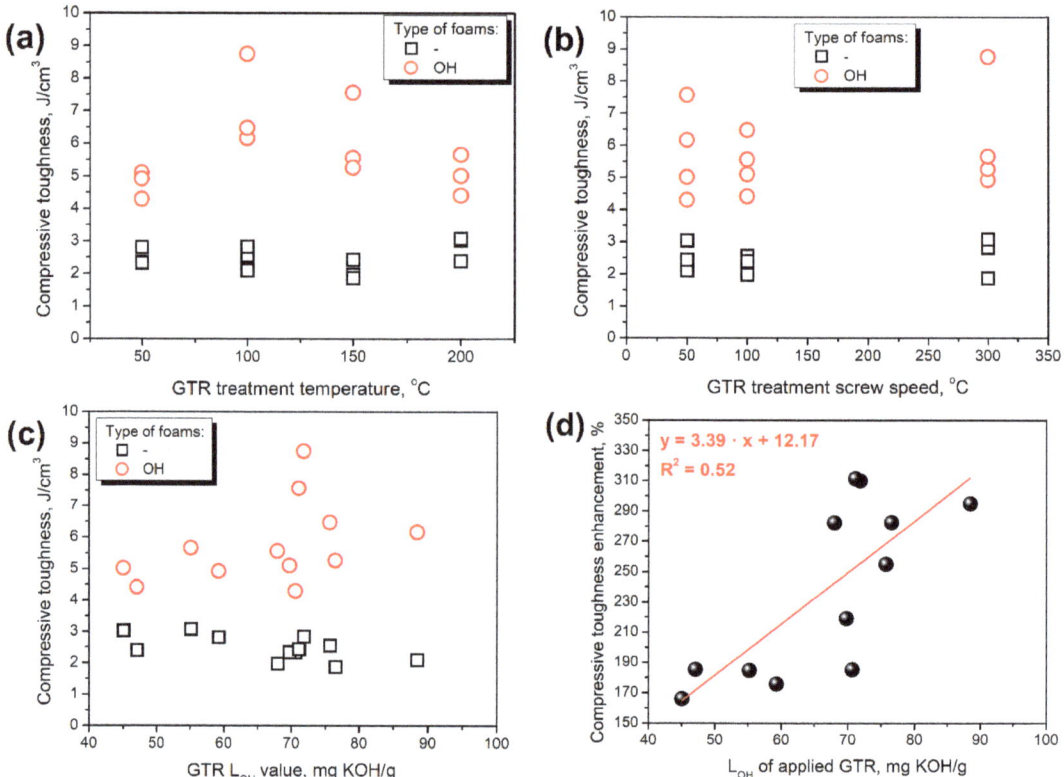

Figure 6. The impact of (**a**) GTR treatment temperature, (**b**) GTR treatment screw speed, and (**c**) GTR L_{OH} value on the compressive toughness of prepared PU/GTR composite foams, and (**d**) the relationship between L_{OH} of applied GTR and compressive toughness enhancement of PU/GTR composite foams after formulation adjustments.

For a more detailed analysis of GTR thermomechanical treatment and formulation adjustments on composites' stiffness, based on the dynamic mechanical analysis results, the composite performance factor (C factor) was calculated according to the following Equation (1):

$$C = ((E'_{g\,c}/E'_{r\,c}) / (E'_{g\,m}/E'_{r\,m})) \tag{1}$$

where: E'_g—storage modulus in the glassy state (−100 °C), MPa; E'_r—storage modulus in the rubbery state (90 °C), MPa; subscripts c and m refer to composite and matrix.

To evaluate the impact of thermomechanical treatment of GTR on the mechanical performance of prepared materials, composite containing unmodified ground tire rubber particles, reported in our previous work [25] reference matrix material. Values of the C factor calculated for prepared composites are presented in Figure 8. They consider the changes in composites' stiffness resulting from the glass transition of polyurethane matrix. The decreasing C factor indicates higher efficiency of reinforcing effect resulting from the thermomechanical treatment of GTR [45].

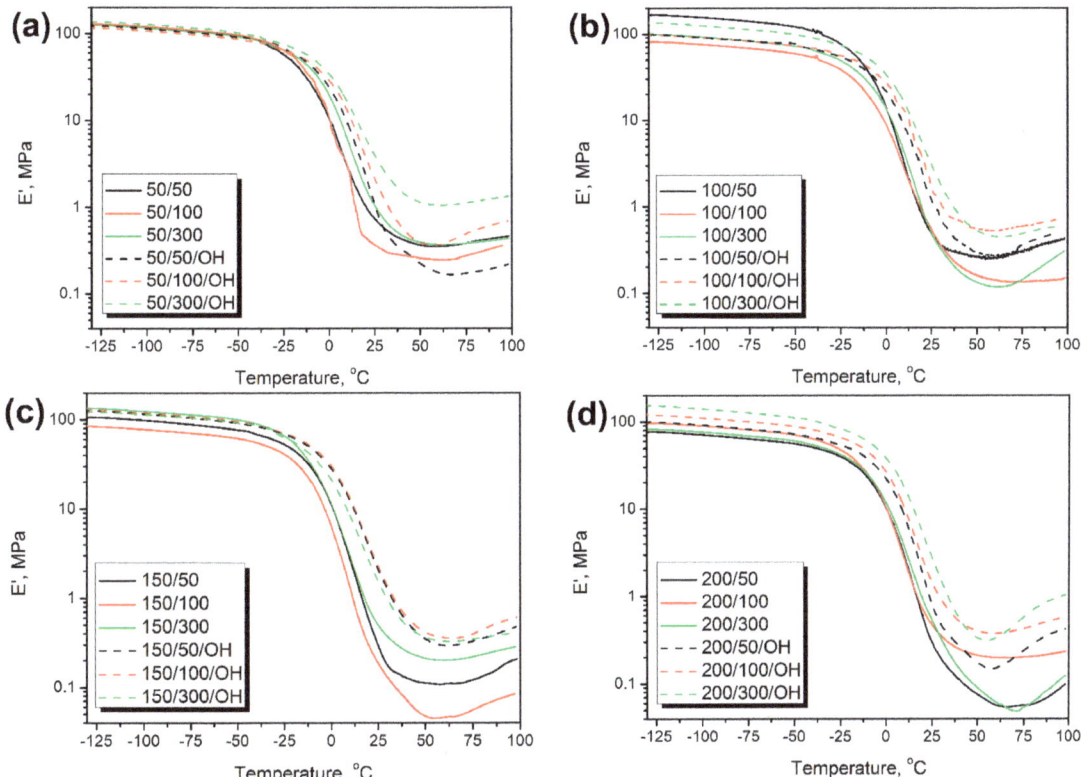

Figure 7. Temperature plots of the storage modulus for PU/GTR foams containing GTR modified at (**a**) 50 °C, (**b**) 100 °C, (**c**) 150 °C, and (**d**) 200 °C.

Noticeably higher values of C factor were noted for foams prepared without consideration of GTR hydroxyl value, which confirms the results of static mechanical tests. The opposite effect was noted only for composites containing 50/50 GTR particles. The drop in C factor values can be attributed to the partial attraction of isocyanate groups by hydroxyls present on the surface of GTR particles resulting in the strengthening of interfacial adhesion [41]. Moreover, increasing the isocyanate loading to match the total L_{OH} of the polyol mixture, taking into account GTR contribution in most of the samples, led to the C factor values below unity, indicating a reinforcing effect compared to the application of unmodified filler [45].

Figure 9 presents the impact of additional chemical interactions on the composite performance factor. For composites prepared without taking into account the impact of GTR hydroxyl value, the C factor was slightly decreasing with L_{OH}, while for the OH-type foams, the trend was hardly noticed. Considering foams without formulation adjustment, increasing hydroxyl values implicate a greater extent of chemical reactions at the interface due to the increasing number of reactive hydroxyls on GTR particles' surface [46]. However, the impact was not very substantial. When the GTR hydroxyl groups were considered in calculating the required isocyanate amount, the hydroxyl value had a very insignificant impact on the C factor due to the formulation adjustments and tailored isocyanate content in the reacting system.

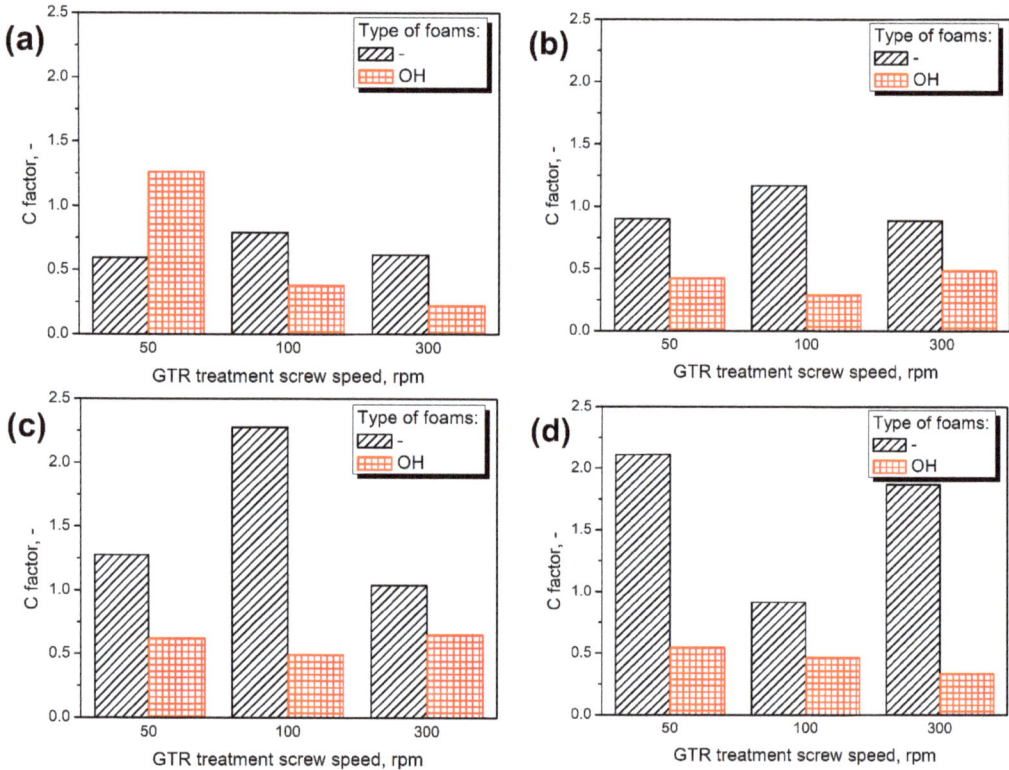

Figure 8. Values of C factor calculated for composites containing GTR modified at (**a**) 50 °C, (**b**) 100 °C, (**c**) 150 °C, and (**d**) 200 °C.

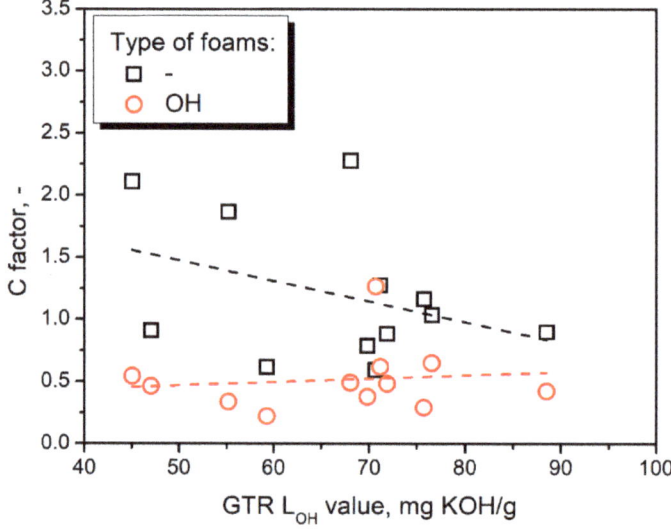

Figure 9. The relationship between GTR L_{OH} values and the C factor of prepared PU/GTR composites.

Temperature plots of loss tangent (tan δ) determined during DMA analysis of prepared composite foams are presented in Figure 10. In addition to the information about materials' stiffness, the dynamic mechanical analysis may provide important insights into their damping behavior. Loss tangent measures a material's ability to absorb and dissipate mechanical energy [47]. Typically, the incorporation of solid fillers into the polymer matrix enhances the storage modulus at the expense of limiting the loss of tangent peak height [48]. Similar effects were noted in the presented work after formulation adjustments, which can be associated with the limited mobility of polymer macromolecules, confirming the enhanced crosslinking and strengthened interfacial adhesion [49].

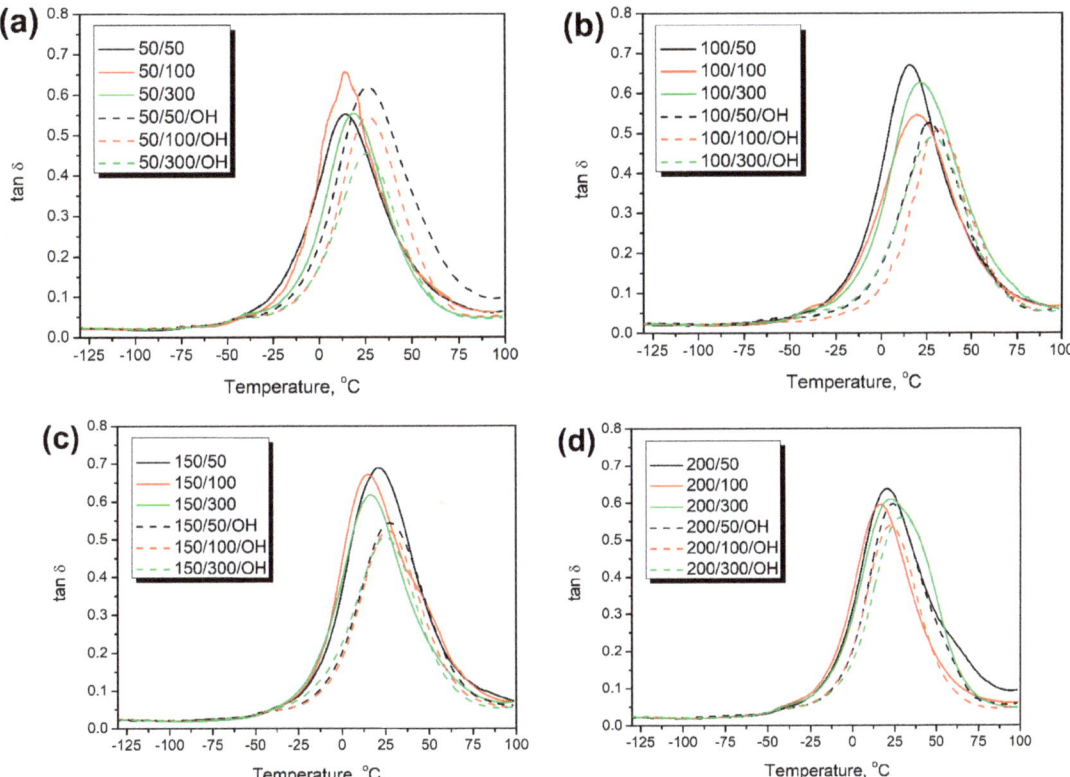

Figure 10. Temperature plots of loss tangent for composites containing GTR modified at (**a**) 50 °C, (**b**) 100 °C, (**c**) 150 °C, and (**d**) 200 °C.

Except for the value of loss tangent, which quantifies the materials' damping ability, the temperature position of the tan δ peak can be used to determine the glass transition temperature (T_g). In the case of polyurethane materials, the T_g value is susceptible to the balance between isocyanate and hydroxyl groups in the system expressed by the isocyanate index [50,51]. Figure 11 shows that the glass transition temperature was significantly more affected by hydroxyl values of introduced GTR particles than conditions of thermomechanical treatment. Similar to other mechanical properties, foams prepared with and without formulation adjustments can be clearly distinguished by the values of T_g. Such an effect is attributed to the differences in the above-mentioned balance between isocyanate and hydroxyl groups in the system. This balance was affected by the L_{OH} of GTR and the adjustments of composites' formulations. Taking the GTR hydroxyl values into account and simultaneously increasing the amount of isocyanate in the system enhanced structural crosslinking by a higher extent of additional reactions leading to allophanates and biurets [12]. As a result, materials require higher energy, so higher temperature, to exceed the Gibbs free energy, increase the free volume and reach the rubbery state. For composites prepared without formulation adjustments, T_g was in the range of 14.0–23.3 °C, while after considering GTR hydroxyl values, its values increased to 22.8–31.3 °C. The shift of T_g was increasing along with the GTR L_{OH} values. However, the relationship was not perfectly proportional. Other researchers have already reported similar changes in T_g induced by increasing the isocyanate index [36,37]. Observed shifts justified the above-mentioned enhancement of composites' tensile and compressive performance.

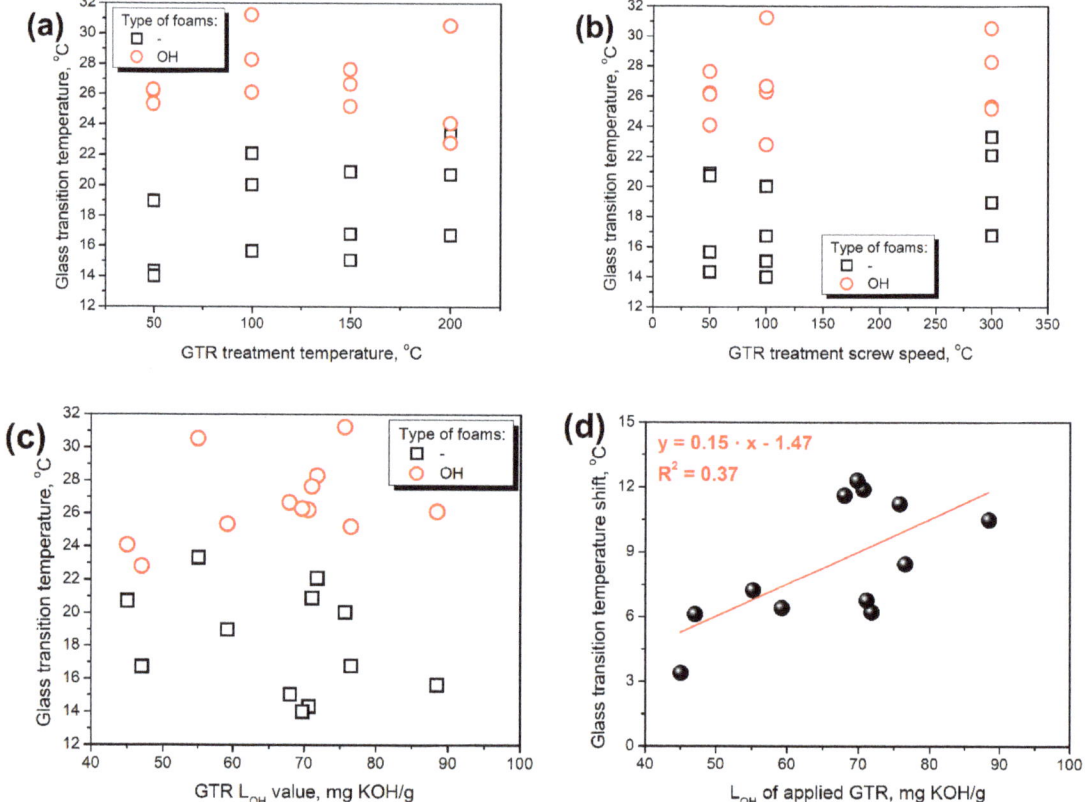

Figure 11. The impact of (**a**) GTR treatment temperature, (**b**) GTR treatment screw speed, and (**c**) GTR L_{OH} value on the glass transition temperature of prepared PU/GTR composite foams, and (**d**) the relationship between L_{OH} of applied GTR and glass transition temperature shift for PU/GTR composite foams after formulation adjustments.

Figure 12 presents the mass loss curves obtained during thermogravimetric analysis of thermomechanically treated GTR samples and PU composite foams. It can be seen that a similar course of thermal decomposition characterized all applied GTR samples despite different modification conditions. Such an effect can be associated with the chemical composition of GTR, particularly the content of natural rubber and styrene-butadiene rubber [52]. For all the analyzed samples, the onset of thermal degradation, determined as a temperature of 2 wt% mass loss, was in the range of 251.1–263.3 °C. Considering the values of char residue, they were in the range of 37.76–39.15 wt%. Figure 13 provides more insights into the impact of GTR treatment parameters and its hydroxyl value on its thermal degradation parameters. It can be seen that contrary to the mechanical performance, temperature and screw speed applied during thermomechanical treatment showed a more direct impact on the thermal stability of GTR. Hardly any relationship between degradation onset and GTR hydroxyl values can be noted. Such an effect can be related to the partial decomposition of GTR particles' surface induced by temperature and shear forces acting on material inside the extruder barrel [53].

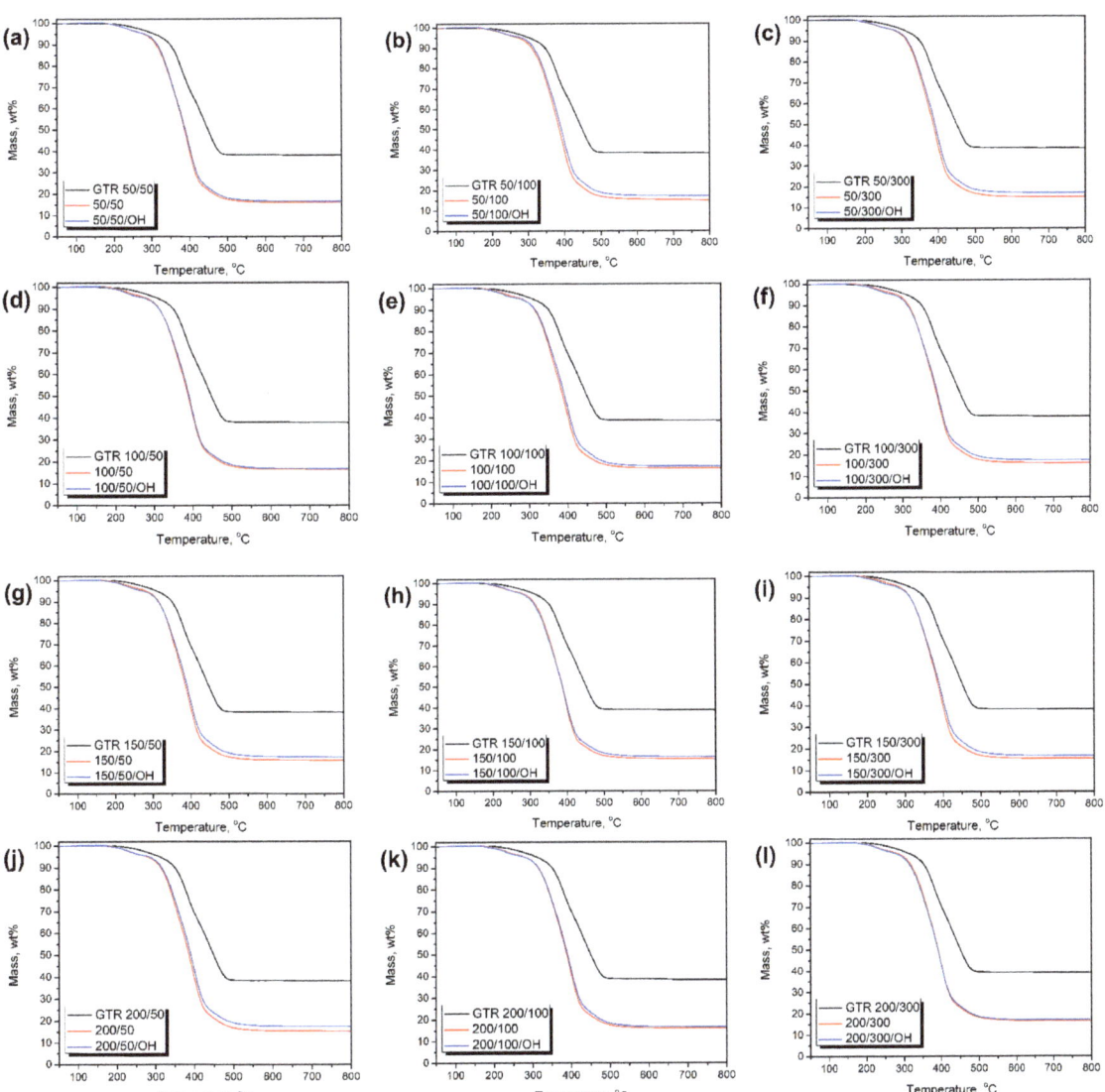

Figure 12. Mass loss curves obtained during thermogravimetric analysis of thermomechanically treated GTR samples and PU composite foams containing different GTR types: (**a**) 50/50, (**b**) 50/100, (**c**) 50/300, (**d**) 100/50, (**e**) 100/100, (**f**) 100/300, (**g**) 150/50, (**h**) 150/100, (**i**) 150/300, (**j**) 200/50, (**k**) 200/100, and (**l**) 200/300.

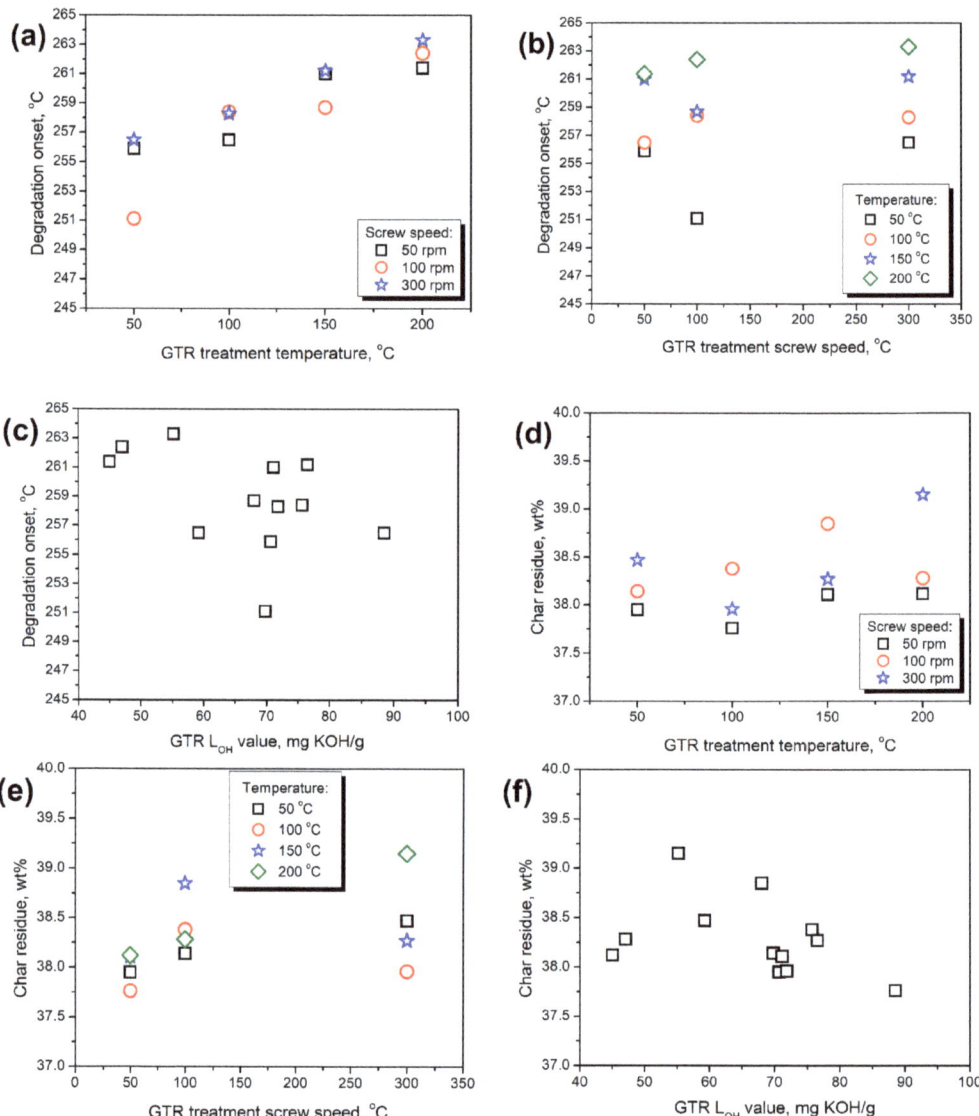

Figure 13. The impact of (**a,d**) GTR treatment temperature, (**b,e**) GTR treatment screw speed, and (**c,f**) GTR L_{OH} value on the (**a–c**) degradation onset, and (**d–f**) char residue of applied GTR samples.

A similar but not as pronounced impact of thermomechanical treatment conditions was noted on char residue. The actions of temperature and shear forces resulted in the decomposition of lower-molecular weight fractions or partially non-crosslinked portions of material characterized by lower thermal stability during treatment. As a result, the remaining part of the material was characterized by superior thermal stability and higher char residue.

Figure 14 presents the impact of GTR treatment parameters and resulting hydroxyl values on the thermal decomposition of prepared composite foams. Despite the noticeable changes in GTR thermal stability induced by varying treatment conditions, the degradation onset of resulting composite foams was hardly affected. Mass loss curves presented in Figure 12 are very similar and for all of the analyzed samples degradation onset was in the range of 221–234 °C, irrespective of the applied GTR type and formulation adjustments. Moreover, no visible trends in thermal stability were noted. Similar to tensile performance, the most noticeable impact was related to the hydroxyl values of GTR. However, the relationships were still far from proportional. Nevertheless, higher hydroxyl values slightly enhanced thermal stability for foams obtained without taking into account the L_{OH} of GTR, while for OH foams, the effect was the opposite. It could be attributed to the formulation adjustments and increased amount of isocyanates in reacting systems leading to the greater extent of reactions yielding urethane, biuret, and allophanate groups, characterized by inferior thermal stability compared to soft segments [54].

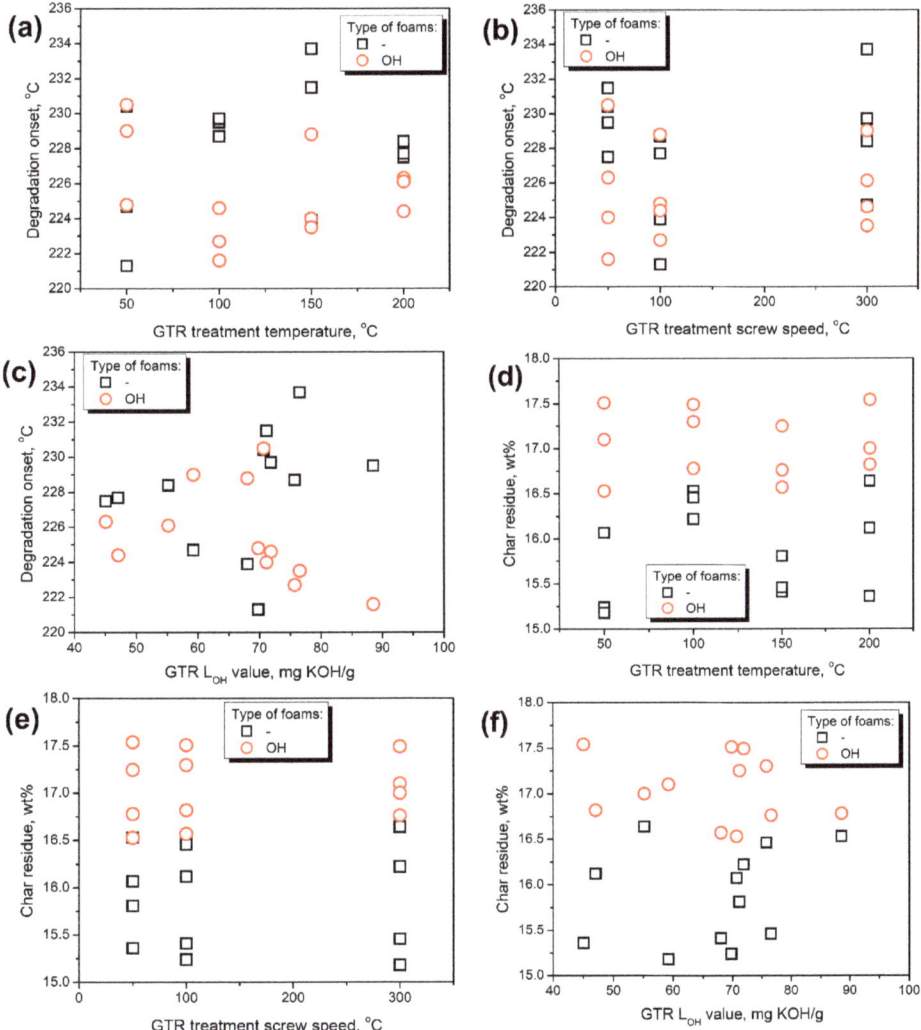

Figure 14. The impact of (**a,d**) GTR treatment temperature, (**b,e**) GTR treatment screw speed, and (**c,f**) GTR L_{OH} value on the (**a–c**) degradation onset, and (**d–f**) char residue of prepared PU/GTR composites.

Considering char residue, for all prepared composites, its values were between 15 and 18 wt%. No direct impact of GTR treatment conditions or L_{OH} of applied GTR particles was noted. Interestingly, OH foams were characterized by higher values of char residue exceeding 16.5 wt%, which can be associated with higher amounts of isocyanates in the system. Similar findings were reported in previous work [55].

4. Conclusions

The presented study was conducted to comprehensively investigate the impact of GTR thermomechanical modification and resulting changes in the L_{OH} values on the stoichiometric balance between isocyanate and hydroxyl groups determining the mechanical, thermomechanical, and thermal performance of foamed PU/GTR composites. The hydroxyl values of applied GTR particles were taken into account during composites' preparation, whose performance was compared to the samples prepared without formulation adjustments. Obtained results indicated that increasing the content of isocyanates in the system led to the additional stiffening of the PU phase, probably induced by the generation of allophanate and biuret groups. As a result, the molecular motions inside composites' were affected, which was expressed by a 3.4–12.3 °C shift of the glass transition temperature towards higher temperatures. Such an effect strengthened composites' structure and increased tensile and compressive strength even by 118 and 243%, respectively. Changes in the composite performance factor calculated from the results of the dynamic mechanical analysis also pointed to the strengthening of materials and the interfacial adhesion between the PU matrix and GTR fillers resulting from formulation adjustments. On the other hand, adjustments reduced composites' thermal stability, especially for GTR fillers with higher L_{OH} values. Such an effect could be attributed to the inferior thermal stability of hard segments comprised of urethane, allophanate, and biuret groups (all generated more extensively for higher isocyanate contents) compared to soft segments. Reported changes in the composites' performance resulting from considering the GTR L_{OH} values during formulation development were proportional to the hydroxyl values of GTR, confirming the reactivity of filler functional groups with isocyanates.

Generally, presented results and their dependence on GTR hydroxyl values and applied formulation adjustments pointed to the partial reactivity of GTR functionalities with isocyanates, confirming the importance of undertaken investigations. Moreover, with careful consideration, the presented study could be extended to other PU materials containing reactive additives or fillers, which affect the isocyanate index. Such an approach may provide important insights into developing flexible foamed PU-based composites on an industrial scale.

Author Contributions: Conceptualization, P.K. and A.H.; methodology, P.K., M.B. and A.H.; validation, P.K., M.B. and A.H.; formal analysis, A.H.; investigation, A.O., W.Ż., M.S., P.W. and Ł.Z.; resources, K.F. and A.H.; data curation, A.H.; writing—original draft preparation, M.B. and A.H.; writing—review and editing, P.K., M.B., K.F. and A.H.; visualization, A.H.; supervision, P.K. and A.H.; project administration, P.K. and A.H.; funding acquisition, A.H. All authors have read and agreed to the published version of the manuscript.

Funding: This work was supported by The National Centre for Research and Development (NCBR, Poland) in the frame of LIDER/3/0013/L-10/18/NCBR/2019 project—Development of technology for the manufacturing of foamed polyurethane-rubber composites for the use as damping materials.

Institutional Review Board Statement: Not applicable.

Informed Consent Statement: Not applicable.

Data Availability Statement: The data presented in this study are available in Insights into stoichiometry adjustments governing the performance of flexible foamed polyurethane/ground tire rubber composites.

Conflicts of Interest: The authors declare no conflict of interest.

References

1. Stachak, P.; Hebda, E.; Pielichowski, K. Foaming extrusion of thermoplastic polyurethane modified by POSS nanofillers. *Compos. Theory Pract.* **2019**, *19*, 23–29.
2. Małysa, T.; Nowacki, K.; Wieczorek, J. Assessment of sound absorbing properties of polyurethane sandwich system. *Compos. Theory Pract.* **2016**, *16*, 244–248.
3. Akindoyo, J.O.; Beg, M.D.H.; Ghazali, S.; Islam, M.R.; Jeyaratnam, N.; Yuvaraj, A.R. Polyurethane types, synthesis and applications—A review. *RSC Adv* **2016**, *6*, 114453–114482. [CrossRef]
4. Das, A.; Mahanwar, P. A brief discussion on advances in polyurethane applications. *Adv. Ind. Eng. Polym. Res.* **2020**, *3*, 93–101. [CrossRef]
5. Somarathna, H.M.C.C.; Raman, S.N.; Mohotti, D.; Mutalib, A.A.; Badri, K.H. The use of polyurethane for structural and infrastructural engineering applications: A state-of-the-art review. *Constr. Build. Mater.* **2018**, *190*, 995–1014. [CrossRef]
6. Lligadas, G.; Ronda, J.C.; Galià, M.; Cádiz, V. Plant oils as platform chemicals for polyurethane synthesis: Current state-of-the-art. *Biomacromolecules* **2010**, *11*, 2825–2835. [CrossRef]
7. Krol, P. Synthesis methods, chemical structures and phase structures of linear polyurethanes. properties and applications of linear polyurethanes in polyurethane elastomers, copolymers and ionomers. *Prog. Mater. Sci.* **2007**, *52*, 915–1015. [CrossRef]
8. Chattopadhyay, D.K.; Raju, K.V.S.N. Structural engineering of polyurethane coatings for high performance applications. *Prog. Polym. Sci.* **2007**, *32*, 352–418. [CrossRef]
9. Kuranska, M.; Polaczek, K.; Auguscik-Krolikowska, M.; Prociak, A.; Ryszkowska, J. Open-cell polyurethane foams based on modified used cooking oil. *Polimery* **2020**, *65*, 216–225. [CrossRef]
10. Gama, N.; Ferreira, A.; Barros-Timmons, A. Polyurethane foams: Past, present, and future. *Materials* **2018**, *11*, 1841. [CrossRef]
11. Kuranska, M.; Prociak, A.; Michalowski, S.; Zawadzinska, K. The influence of blowing agents type on foaming process and properties of rigid polyurethane foams. *Polimery* **2018**, *63*, 672–678. [CrossRef]
12. Lapprand, A.; Boisson, F.; Delolme, F.; Méchin, F.; Pascault, J.-P. Reactivity of isocyanates with urethanes: Conditions for allophanate formation. *Polym Degrad Stab* **2005**, *90*, 363–373. [CrossRef]
13. Vilar, W.D. *Química e Tecnologia Dos Poliuretanos*, 2nd ed.; Vilar Consultoria Técnica Ltda: Rio de Janeiro, Brazil, 1998.
14. Zieleniewska, M.; Ryszkowska, J.; Bryskiewicz, A.; Auguscik, M.; Szczepkowski, L.; Swiderski, A.; Wrzesniewska-Tosik, K. The structure and properties of viscoelastic polyurethane foams with fyrol and keratin fibers. *Polimery* **2017**, *62*, 127–135. [CrossRef]
15. Członka, S.; Bertino, M.F.; Strzelec, K.; Strąkowska, A.; Masłowski, M. Rigid polyurethane foams reinforced with solid waste generated in leather industry. *Polym Test* **2018**, *69*, 225–237. [CrossRef]
16. Bryśkiewicz, A.; Zieleniewska, M.; Przyjemska, K.; Chojnacki, P.; Ryszkowska, J. Modification of flexible polyurethane foams by the addition of natural origin fillers. *Polym. Degrad. Stab.* **2016**, *132*, 32–40. [CrossRef]
17. Dang, X.; Yang, M.; Zhang, B.; Chen, H.; Wang, Y. Recovery and utilization of collagen protein powder extracted from chromium leather scrap waste. *Environ. Sci. Pollut. Res.* **2019**, *26*, 7277–7283. [CrossRef] [PubMed]
18. Barczewski, M.; Szostak, M.; Nowak, D.; Piasecki, A. Effect of wood flour addition and modification of its surface on the properties of rotationally molded polypropylene composites. *Polimery* **2018**, *63*, 772–784. [CrossRef]
19. Szefer, E.; Leszczyńska, A.; Pielichowski, K. Modification of microcrystalline cellulose filler with succinic anhydride—Effect of microwave and conventional heating. *Compos. Theory Pract.* **2018**, *18*, 25–31.
20. Prut, E.V.; Zhorina, L.A.; Kompaniets, L.V.; Novikov, D.D.; Gorenberg, A.Y. The role of functional polymers in rubber powder/thermoplastic composites. *Polimery* **2017**, *62*, 548–555. [CrossRef]
21. Zedler, Ł.; Kosmela, P.; Olszewski, A.; Burger, P.; Formela, K.; Hejna, A. Recycling of waste rubber by thermo-mechanical treatment in a twin-screw extruder. In Proceedings of the First International Conference on "Green" Polymer Materials 2020, Basel, Switzerland, 4 November 2020; p. 10.
22. Olszewski, A.; Kosmela, P.; Zedler, Ł.; Formela, K.; Hejna, A. Optimization of foamed polyurethane/ground tire rubber composites manufacturing. In Proceedings of the 2nd International Online Conference on Polymer Science, Polymers and Nanotechnology for Industry 4.0, Basel, Switzerland, 30 October 2021; p. 12.
23. Formela, K.; Klein, M.; Colom, X.; Saeb, M.R. Investigating the combined impact of plasticizer and shear force on the efficiency of low temperature reclaiming of Ground Tire Rubber (GTR). *Polym. Degrad. Stab.* **2016**, *125*, 1–11. [CrossRef]
24. Kosmela, P.; Olszewski, A.; Zedler, Ł.; Burger, P.; Piasecki, A.; Formela, K.; Hejna, A. Ground tire rubber filled flexible polyurethane foam—effect of waste rubber treatment on composite performance. *Materials* **2021**, *14*, 3807. [CrossRef] [PubMed]
25. Kosmela, P.; Olszewski, A.; Zedler, Ł.; Hejna, A.; Burger, P.; Formela, K. Structural changes and their implications in foamed flexible polyurethane composites filled with rapeseed oil-treated Ground Tire Rubber. *J. Compos. Sci.* **2021**, *5*, 90. [CrossRef]
26. Simon, D.Á.; Pirityi, D.Z.; Bárány, T. Devulcanization of Ground Tire Rubber: Microwave and thermomechanical approaches. *Sci. Rep.* **2020**, *10*, 16587. [CrossRef]
27. Hejna, A.; Olszewski, A.; Zedler, Ł.; Kosmela, P.; Formela, K. The impact of Ground Tire Rubber oxidation with H_2O_2 and $KMnO_4$ on the structure and performance of flexible polyurethane/Ground Tire Rubber composite foams. *Materials* **2021**, *14*, 499. [CrossRef] [PubMed]
28. Yeh, A.; Hwang, S.; Guo, J. Effects of screw speed and feed rate on residence time distribution and axial mixing of wheat flour in a twin-screw extruder. *J. Food Eng.* **1992**, *17*, 1–13. [CrossRef]

29. Suparno, M.; Dolan, K.D.; NG, P.K.W.; Steffe, J.F. average shear rate in a twin-screw extruder as a function of degree of fill, flow behavior index, screw speed and screw configuration. *J. Food Process. Eng.* **2011**, *34*, 961–982. [CrossRef]
30. Altomare, R.E.; Ghossi, P. An analysis of residence time distribution patterns in a twin screw cooking extruder. *Biotechnol. Prog.* **1986**, *2*, 157–163. [CrossRef]
31. Hejna, A.; Barczewski, M.; Skórczewska, K.; Szulc, J.; Chmielnicki, B.; Korol, J.; Formela, K. Sustainable upcycling of brewers' spent grain by thermo-mechanical treatment in twin-screw extruder. *J. Clean Prod.* **2021**, *285*, 124839. [CrossRef]
32. Formela, K.; Cysewska, M.; Haponiuk, J. The influence of screw configuration and screw speed of co-rotating twin screw extruder on the properties of products obtained by thermomechanical reclaiming of Ground Tire Rubber. *Polimery* **2014**, *59*, 170–177. [CrossRef]
33. Członka, S.; Strąkowska, A.; Pospiech, P.; Strzelec, K. Effects of chemically treated eucalyptus fibers on mechanical, thermal and insulating properties of polyurethane composite foams. *Materials* **2020**, *13*, 1781. [CrossRef]
34. Paukszta, D.; Szostak, M.; Rogacz, M. Mechanical properties of polypropylene copolymers composites filled with rapeseed straw. *Polimery* **2014**, *59*, 165–169. [CrossRef]
35. Duque, A.; Manzanares, P.; Ballesteros, M. Extrusion as a pretreatment for lignocellulosic biomass: Fundamentals and applications. *Renew. Energy* **2017**, *114*, 1427–1441. [CrossRef]
36. Prociak, A.; Malewska, E.; Bąk, S. Influence of isocyanate index on selected properties of flexible polyurethane foams modified with various bio-components. *J. Renew. Mater.* **2016**, *4*, 78–85. [CrossRef]
37. Rojek, P.; Prociak, A. Effect of different rapeseed-oil-based polyols on mechanical properties of flexible polyurethane foams. *J. Appl. Polym. Sci.* **2012**, *125*, 2936–2945. [CrossRef]
38. Hejna, A.; Kopczyńska, M.; Kozłowska, U.; Klein, M.; Kosmela, P.; Piszczyk, Ł. Foamed polyurethane composites with different types of ash—Morphological, mechanical and thermal behavior assessments. *Cell. Polym.* **2016**, *35*, 213–220. [CrossRef]
39. Arnold, R.G.; Nelson, J.A.; Verbanc, J.J. Recent advances in isocyanate chemistry. *Chem. Rev.* **1957**, *57*, 47–76. [CrossRef]
40. Bukowczan, A.; Hebda, E.; Michałowski, S.; Pielichowski, K. Modification of polyurethane viscoelastic foams by functionalized Polyhedral Oligomeric Silsesquioxanes (POSS). *Compos. Theory Pract.* **2018**, *18*, 77–81.
41. Gómez-Fernández, S.; Ugarte, L.; Calvo-Correas, T.; Peña-Rodríguez, C.; Corcuera, M.A.; Eceiza, A. Properties of flexible polyurethane foams containing isocyanate functionalized kraft lignin. *Ind. Crops Prod.* **2017**, *100*, 51–64. [CrossRef]
42. Smoleń, J.; Olszowska, K.; Godzierz, M. Composites of rigid polyurethane foam and shredded car window glass particles—Structure and mechanical properties. *Compos. Theory Pract.* **2021**, *21*, 135–140.
43. Caban, R. Examinations of structure and properties of polymer composite with glass fiber. *Compos. Theory Pract.* **2019**, *19*, 150–156.
44. Wolska, A.; Goździkiewicz, M.; Ryszkowska, J. Thermal and mechanical behaviour of flexible polyurethane foams modified with graphite and phosphorous fillers. *J. Mater. Sci.* **2012**, *47*, 5627–5634. [CrossRef]
45. Andrzejewski, J.; Szostak, M.; Barczewski, M.; Łuczak, P. Cork-wood hybrid filler system for polypropylene and poly(lactic acid) based injection molded composites. structure evaluation and mechanical performance. *Compos. B Eng.* **2019**, *163*, 655–668. [CrossRef]
46. Zedler, Ł.; Kowalkowska-Zedler, D.; Colom, X.; Cañavate, J.; Saeb, M.R.; Formela, K. Reactive sintering of Ground Tire Rubber (GTR) modified by a trans-polyoctenamer rubber and curing additives. *Polymers* **2020**, *12*, 3018. [CrossRef] [PubMed]
47. Hejna, A.; Barczewski, M.; Kosmela, P.; Mysiukiewicz, O.; Kuzmin, A. Coffee silverskin as a multifunctional waste filler for high-density polyethylene green composites. *J. Compos. Sci.* **2021**, *5*, 44. [CrossRef]
48. Russo, P.; Acierno, D.; Corradi, A.; Leonelli, C. RETRACTED: Dynamic-mechanical behavior and morphology of polystyrene/perovskite composites: Effects of filler size. *Procedia Eng.* **2011**, *10*, 1017–1022. [CrossRef]
49. Ma, P.-C.; Siddiqui, N.A.; Marom, G.; Kim, J.-K. Dispersion and functionalization of carbon nanotubes for polymer-based nanocomposites: A review. *Compos. Part A Appl. Sci. Manuf.* **2010**, *41*, 1345–1367. [CrossRef]
50. Hejna, A.; Haponiuk, J.; Piszczyk, Ł.; Klein, M.; Formela, K. Performance properties of rigid polyurethane-polyisocyanurate/Brewers' spent grain foamed composites as function of isocyanate index. *e-Polymers* **2017**, *17*, 427–437. [CrossRef]
51. Ivdre, A.; Abolins, A.; Sevastyanova, I.; Kirpluks, M.; Cabulis, U.; Merijs-Meri, R. Rigid polyurethane foams with various isocyanate indices based on polyols from rapeseed oil and waste PET. *Polymers* **2020**, *12*, 738. [CrossRef]
52. Nadal Gisbert, A.; Crespo Amorós, J.E.; López Martínez, J.; Garcia, A.M. Study of thermal degradation kinetics of elastomeric powder (Ground Tire Rubber). *Polym. Plast. Technol. Eng.* **2007**, *47*, 36–39. [CrossRef]
53. Zedler, Ł.; Kowalkowska-Zedler, D.; Vahabi, H.; Saeb, M.R.; Colom, X.; Cañavate, J.; Wang, S.; Formela, K. Preliminary investigation on auto-thermal extrusion of Ground Tire Rubber. *Materials* **2019**, *12*, 2090. [CrossRef]
54. Levchik, S.V.; Weil, E.D. Thermal decomposition, combustion and fire-retardancy of polyurethanes—A review of the recent literature. *Polym. Int.* **2004**, *53*, 1585–1610. [CrossRef]
55. Mizera, K.; Ryszkowska, J. Thermal properties of polyurethane elastomers from soybean oil-based polyol with a different isocyanate index. *J. Elastomers Plast.* **2019**, *51*, 157–174. [CrossRef]

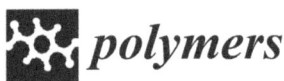

Article

Dual X-ray- and Neutron-Shielding Properties of Gd$_2$O$_3$/NR Composites with Autonomous Self-Healing Capabilities

Worawat Poltabtim [1,2,3], Arkarapol Thumwong [3,4], Ekachai Wimolmala [5], Chanis Rattanapongs [1,3], Shinji Tokonami [2], Tetsuo Ishikawa [6] and Kiadtisak Saenboonruang [1,3,7,8,*]

1. Department of Applied Radiation and Isotopes, Faculty of Science, Kasetsart University, Bangkok 10900, Thailand
2. Institute of Radiation Emergency Medicine, Hirosaki University, Aomori 0368564, Japan
3. Special Research Unit of Radiation Technology for Advanced Materials (RTAM), Faculty of Science, Kasetsart University, Bangkok 10900, Thailand
4. Department of Materials Science, Faculty of Science, Kasetsart University, Bangkok 10900, Thailand
5. Polymer PROcessing and Flow (P-PROF) Research Group, Division of Materials Technology, School of Energy, Environment and Materials, King Mongkut's University of Technology Thonburi, Bangkok 10140, Thailand
6. Department of Radiation Physics and Chemistry, Fukushima Medical University, Fukushima 9601295, Hikarigaoka, Japan
7. Kasetsart University Research and Development Institute (KURDI), Kasetsart University, Bangkok 10900, Thailand
8. Specialized Center of Rubber and Polymer Materials in Agriculture and Industry (RPM), Faculty of Science, Kasetsart University, Bangkok 10900, Thailand
* Correspondence: kiadtisak.s@ku.th; Tel.: +66-2-562-5555 (ext. 646219)

Abstract: The neutron- and X-ray-shielding, morphological, physical, mechanical, and self-healing properties were investigated for natural rubber (NR) composites containing varying gadolinium oxide (Gd$_2$O$_3$) contents (0, 25, 50, 75, and 100 parts per hundred parts of rubber; phr) to investigate their potential uses as self-healing and flexible neutron- and X-ray-shielding materials. Gd$_2$O$_3$ was selected as a radiation protective filler in this work due to its preferable properties of having relatively high neutron absorption cross-section (σ_{abs}), atomic number (Z), and density (ρ) that could potentially enhance interaction probabilities with incident radiation. The results indicated that the overall neutron-shielding and X-ray-shielding properties of the NR composites were enhanced with the addition of Gd$_2$O$_3$, as evidenced by considerable reductions in the half-value layer (HVL) values of the samples containing 100 phr Gd$_2$O$_3$ to just 1.9 mm and 1.3 mm for thermal neutrons and 60 kV X-rays, respectively. Furthermore, the results revealed that, with the increase in Gd$_2$O$_3$ content, the mean values (± standard deviations) of the tensile strength and elongation at break of the NR composites decreased, whereas the hardness (Shore A) increased, for which extreme values were found in the sample with 100 phr Gd$_2$O$_3$ (3.34 ± 0.26 MPa, 411 ± 9%, and 50 ± 1, respectively). In order to determine the self-healing properties of the NR composites, the surfaces of the cut samples were gently pressed together, and they remained in contact for 60 min; then, the self-healing properties (the recoverable strength and the %Recovery) of the self-healed samples were measured, which were in the ranges of 0.30–0.40 MPa and 3.7–9.4%, respectively, for all the samples. These findings confirmed the ability to autonomously self-heal damaged surfaces through the generation of a reversible ionic supramolecular network. In summary, the outcomes from this work suggested that the developed Gd$_2$O$_3$/NR composites have great potential to be utilized as effective shielding materials, with additional dual shielding and self-healing capabilities that could prolong the lifetime of the materials, reduce the associated costs of repairing or replacing damaged equipment, and enhance the safety of all users and the public.

Keywords: natural rubber; Gd$_2$O$_3$; self-healing; shielding; mechanical properties; X-rays; neutrons

Citation: Poltabtim, W.; Thumwong, A.; Wimolmala, E.; Rattanapongs, C.; Tokonami, S.; Ishikawa, T.; Saenboonruang, K. Dual X-ray- and Neutron-Shielding Properties of Gd$_2$O$_3$/NR Composites with Autonomous Self-Healing Capabilities. *Polymers* **2022**, *14*, 4481. https://doi.org/10.3390/polym14214481

Academic Editor: Kamila Sałasińska

Received: 29 September 2022
Accepted: 19 October 2022
Published: 22 October 2022

Publisher's Note: MDPI stays neutral with regard to jurisdictional claims in published maps and institutional affiliations.

Copyright: © 2022 by the authors. Licensee MDPI, Basel, Switzerland. This article is an open access article distributed under the terms and conditions of the Creative Commons Attribution (CC BY) license (https://creativecommons.org/licenses/by/4.0/).

1. Introduction

As the demand for greener technologies has rapidly increased in recent years following the Sustainable Development Goals (SDGs) introduced by the United Nations

(UN) [1], radiation technologies have become one of the most sought-after tools to satisfy such demands due to their reduced use of hazardous chemicals during irradiation and procedures, adaptability to large-scale production, and vast range of applications, such as the determination of transfer mechanisms for minerals and radionuclides in plants [2,3], non-destructive imaging for cultural heritage artifacts [4], diagnostic and radiotherapy purposes for brain and breast cancers [5,6], measurement of moisture in soils [7], and gemstone modification [8]. However, despite their acknowledged benefits, excessive exposure to different types of radiation, especially those from neutrons and X-rays, can harmfully affect users and the public, possibly resulting in permanent injuries or deaths [9].

To minimize the risk of potential adverse effects from excessive radiation exposure, suitable and effective radiation-shielding equipment must be implemented in all nuclear-related facilities following a radiation safety concept, namely As Low As Reasonably Achievable, or ALARA [10]. Generally, the selection of the main materials and radiation-protective fillers used to produce radiation-shielding equipment depends on several factors, such as the type and energy of the incident radiation, as well as the physical and mechanical requirements for the intended applications. For example, to attenuate thermal neutrons (neutrons with an energy of 0.025 eV), compounds containing elements with a high neutron absorption cross-section (σ_{abs}), such as boron (B), boron carbide (B_4C), and boron oxide (B_2O_3), are often used due to the relatively high σ_{abs} value of B (^{10}B has a σ_{abs} value of 3840 barns, while ^{nat}B has a value of 768 barns) [11], which considerably enhances the absorption probabilities between incident thermal neutrons and the material. On the other hand, for X-ray attenuation, materials consisting of heavy elements or compounds, such as lead (Pb), lead oxide (PbO), bismuth oxide (Bi_2O_3), tungsten oxide (WO_3), and barium sulfate ($BaSO_4$), are commonly implemented due to the relatively high atomic numbers (Z) of Pb, Bi, W, and Ba (Z = 82, 83, 74, and 56, respectively), as well as the high densities (ρ) of Pb, PbO, Bi_2O_3, WO_3, and $BaSO_4$ (ρ = 11.3, 9.5, 8.9, 7.2, and 4.5 g/cm^3, respectively) [12–14], which considerably enhance the interaction probabilities between incident X-rays and the material through two main mechanisms, namely photoelectric absorption and Compton scattering, subsequently resulting in improved X-ray-shielding properties of the composites [15].

While the use of these fillers can noticeably improve the radiation attenuation capabilities of materials, the lack of dual shielding properties (that is, the ability to effectively and simultaneously attenuate both thermal neutrons and X-rays) has resulted in the need to either acquire two distinct types of shielding materials or to mix two different fillers in the same material [16,17]. While these methods are possible, they could potentially increase the cost and space requirements to accommodate thicker materials, as well as possibly reducing desirable mechanical and physical properties of the shielding materials due to particle agglomeration from having filler contents that are too high [13]. To alleviate such drawbacks, gadolinium oxide (Gd_2O_3), which is a rare-earth compound, has drawn much attention from researchers and product developers in radiation safety due to the high values of σ_{abs} (49,700 barns) and Z (64) for Gd, as well as the high ρ of Gd_2O_3 (7.4 g/cm^3), which result in its ability to simultaneously attenuate both thermal neutrons and X-rays. Some examples of Gd_2O_3 used as radiation protective filler are the development of neutron-shielding hydrogels from poly(vinyl) alcohol (PVA), which indicated substantial enhancements in the ability of the hydrogels to attenuate thermal neutrons after the addition of Gd_2O_3. This was evidenced by the half-value layer (HVL; the thickness of a material that can attenuate 50% of the initial intensity of radiation), which were reduced from 146.3 mm in a pristine PVA hydrogel to just 3.6 mm in a 10.5 wt% Gd_2O_3/PVA hydrogel. Subsequently, this shielding improvement reduced space requirements to accommodate the materials by almost 40-fold [18]. Another work on the use of Gd_2O_3 by Kaewnuam et al. investigated the gamma-shielding properties of WO_3-Gd_2O_3-B_2O_3 glass and showed that the HVL values of the glasses were reduced from 1.424 cm in a sample with 17.5 wt% Gd_2O_3 to 1.326 cm in a sample with 27.5 wt% Gd_2O_3 determined based on 662 keV gamma rays emitted from ^{137}Cs [19]. These two examples clearly show the shielding effectiveness of Gd_2O_3 for both thermal neutron and high-energy photon attenuations and present the advantages of Gd_2O_3 as a radiation-protective filler in comparison to common Pb, Bi, and B compounds.

Another important factor to consider in producing radiation-shielding materials is the selection of the main matrix, for which the selection largely depends on the requirements of the intended applications. For example, applications requiring high flexibility, strength, and elongation usually rely on natural rubber (NR) or synthetic rubber (SR). For example, B_2O_3/NR [11], B_4C/NR [20], and H_3BO_3/ethylene-propylene diene monomer (EPDM) [21] composites have been developed for use as flexible, neutron-shielding materials, while Bi_2O_3/EPDM [13], WO_3/EPDM [13], $BaSO_4$/NR [14], and Pb/NR [22] composites have been utilized as flexible, X-ray-shielding and gamma-shielding materials. While these composites could serve their mandatory purpose (the ability to attenuate incident thermal neutrons or X-rays (depending on filler type) with high flexibility and strength), the lack of self-healing capabilities in most common NR and SR composites has resulted in extra procedures or new materials needed to restore full function once the materials are damaged, inevitably shortening their lifetimes and increasing operational costs. To resolve these shortcomings, Xu et al. successfully developed autonomously self-healing NR composites by introducing controlled peroxide-induced vulcanization to generate ionic cross-links to NR networks via the polymerization of zinc dimethacrylate (ZDMA), which slowed the formation of non-reversible covalent cross-links while generating a reversible ionic supramolecular network to NR, enabling the ability to autonomously heal after damage [23,24]. Hence, to expand their usefulness to other applications, the concept of autonomously self-healing NR materials can be adapted for the production of radiation-shielding materials, which could not only present the mentioned benefits but also improve safety for radiation users from damaged equipment.

Therefore, this current work investigates the properties of flexible Gd_2O_3/NR composites for their potential use as dual thermal-neutron- and X-ray-shielding materials with autonomously self-healing capabilities by introducing reversible ionic supramolecular cross-links to NR networks. In order to understand the effects of the Gd_2O_3 fillers on the properties of the composites, the Gd_2O_3 contents were varied with values of 0, 25, 50, 75, and 100 parts per hundred parts of rubber (phr) by weight to thoroughly investigate the properties of interest, which consisted of thermal-neutron and X-ray-shielding properties (based on the linear attenuation coefficient (μ), the mass attenuation coefficient (μ_m), the half-value layer (HVL), the tenth-value layer (TVL), and the Pb equivalence (Pb_{eq})), as well as mechanical (based on tensile strength, elongation at break, and hardness (Shore A) both before and after self-healing), morphological, and physical (based on density) properties. The outcomes of this work can not only present valuable information on the dual neutron- and X-ray-shielding properties of the developed Gd_2O_3/NR composites, but may also offer a novel procedure to obtain self-healing NR composites that is beneficial for the future development of other radiation-shielding products.

2. Experimental

2.1. Materials and Chemicals

Natural rubber (STR 5CV) with a Mooney viscosity of 60.8 (at 100 °C) was supplied by Hybrid Post Co., Ltd. (Bangkok, Thailand). The names, contents, roles, and suppliers of the chemicals used for sample preparation are shown in Table 1. An image of Gd_2O_3 powder captured using a scanning electron microscope (SEM; Quanta 450 FEI: JSM-6610LV, Eindhoven, the Netherlands) is shown in Figure 1, which indicates that the average particle size of the Gd_2O_3 powder was 3.4 ± 0.4 μm, as determined using ImageJ software version 1.50i (Bethesda, MD, USA).

Table 1. Material formulations of Gd_2O_3/NR composites and their chemical names, contents, roles, and suppliers.

Chemical	Content (phr)	Role	Supplier
Natural rubber (NR: STR 5CV)	100	Main matrix	Hybrid Post Co., Ltd. (Bangkok, Thailand)
Zinc dimethacrylate (ZDMA)	40	Accelerator	Shanghai Ruizheng Chemical Technology Co., Ltd. (Shanghai, China)
Dicumyl peroxide (DCP)	1	Curing agent	Shanghai Ruizheng Chemical Technology Co., Ltd. (Shanghai, China)
Gadolinium oxide (Gd_2O_3)	0, 25, 50, 75, and 100	Radiation-protective filler	Shanghai Ruizheng Chemical Technology Co., Ltd. (Shanghai, China)

Figure 1. SEM image of Gd_2O_3 particles used in this work.

2.2. Sample Preparation

The NR samples were prepared using two steps: mastication and then compounding. Initially, the NR was masticated on a two-roll mill (R11-3FF, Kodaira Seisakusho Co., Ltd., Tokyo, Japan) for 5 min. Then, the masticated NR was compounded with the chemicals (Table 1) for a further 15–20 min. Notably, although the content of Gd_2O_3 was as high as 100 phr, the much higher density of Gd_2O_3 (ρ = 7.4 g/cm^3) than that of NR (approximately 0.93–0.97 g/cm^3 [25]) resulted in the volume of Gd_2O_3 powder used during the compounding being much less than that of NR, making the mixing of all the chemicals on a two-roll mill possible. After the compounding, the NR samples were vulcanized using hot compression molding (CC-HM-2060, Chaicharoen Karnchang Co., Ltd., Bangkok, Thailand) at 150 °C and a pressure of 160 kg/cm^2 for 150 secs in a mold with dimensions of either 15 cm × 15 cm × 0.2 cm or 10 cm × 10 cm × 0.2 cm. Notably, the procedure for sample preparation was mainly based on the published works of Xu et al. [23,24], while the cure time of 150 secs was selected following preliminary studies for optimized cure times, for which shorter or longer cure times resulted in the samples being too soft or too hard, respectively, which limited their useability and prevented the initiation of self-healing mechanisms from occurring [26].

2.3. Characterization

2.3.1. Neutron-Shielding Properties

The neutron shielding properties of the Gd_2O_3/NR composites were investigated at the Thailand Institute of Nuclear Technology (Public Organization), Bangkok, Thailand. The neutron-shielding parameters investigated in this work were the neutron transmission (I/I_0), the linear attenuation coefficient (μ), the half-value layer (HVL), and the tenth-value layer (TVL), and their relationships are shown in Equations (1)–(4) [14]:

$$I/I_0 = e^{-\mu x} \quad (1)$$

$$HVL = \frac{\ln(2)}{\mu} \quad (2)$$

$$\mu_m = \frac{\mu}{\rho} \quad (3)$$

$$TVL = \frac{\ln(10)}{\mu} \quad (4)$$

where I_0 is the initial intensity of the incident neutrons, I is the final intensity of the transmitted neutrons, x is the thickness of the sample, and ρ is the density of the sample.

The setup for neutron-shielding measurement is schematically shown in Figure 2, with a ^{241}Am/Be used as a thermal neutron source. The values of I and I_0 were recorded using a ^{3}He neutron detector that was connected to a high-voltage supplier (Model 659, ORTEC, CA, USA), an amplifier (Model 2022, Canberra, CT, USA), and a time counter (Model TC 535P, Tennelec, TN, USA). The neutron source was positioned such that it was 0.89 m away from the NR sample and 1.00 m away from the detector. Notably, to investigate the effects of sample thickness on the neutron-shielding properties, the total thickness values of the NR samples were also varied (2, 4, 6, 8, and 10 mm).

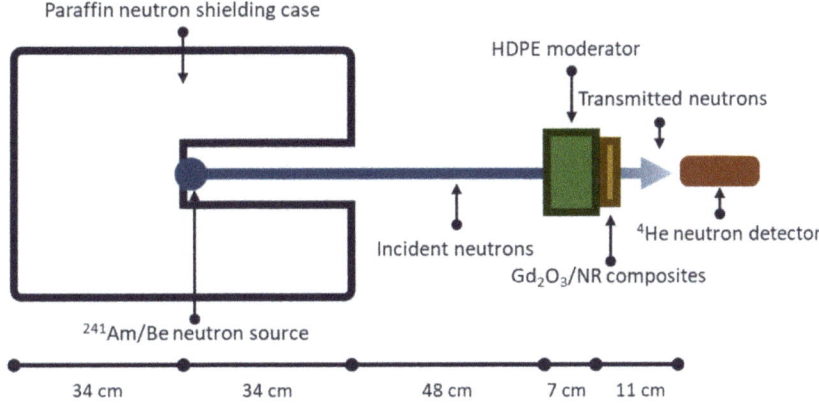

Figure 2. Schematic setup for neutron-shielding measurement.

2.3.2. X-ray-Shielding Properties

The schematic setup for X-ray-shielding measurement is shown in Figure 3. The measurement was carried out at the Secondary Standard Dosimetry Laboratory (SSDL), the Office of Atoms for Peace (OAP), Bangkok, Thailand. The X-ray-shielding parameters investigated in this work were X-ray transmission (I/I_0), the linear attenuation coefficient (μ), the mass attenuation coefficient (μ_m), the half-value layer (HVL), the tenth-value layer (TVL), and the Pb equivalence (Pb_{eq}), for which Pb_{eq} could be determined using Equation (5) [14]:

$$Pb_{eq} = \frac{\mu x}{\mu_{Pb}} \quad (5)$$

where μ_{Pb} is the linear attenuation coefficient of a pure Pb sheet. It should be noted that the values of μ_{Pb} were 63.06 cm^{-1} and 25.99 cm^{-1} for the incident X-ray energies of 45 keV and 80 keV, respectively, and were numerically determined using XCOM software (National Institute of Standards and Technology, Gaithersburg, MD, USA) [14,27]. The X-ray energies of 45 keV and 80 keV were selected for the determination of Pb_{eq} due to being the average energies of X-rays generated from an X-ray tube (YXLON MGC41, NY, USA) with the supplied voltages of 60 and 100 kV (Keithley 651B, OH, USA), respectively, used in this work. The emitted X-ray beam was collimated using a Pb collimator with a 1 mm pinhole, and the transmitted X-rays were detected and counted using a free-air ionization chamber (Korea Research Institute of Standards and Science; KRISS, Daejeon, Korea). More details for the setup of the neutron-shielding measurement are available in [14]. Similar to the neutron measurement, the total thickness values of the NR samples varied from 2 to 10 mm in 2 mm increments to investigate effects of material thickness on X-ray-shielding abilities.

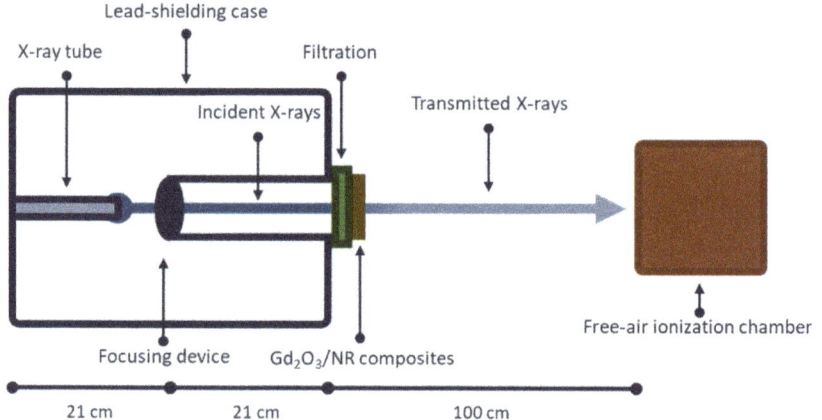

Figure 3. Schematic setup for X-ray-shielding measurement.

2.3.3. Mechanical Properties

The mechanical properties of tensile strength and elongation at break for all the Gd_2O_3/NR composites were determined using a universal testing machine (Auto-graph AG-I 5kN, Shimadzu, Kyoto, Japan) following ASTM D412-06 standard testing. The tensile testing speed used for all the samples was 50 mm/min. The surface hardness (Shore A) was determined using a hardness durometer (Teclock GS-719G, Japan) following the ASTM D2240-05 standard testing method.

For the determination of the self-healing capabilities of the developed Gd_2O_3/NR composites, samples having shapes and sizes based on ASTM D412-06 standard testing were cut into two equal pieces using a surgical knife and were immediately brough into contact. Then, after 60 min of contact, the samples were installed in a universal testing machine (Auto-graph AG-I 5kN, Shimadzu, Kyoto, Japan) to determine their tensile strength and elongation at break, following the same testing procedures as those for the uncut samples. Then, the tensile strength values of the self-healed samples were used for the calculation of the percentage of recoverable strength (%Recovery) using Equation (6) [18]:

$$\%\text{Recovery} = \frac{TS_{self-healing}}{TS_{uncut}} \times 100\% \quad (6)$$

where $TS_{self-healing}$ and TS_{uncut} are the tensile strengths of the self-healed and uncut samples, respectively [28].

2.3.4. Density Measurement

The densities for all the Gd_2O_3/NR composites were determined using a densitometer (MH-300A, Shanghai, China) with a precision of $0.01\ g/cm^3$, and the determination was carried out based on the Archimedes principle [29]. Additionally, to verify the correctness of the density measurement, theoretical densities (ρ_{th}) for all the samples were calculated using Equation (7) [14]:

$$\rho_{th} = \frac{C_{NR} + C_{Gd2O3}}{\frac{C_{NR}}{\rho_{NR}} + \frac{C_{Gd2O3}}{\rho_{Gd2O3}}} \quad (7)$$

where C_{NR} is the content of NR, C_{Gd2O3} is the content of Gd_2O_3, ρ_{NR} is the density of NR ($0.95\ g/cm^3$), and ρ_{Gd2O3} is the density of Gd_2O_3 ($7.4\ g/cm^3$).

2.3.5. Morphological Studies

The morphology, dispersion of Gd_2O_3 particles, and dispersion of Gd elements were determined using scanning electron microscopy (SEM) with energy-dispersive X-ray (EDX) spectroscopy (Quanta 450 FEI: JSM-6610LV, Eindhoven, the Netherlands) at a 10 kV accelerating voltage. Prior to the SEM-EDX studies, all specimens were coated with gold using a sputter coater (Quorum SC7620: Mini Sputter Coater/Glow Discharge System, Nottingham, UK) at a power voltage of 10 kV and a current of 10 mA for 120 secs.

3. Results and Discussion

3.1. Density

Table 2 indicates the experimental and theoretical densities, as well as the differences between these two values, of all the Gd_2O_3/NR composites investigated in this work. The results showed that the densities of the NR samples increased with increasing Gd_2O_3 content, while the differences between the experimental and theoretical values were below 5.0% for all the samples, clearly verifying the correctness and reliability of the experimental values for further use. The positive relationship between density and filler content was due to the much higher density of Gd_2O_3 than NR, resulting in a greater sample mass (determined at the same total volume) and, subsequently, greater overall density of the NR composites containing higher Gd_2O_3 contents [30].

Table 2. Experimental and theoretical densities, as well as the differences between the methods, of Gd_2O_3/NR composites with varying Gd_2O_3 contents of 0, 25, 50, 75, and 100 phr. Experimental densities shown as mean ± standard deviation.

Gd_2O_3 Content (phr)	Experimental Density (g/cm^3)	Theoretical Density (g/cm^3)	Difference (%)
0 (Control)	0.99 ± 0.01	0.95	4.0
25	1.19 ± 0.01	1.15	3.4
50	1.35 ± 0.01	1.33	1.5
75	1.49 ± 0.01	1.51	1.3
100	1.60 ± 0.01	1.68	5.0

3.2. Neutron-Shielding Properties

The results for the neutron-shielding properties, consisting of I/I_0, μ, μ_m, HVL, and TVL, for all the Gd_2O_3/NR composites are shown in Figure 4, which indicates that the overall neutron-shielding properties of the samples increased with increasing Gd_2O_3 content, as evidenced by the lower values of I/I_0, HVL, and TVL and the higher values of μ and μ_m in the samples containing higher contents of Gd_2O_3. This shielding enhancement from the addition of Gd_2O_3 was mainly due to Gd having a much higher σ_{abs} value than the C and H in NR (σ_{abs} values for Gd, C, and H are 49,700 barns, 0.0035 barns, and 0.3326 barns, respectively [31]), resulting in considerably increased chances for incident thermal neutrons to be absorbed and attenuated by the composites and, consequently, leading to superior neutron-shielding properties for the Gd_2O_3/NR composites [32]. Figure 5 shows the dispersion of Gd in the NR composites based on SEM-EDX and reveals that the highest elemental density of Gd was in the sample containing 100 phr Gd_2O_3 (Figure 5d), confirming the rationale for the improved neutron-shielding properties of the Gd_2O_3/NR composites by the addition of Gd_2O_3.

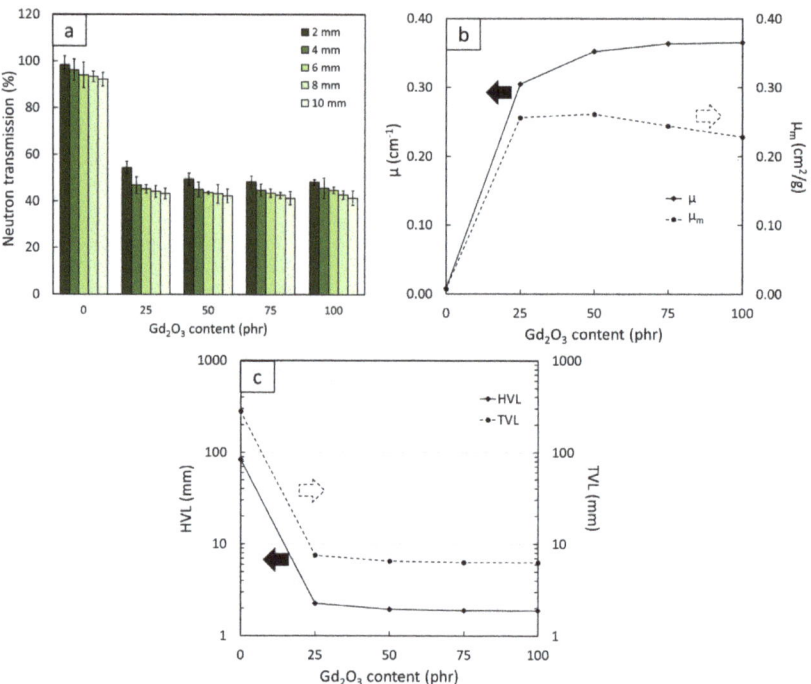

Figure 4. Neutron-shielding properties: (**a**) the neutron transmission, (**b**) linear attenuation coefficient (μ), mass attenuation coefficient (μ_m), (**c**) half-value layer (HVL) and tenth-value layer (TVL) of Gd_2O_3/NR composites containing varying Gd_2O_3 contents of 0, 25, 50, 75, and 100 phr, where error bars indicate ± standard error.

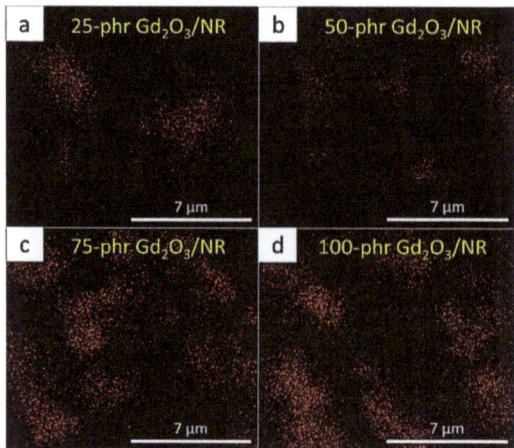

Figure 5. Dispersion of Gd_2O_3/NR composites containing varying Gd_2O_3 contents captured using SEM-EDX: (**a**) 25 phr, (**b**) 50 phr, (**c**) 75 phr, and (**d**) 100 phr.

In addition, Figure 4 indicates that a small addition of 25 phr Gd_2O_3 to pristine NR could sharply increase the neutron-shielding properties of the samples, as seen by the sharp decreases in I/I_0 and HVL from 97% and 84 mm in pristine NR to just 55% and 2.3 mm, respectively, in the 25 phr Gd_2O_3/NR sample (values of I/I_0 and HVL were compared using 2 mm thick samples). This notable improvement in the neutron-shielding properties was mainly due to the sudden change in dominant neutron interactions from elastic scattering in the pristine NR to neutron absorption in the Gd_2O_3/NR composites, for which the latter mechanism was relatively more effective in neutron attenuation than the former [33]. However, as more Gd_2O_3 powder was added to the composites, only a slight improvement was observed, perhaps because the samples already relied on the preferable absorption mechanism such that further addition of Gd_2O_3 could only slightly increase the probabilities of neutrons being absorbed [34].

Another point worth mentioning is that the ability to attenuate neutrons increased with increasing sample thickness. The dependence of neutron-shielding properties on sample thickness, as illustrated in the determination of I/I_0 and shown in Figure 4a, was mostly due to more materials being available to elastically scatter (in the case of a pristine NR sample) or absorb (in the case of Gd_2O_3/NR samples) incident neutrons in thicker samples, subsequently reducing the transmitted neutrons (lower I/I_0). This relationship could also be mathematically explained using Equation (1) when re-arranged as shown in Equation (8), which depicts that the value of $\ln(I/I_0)$ was inversely proportional to x (thickness of the sample); hence, I/I_0 was negatively related to x:

$$\ln(I/I_0) = -\mu x \tag{8}$$

3.3. X-ray-Shielding Properties

Figure 6 shows the results for the percentage of X-ray transmission for 60 kV and 100 kV X-rays, respectively, through varying thicknesses (2, 4, 6, 8, and 10 mm) of Gd_2O_3/NR composites. Similar to the results from the neutron-shielding measurement, the X-ray transmission decreased with increasing Gd_2O_3 content and sample thickness. The dependence of the X-ray transmission on Gd_2O_3 content was mainly due to the increased interaction probabilities between the incident X-rays and the materials through photoelectric absorption and Compton scattering with the addition of Gd_2O_3, for which their cross-sections were positively correlated to the Z and ρ values of the materials, hence improving the attenuation ability of the composites [35,36]. In addition, increasing the thickness of the samples could decrease X-ray transmission due to more Gd_2O_3 being available to interact with the incident X-rays, resulting in fewer X-rays being transmitted through the samples.

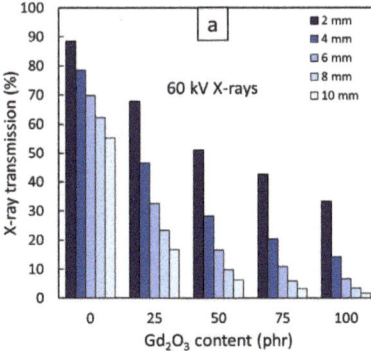
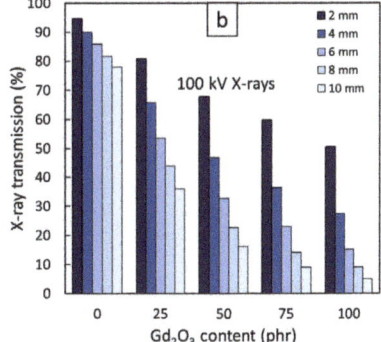

Figure 6. X-ray transmission of Gd_2O_3/NR composites with varying thicknesses from 2 to 10 mm and varying Gd_2O_3 contents from 0 to 100 phr determined at X-ray supplied voltages of (**a**) 60 kV and (**b**) 100 kV.

Figure 7 shows the results of the μ, μ_m, HVL, TVL, and Pb_{eq} values of Gd_2O_3/NR composites containing varying Gd_2O_3 content determined using 60 kV and 100 kV X-rays (common supplied voltages used for medical diagnostics [37]). The results imply that the overall X-ray-shielding properties of the composites generally increased with increasing Gd_2O_3 content, as evidenced by the lowest values of HVL and TVL and the highest values of μ, μ_m, and Pb_{eq} being found in the sample with 100 phr Gd_2O_3. For example, the values of HVL (Pb_{eq}) of the NR composites were reduced from 1.12 cm and 2.50 cm, respectively, (0.03 mmPb and 0.03 mmPb) for pristine NR to 0.36 and 0.65 cm (0.09 mmPb and 0.12 mmPb) for 25 phr Gd_2O_3/NR composites, determined at 60 kV and 100 kV X-rays, respectively, exhibiting approximately a 3–4-fold improvement in X-ray attenuation ability with the addition of just 25 phr Gd_2O_3.

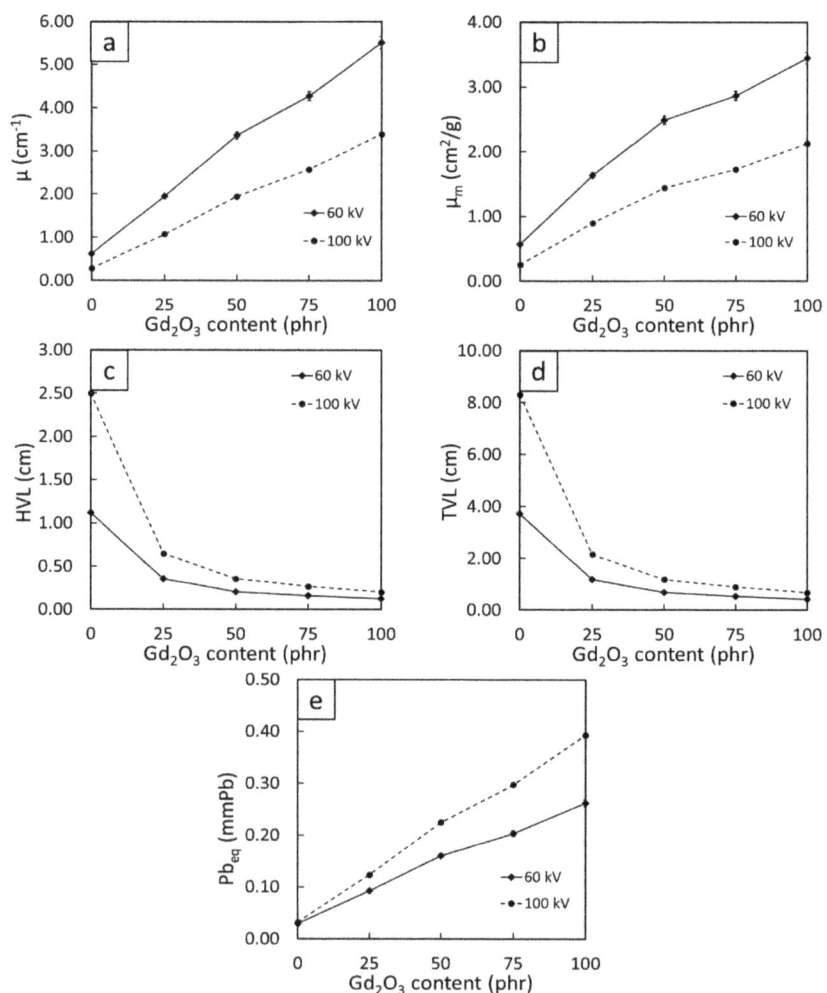

Figure 7. X-ray-shielding properties: (**a**) linear attenuation coefficient (μ), (**b**) mass attenuation coefficient (μ_m), (**c**) half-value layer (HVL), (**d**) tenth-value layer (TVL), and (**e**) lead equivalence (Pb_{eq}) of 3 mm thick Gd_2O_3/NR composites containing varying Gd_2O_3 contents of 0, 25, 50, 75, and 100 phr.

Another interesting result from Figure 7 was that the X-ray-shielding properties of the samples determined at the 60 kV supplied voltage were higher than those at the 100 kV supplied voltage due to the cross-section of the dominant photoelectric absorption (σ_{pe}) being inversely related to the cube of incident X-ray energy (E), as shown in Equation (9) [38]:

$$\sigma_{pe} \propto \frac{Z^n}{E^3} \quad (9)$$

Consequently, the incident X-rays emitted from an X-ray tube with higher supplied voltages could have fewer chances of interaction with the materials, resulting in more X-rays being transmitted and, subsequently, lower overall X-ray attenuation properties than lower-energy X-rays [39]. Notably, the values of Pb_{eq} at a specific Gd_2O_3 content could be tailored according to the shielding requirements for intended applications by lowering or increasing the thickness of the sample (the sample thickness for Figure 7e was 3 mm) using Equation (5).

3.4. Comparative Neutron- and X-ray-Shielding Properties between Current and Other Similar Materials

Table 3 shows a comparison of neutron- and X-ray-shielding properties (based on the values of HVL) from this work with other similar materials, which indicates that the current materials exhibited comparable or better neutron and X-ray attenuation capabilities than those from other reports. As a result, this comparison clearly confirmed the useability and potential of the NR composites for utilization as effective neutron- and X-ray-shielding materials with potential self-healing capabilities. It should be noted that the differences in the HVL values from all the materials in Table 3 could be due to several factors, such as differences in the filler types and contents used, as well as various energies of the incident radiation during measurement.

Table 3. Comparison of neutron- and X-ray-shielding properties based on the half-value layer (HVL) between the results from this work and those from similar materials. Numbers in parentheses represent supplied voltages used for X-ray-shielding measurement.

Main Matrix	Filler	Filler Content	Half-Value Layer (mm)		Reference
			Neutrons	X-rays	
NR	Gd_2O_3	50 phr	2.0	2.1 (60 kV)/ 3.6 (100 kV)	This work
NR	Gd_2O_3	75 phr	1.9	1.6 (60 kV)/ 2.7 (100 kV)	This work
NR	B_2O_3	80 phr	3.2	–	[11]
PVA	Sm_2O_3	10.5 wt%	4.2	–	[18]
PVA	Gd_2O_3	10.5 wt%	3.6	–	[18]
EPDM	B_2O_3	42.6 phr	3.7	–	[40]
NR	Bi_2O_3	50 phr	–	6.0 (60 kV)	[14]
NR	$BaSO_4$	50 phr	–	6.0 (60 kV)	[14]
NR	Bi_2O_3	40 phr	–	~3.5 (120 kV)	[41]
NR	Bi_2O_3	80 phr	–	~3.0 (120 kV)	[41]

3.5. Mechanical Properties

Table 4 shows the mechanical properties—tensile strength, elongation at break, and hardness (Shore A) of Gd_2O_3/NR composites containing varying Gd_2O_3 contents (0, 25, 50, 75, and 100 phr). The results indicate that the values of tensile strength and elongation at break generally decreased, while the hardness (Shore A) increased with increasing Gd_2O_3 content. The decreases in tensile properties after the addition of Gd_2O_3 could be due to poor surface compatibility between Gd_2O_3 and the NR matrix, which possibly resulted in the formation of voids and discontinuities in the matrix, subsequently obstructing the transfer of external forces and reducing the overall strength and elongation of the materials [42]. Furthermore, the addition of high levels of Gd_2O_3 contents, especially at 100 phr, led to high agglomeration of the Gd_2O_3 particles due to filler–filler interactions that prevented more preferable and stronger rubber–filler interactions from occurring [43]. The SEM images depicting the dispersion of Gd_2O_3 particles in the NR matrix are shown in Figure 8 and reveal that, while the particles were fairly evenly distributed throughout the NR matrix, some particle agglomeration was found in samples with higher filler contents, especially at 50, 75, and 100 phr (Figure 8c–e), compared to pristine NR (Figure 8a) and the sample with 25 phr filler content (Figure 8b). On the other hand, hardness (Shore A) had a strong positive relationship with Gd_2O_3 content due to the high rigidity of the Gd_2O_3 particles, which enhanced the overall rigidity and, hence, the hardness of the composites [44]. These findings are consistent with other reports, where the mechanical properties of materials generally decrease with the addition of high filler content, especially those developed for use in radiation protection [13,14].

Table 4. Mechanical properties of tensile strength, elongation at break, and hardness (Shore A) of Gd_2O_3/NR composites containing varying Gd_2O_3 contents of 0, 25, 50, 75, and 100 phr. Values are shown as mean ± standard deviation.

Gd_2O_3 Content (phr)	Tensile Strength (MPa)	Elongation at Break (%)	Hardness (Shore A)
0 (Control)	8.29 ± 0.83	555 ± 53	38 ± 1
25	5.02 ± 0.79	387 ± 33	41 ± 1
50	4.07 ± 0.14	515 ± 10	45 ± 1
75	4.08 ± 0.77	463 ± 16	46 ± 1
100	3.34 ± 0.26	411 ± 9	50 ± 1

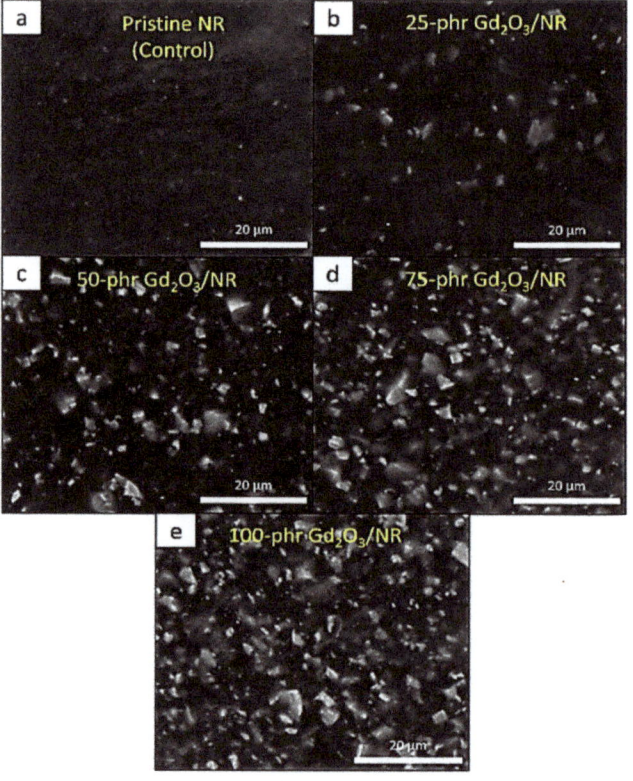

Figure 8. SEM images showing morphology and particle dispersion of NR composites containing varying Gd_2O_3 contents: (**a**) 0, (**b**) 25 phr, (**c**) 50 phr, (**d**) 75 phr, and (**e**) 100 phr.

Comparing the tensile properties obtained from this work with another work by Xu et al. indicated that the current pristine NR samples were approximately four times higher in tensile strength, as evidenced by the values reported in [23,24] being lower than 2 MPa for all the formulations. The differences in mechanical properties between these two works could be due to several factors, such as different formulation and cure times, as well as the types of NR used for sample preparation, which can affect the degree of cross-linking and, hence, the strength of the composites [45].

3.6. Self-Healing Properties

Figure 9 shows the comparative values of strength and elongation at break, as well as the percentage of recoverable strength (%Recovery), of the original and 60 min self-healed NR composites containing varying Gd_2O_3 contents (0, 25, 50, 75, and 100 phr). The results indicate that the values of tensile strength and elongation at break for all the self-healed samples (Figure 9a,b) were lower than the original ones, with values for recoverable strength and elongation at break in the ranges of 0.30–0.40 MPa and 22.6–36.4%, respectively, leading to values of %Recovery in the range of 3.7–9.4% (Figure 9c).

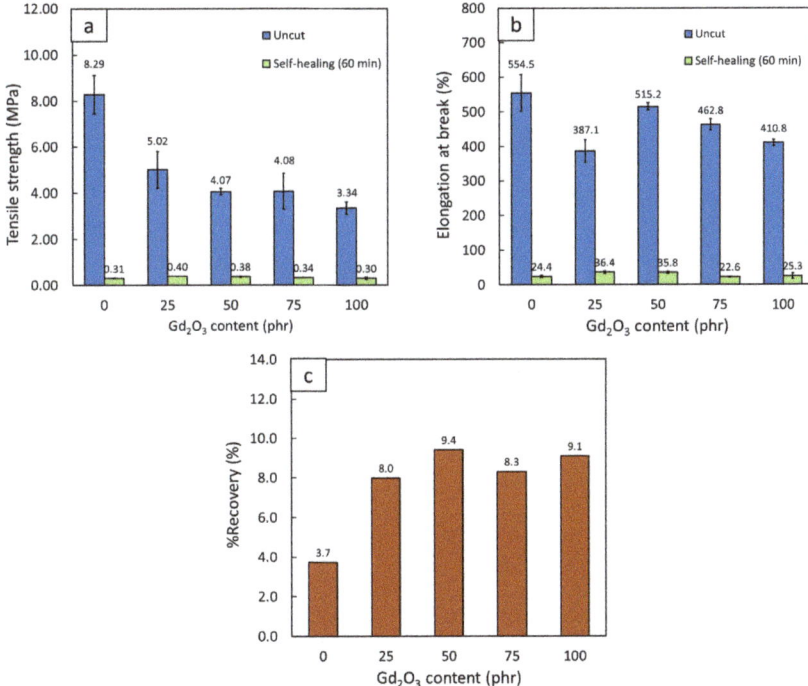

Figure 9. Comparison of (**a**) tensile strength, (**b**) elongation at break, and (**c**) percentage of recovery (%Recovery) for uncut and self-healed Gd_2O_3/NR composites containing varying Gd_2O_3 contents (0, 25, 50, 75, and 100 phr), where error bars indicate ± standard deviation.

The reduction in the tensile strength and elongation at break of the self-healed NR composites in comparison to the original samples could be explained by the NR molecular chains in the uncut samples being originally cross-linked with a combination of covalent and ionic bonds, which resulted in relatively high tensile strength and elongation at break before the cut [46]. However, as the two damaged surfaces were gently pressed together and remained in contact for 60 min at room temperature (approximately 25 °C), a reversible ionic supramolecular network via the polymerization of ZDMA was able to recreate the sample through the mobility of NR molecular chains and, subsequently, restore some recoverable strength to the self-healed surfaces [47]. Nonetheless, the overall strengths of the self-healed samples were considerably lower than the original samples, with %Recovery values in the range of 3.7–9.1% (Figure 9c). This could be due to the reduced level of cross-link density in the samples after self-healing that could only be recreated by ionic bonds. On the other hand, the covalent bonds, which also initially presented and played major roles in providing exceptional strength to the original samples, were irreversible and, consequently, absent in the self-healed contact, resulting in much-reduced levels of recoverable strength and elongation at break for the samples [48]. Another factor that affected the self-healing mechanism was the addition of the Gd_2O_3 particles to the NR matrix, for which the fillers were not a part of the reversible supramolecular network and, hence, hindered or blocked the initiation of self-healing [24].

Nonetheless, despite having Gd_2O_3 contents of up to 100 phr, the recoverable strengths of the Gd_2O_3/NR composites were in the range of 0.30–0.40 MPa, which were in the same order of magnitude as that of pristine NR reported by Xu et al. (being in the range of 0.5–0.7 MPa, depending on self-healing times [23]), implying the useability and potential of the current self-healing materials for applications in radiation protection. Furthermore, the success of the current work could promote further attempts to develop 'smart' and more effective materials for use in radiation shielding, along with the previously reported composites of Bi_2O_3/PVA, Sm_2O_3/PVA, Gd_2O_3/PVA, graphene/PVA, and PbO_2/acrylamide [18,28,49,50]. It should be noted that the %Recovery values of the NR samples with the addition of Gd_2O_3 were higher than that of the pristine NR because the original Gd_2O_3/NR composite had 2–3 times lower tensile strength than pristine NR, while having similar recoverable strength after self-healing, which resulted in considerably higher %Recovery values for the Gd_2O_3/NR composites.

4. Conclusions

This work developed dual neutron-shielding and X-ray-shielding NR composites containing varying contents of Gd_2O_3 (0, 25, 50, 75, and 100 phr) with autonomously self-healing capabilities. The results showed that the added Gd_2O_3 acted as an effective protective filler against neutrons and X-rays, as evidenced by the decreases in I/I_0, HVL, and TVL and the increases in μ, μ_m, and Pb_{eq} of the NR composites after being added to the matrix. In addition, the results indicated that the increased filler content led to decreased tensile strength and elongation at break, whereas the hardness (Shore A) increased, mainly due to the initiation of particle agglomeration at high filler contents and poor surface compatibility between the NR matrix and the filler. The developed NR composites also offered self-healing capabilities at the fractured surfaces through a reversible ionic supramolecular network, with recoverable strength and %Recovery values in the ranges 0.30–0.4 MPa and 3.7–9.4%, respectively (after self-healing for 60 min). Based on the overall results obtained, the developed Gd_2O_3/NR composites showed great potential for use as novel and self-healing radiation-shielding materials that could effectively attenuate both neutrons and X-rays, thus prolonging the lifetime of the protective material and enhancing the safety for users, as well as being a basis for the future development of 'smart' shielding materials.

Author Contributions: Conceptualization, K.S.; formal analysis, W.P., A.T., E.W., C.R., S.T., T.I. and K.S.; funding acquisition, K.S.; investigation, W.P., A.T., E.W., C.R., S.T., T.I. and K.S.; methodology, W.P., A.T., E.W., C.R., S.T., T.I. and K.S.; supervision, K.S.; validation, W.P., A.T., E.W., C.R., S.T., T.I. and K.S.; visualization, K.S.; writing—original draft, K.S.; writing—review and editing, W.P., A.T., E.W., C.R., S.T., T.I. and K.S. All authors have read and agreed to the published version of the manuscript.

Funding: This research was financially supported by the Office of the Ministry of Higher Education, Science, Research, and Innovation; the Thailand Science Research and Innovation through the Kasetsart University Reinventing University Program 2021; and the Kasetsart University Research and Development Institute (KURDI), Bangkok, Thailand (grant number FF(KU)25.64).

Institutional Review Board Statement: Not applicable.

Informed Consent Statement: Not applicable.

Data Availability Statement: The data presented in this study are available on request from the corresponding author.

Acknowledgments: The Kasetsart University Research and Development Institute (KURDI) and the Specialized Center of Rubber and Polymer Materials in Agriculture and Industry (RPM) provided publication support.

Conflicts of Interest: The authors declare no conflict of interest.

References

1. Pradhan, P.; Costa, L.; Rybski, D.; Lucht, W.; Kropp, J.P. A systematic study of sustainable development goal (SDG) interactions. *Earth's Future* **2017**, *5*, 1169–1179. [CrossRef]
2. Courbet, G.; Gallardo, K.; Vigani, G.; Brunel-Muguet, S.; Trouverie, J.; Salon, C.; Ourry, A. Disentangling the complexity and diversity of crosstalk between sulfur and other mineral nutrients in cultivated plants. *J. Exp. Bot.* **2019**, *70*, 4183–4196. [CrossRef] [PubMed]
3. Saenboonruang, K.; Phonchanthuek, E.; Prasandee, K. Soil-to-plant transfer factors of natural radionuclides (^{226}Ra and ^{40}K) in selected Thai medicinal plants. *J. Environ. Radioact.* **2018**, *184–185*, 1–5. [CrossRef] [PubMed]
4. Jassens, K.; Dik, J.; Cottee, M.; Susini, J. Photon-based techniques for nondestructive subsurface analysis of painted cultural heritage artifacts. *Acc. Chem. Res.* **2010**, *43*, 814–825. [CrossRef]
5. Kut, C.; Chaichana, K.; Xi, J.; Raza, S.M.; Ye, X.; McVeigh, E.R.; Rodriguez, F.J.; Quinones-Hinojosa, A.; Li, X. Detection of human brain cancer infiltration ex vivo and in vivo using quantitative optical coherence tomography. *Sci. Transl. Med.* **2015**, *7*, 292ra100. [CrossRef]
6. Taylor, C.W.; Kirby, A.M. Cardiac side-effects from breast cancer radiotherapy. *Clin. Oncol.* **2015**, *27*, 621–629. [CrossRef]
7. Lekshmi, S.; Singh, D.N.; Baghini, M.S. A critical review of soil moisture measurement. *Measurement* **2014**, *54*, 92–105.
8. Parejo Calvo, W.A.; Duarte, C.L.; Machado, L.D.B.; Manzoli, J.E.; Geraldo, A.B.C.; Kodama, Y.; Silva, L.G.A.; Pino, E.S.; Somessari, E.S.R.; Silveira, C.G.; et al. Electron beam accelerators—trends in radiation processing technology for industrial and environmental applications in Latin America and the Caribbean. *Radiat. Phys. Chem.* **2012**, *81*, 1276–1281. [CrossRef]
9. Leuraud, K.; Richardson, D.B.; Cardis, E.; Daniels, R.D.; Gillies, M.; O'Hagan, J.A.; Hamra, G.B.; Haylock, R.; Laurier, D.; Moissonnier, M.; et al. Ionising radiation and risk of death from leukaemia and lymphoma in radiation-monitored workers (INWORKS): An international cohort study. *Lancet Haematol.* **2015**, *2*, e276–e281. [CrossRef]
10. Yeung, A.W.K. The "As Low As Reasonably Achievable" (ALARA) principle: A brief historical overview and a bibliometric analysis of the most cited publications. *Radioprotection* **2019**, *54*, 103–109. [CrossRef]
11. Ninyong, K.; Wimolmala, E.; Sombatsompop, N.; Saenboonruang, K. Potential use of NR and wood/NR composites as thermal neutron shielding materials. *Polym. Test.* **2017**, *59*, 336–343. [CrossRef]
12. Kumar, A. Gamma ray shielding properties of PbO-Li$_2$O-B$_2$O$_3$ glasses. *Radiat. Phys. Chem.* **2017**, *136*, 50–53. [CrossRef]
13. Poltabtim, W.; Wimolmala, E.; Saenboonruang, K. Properties of lead-free gamma-ray shielding materials from metal oxide/EPDM rubber composites. *Radiat. Phys. Chem.* **2018**, *153*, 1–9. [CrossRef]
14. Thumwong, A.; Chinnawet, M.; Intarasena, P.; Rattanapongs, C.; Tokonami, S.; Ishikawa, T.; Saenboonruang, K. A comparative study on X-ray shielding and mechanical properties of natural rubber latex nanocomposites containing Bi$_2$O$_3$ or BaSO$_4$: Experimental and numerical determination. *Polymers* **2022**, *14*, 3654. [CrossRef]
15. Singh, V.P.; Ali, A.M.; Badiger, N.M.; Al-Khayatt, A.M. Monte Carlo simulation of gamma ray shielding parameters of concretes. *Nucl. Eng. Des.* **2013**, *265*, 1071–1077. [CrossRef]
16. Hu, G.; Shi, G.; Hu, H.; Yang, Q.; Yu, B.; Sun, W. Development of gradient composite shielding material for shielding neutrons and gamma rays. *Nucl. Eng. Technol.* **2020**, *52*, 2387–2393. [CrossRef]
17. Tekin, H.O.; Altunsoy, E.E.; Kavaz, E.; Sayyed, M.I.; Agar, O.; Kamislioglu, M. Photon and neutron shielding performance of boron phosphate glasses for diagnostic radiology facilities. *Results Phys.* **2019**, *12*, 1457–1464. [CrossRef]
18. Tiamduantawan, P.; Wimolmala, E.; Meesat, R.; Saenboonruang, K. Effects of Sm$_2$O$_3$ and Gd$_2$O$_3$ in poly (vinyl alcohol) hydrogels for potential use as self-healing thermal neutron shielding materials. *Radiat. Phys. Chem.* **2020**, *172*, 108818. [CrossRef]
19. Kaewnuam, E.; Wantana, N.; Tanusilp, S.; Kurosaki, K.; Limkitjaroenporn, P.; Kaewkhao, J. The influence of Gd$_2$O$_3$ on shielding, thermal and luminescence properties of WO$_3$-Gd$_2$O$_3$-B$_2$O$_3$ glass for radiation shielding and detection material. *Radiat. Phys. Chem.* **2022**, *190*, 109805. [CrossRef]
20. Gwaily, S.E.; Badawy, M.M.; Hassan, H.H.; Madani, M. Natural rubber composites as thermal neutron radiation shields: I. B$_4$C/NR composites. *Polym. Test.* **2002**, *21*, 129–133. [CrossRef]
21. Ozdemir, T.; Akbay, I.K.; Uzun, H.; Reyhancan, I.A. Neutron shielding of EPDM rubber with boric acid: Mechanical, thermal properties and neutron absorption tests. *Prog. Nucl. Energy* **2016**, *89*, 102–109. [CrossRef]
22. El-Khatib, A.M.; Doma, A.S.; Badawi, M.S.; Abu-Rayan, A.E.; Aly, N.S.; Alzahrani, J.S.; Abbas, M.I. Conductive natural and waste rubbers composites-loaded with lead powder as environmental flexible gamma radiation shielding material. *Mater. Res. Express* **2020**, *7*, 105309. [CrossRef]
23. Xu, C.; Cao, L.; Lin, B.; Liang, X.; Chen, Y. Design of self-healing supramolecular rubbers by introducing ionic cross-links into natural rubber via a controlled vulcanization. *ACS Appl. Mater. Interfaces* **2016**, *8*, 17728–17737. [CrossRef]

24. Xu, C.; Cao, L.; Huang, X.; Chen, Y.; Lin, B.; Fu, L. Self-healing natural rubber with tailorable mechanical properties based on ionic supramolecular hybrid network. *ACS Appl. Mater. Interfaces* **2017**, *9*, 29363–29373. [CrossRef] [PubMed]
25. Norhazariah, S.; Azura, A.R.; Sivakumar, R.; Azahari, B. Effect of different preparation methods on crosslink density and mechanical properties of Carrageenan filled natural rubber (NR) latex films. *Procedia Chem.* **2016**, *19*, 986–992. [CrossRef]
26. Boden, J.; Bowen, C.R.; Buchard, A.; Davidson, M.G.; Norris, C. Understanding the effects of cross-linking density on the self-healing performance of epoxidized natural rubber and natural rubber. *ACS Omega* **2022**, *7*, 15098–15105. [CrossRef]
27. Gerward, L.; Guilbert, N.; Bjorn Jensen, K.; Levring, H. X-ray absorption in matter. Reengineering XCOM. *Radiat. Phys. Chem.* **2001**, *60*, 23–24. [CrossRef]
28. Tiamduangtawan, P.; Kamkaew, C.; Kuntonwatchara, S.; Wimolmala, E.; Saenboonruang, K. Comparative mechanical, self-healing, and gamma attenuation properties of PVA hydrogels containing either nano- or micro-sized Bi_2O_3 for use as gamma-shielding materials. *Radiat. Phys. Chem.* **2020**, *177*, 109164. [CrossRef]
29. Kires, M. Archimedes' principle in action. *Phys. Educ.* **2007**, *42*, 484. [CrossRef]
30. Moonart, U.; Utara, S. Effect of surface treatments and filler loading on the properties of hemp fiber/natural rubber composites. *Cellulose* **2019**, *26*, 7271–7295. [CrossRef]
31. Sears, V.F. Neutron scattering lengths and cross sections. *Neutron News* **1992**, *3*, 26–37. [CrossRef]
32. Piotrowski, T. Neutron shielding evaluation of concretes and mortars: A review. *Const. Build Mater.* **2021**, *277*, 122238. [CrossRef]
33. Celli, M.; Grazzi, F.; Zoppi, M. A new ceramic material for shielding pulsed neutron scattering instruments. *Nucl. Instrum. Methods Phys. Res. A* **2006**, *565*, 861–863. [CrossRef]
34. Xu, Z.G.; Jiang, L.T.; Zhang, Q.; Qiao, J.; Gong, D.; Wu, G.H. The design of a novel neutron shielding B_4C/Al composite containing Gd. *Mater. Des.* **2016**, *111*, 375–381. [CrossRef]
35. Manohara, S.R.; Hanagodimath, S.M.; Thind, K.S.; Gerward, L. On the effective atomic number and electron density: A comprehensive set of formulas for all types of materials and energies above 1 keV. *Nucl. Instrum. Methods Phys. Res. Sec. B* **2008**, *266*, 3906–3912. [CrossRef]
36. Singh, V.P.; Badiger, N.M. Study of mass attenuation coefficients, effectiv1e atomic numbers and electron densities of carbon steel and stainless steels. *Radioprotection* **2013**, *48*, 431–443. [CrossRef]
37. Aghaz, A.; Faghihi, R.; Mortazavi, S.M.J.; Haghparast, A.; Mehdizadeh, S.; Sina, S. Radiation attenuation properties of shields containing micro and nano WO_3 in diagnostic X-ray energy range. *Int. J. Radiat. Res.* **2016**, *14*, 127–131. [CrossRef]
38. Chantler, C.T. Detailed tabulation of atomic form factors, photoelectric absorption and scattering cross section, and mass attenuation coefficients in the vicinity of absorption edges in the soft X-ray (Z=30–36, Z=60–89, E=0.1 keV–10 keV), addressing convergence issues of earlier work. *J. Phys. Chem. Ref. Data* **2000**, *29*, 597.
39. McCaffrey, J.P.; Shen, H.; Downton, B.; Mainegra-Hing, E. Radiation attenuation by lead and nonlead materials used in radiation shielding garments. *Med. Phys.* **2007**, *34*, 530–537. [CrossRef]
40. Ozdemir, T.; Gungor, A.; Reyhancan, I.A. Flexible neutron shielding composite material of EPDM rubber with boron trioxide: Mechanical, thermal investigations and neutron shielding tests. *Radiat. Phys. Chem.* **2017**, *131*, 7–12. [CrossRef]
41. Intom, S.; Kalkornsurapranee, E.; Johns, J.; Kaewjaeng, S.; Kothan, S.; Hongtong, W.; Chaiphaksa, W.; Kaewkhao, J. Mechanical and radiation shielding properties of flexible material based on natural rubber/ Bi_2O_3 composites. *Radiat. Phys. Chem.* **2020**, *172*, 108772. [CrossRef]
42. Kong, S.M.; Mariatti, M.; Busfield, J.J.C. Effects of types of fillers and filler loading on the properties of silicone rubber composites. *J. Reinf. Plast. Compos.* **2011**, *30*, 1087–1096. [CrossRef]
43. Frohlich, J.; Niedermeier, W.; Luginsland, H.D. The effect of filler–filler and filler–elastomer interaction on rubber reinforcement. *Compos. Part A Appl. Sci. Manuf.* **2005**, *36*, 449–460. [CrossRef]
44. Sambhudevan, S.; Shankar, B.; Appukuttan, S.; Joseph, K. Evaluation of kinetics and transport mechanism of solvents through natural rubber composites containing organically modified gadolinium oxide. *Plast. Rubber Compos.* **2016**, *45*, 216–223. [CrossRef]
45. Fan, R.L.; Zhang, Y.; Li, F.; Zhang, Y.X.; Sun, K.; Fan, Y.Z. Effect of high-temperature curing on the crosslink structures and dynamic mechanical properties of gum and N330-filled natural rubber vulcanizates. *Polym. Test.* **2001**, *20*, 925–936. [CrossRef]
46. Liu, Y.; Li, Z.; Liu, R.; Liang, Z.; Yang, J.; Zhang, R.; Zhou, Z.; Nie, Y. Design of self-healing rubber by introducing ionic interaction to construct a network composed of ionic and covalent cross-linking. *Ind. Eng. Chem. Res.* **2019**, *58*, 14848–14858. [CrossRef]
47. Zhao, M.; Chen, H.; Yuan, J.; Wu, Y.; Li, S.; Liu, R. The study of ionic and entanglements self-healing behavior of zinc dimethacrylate enhanced natural rubber and natural rubber/butyl rubber composite. *J. Appl. Polym. Sci.* **2022**, *139*, 52048. [CrossRef]
48. Wu, M.; Yang, L.; Shen, Q.; Zheng, Z.; Xu, C. Endeavour to balance mechanical properties and self-healing of nature rubber by increasing covalent crosslinks via a controlled vulcanization. *Eur. Polym. J.* **2011**, *161*, 110823. [CrossRef]
49. Peymanfar, R.; Selseleh-Zakerin, E.; Ahmadi, A.; Saeidi, A.; Tavassoli, S.H. Preparation of self-healing hydrogel toward improving electromagnetic interference shielding and energy efficiency. *Sci. Rep.* **2021**, *11*, 16361. [CrossRef]
50. Park, J.; Kim, M.; Choi, S.; Sun, J. Self-healable soft shield for γ-ray radiation based on polyacrylamide hydrogel composites. *Sci. Rep.* **2020**, *10*, 21689. [CrossRef]

Article

Detection of Leachable Components from Conventional and Dental Bulk-Fill Resin Composites (High and Low Viscosity) Using Liquid Chromatography-Tandem Mass Spectrometry (LC-MS/MS) Method

Matea Lapaš Barišić [1], Hrvoje Sarajlija [2], Eva Klarić [3,*], Alena Knežević [4], Ivan Sabol [5] and Vlatko Pandurić [3,*]

1. Private Dental Office, 10000 Zagreb, Croatia
2. Forensic Science Center, Ivan Vucetic, 10000 Zagreb, Croatia
3. Department of Endodontics and Restorative Dentistry, School of Dental Medicine University of Zagreb, 10000 Zagreb, Croatia
4. Division of Restorative Sciences, Herman Ostrow School of Dentistry, University of Southern California, Los Angeles, CA 90089, USA
5. Division of Molecular Medicine, Ruđer Boskovic Institute, 10000 Zagreb, Croatia
* Correspondence: eklaric@sfzg.hr (E.K.); vpanduric@sfzg.hr (V.P.)

Abstract: The aim of this study was to investigate leachable components (monomers) in high and low viscosity dental bulk-fill resin composites and conventional resin composite materials after polymerization. Six bulk-fill and six conventional dental resin composite materials were used in this study. The samples of each material (three sets of triplicates) were cured for 20 s with irradiance of 1200 mW/cm^2 with a LED curing unit and immersed in a 75% ethanol solution at 37 °C. The eluates from each triplicate set were analyzed after 24 h, 7 days or 28 days using liquid chromatography coupled with triple quadrupole tandem mass spectrometry (LC-MS/MS). Detectable amounts of 2-Hydroxyethyl methacrylate (HEMA) were found in both Gradia materials and the amount observed across different time points was statistically different ($p < 0.05$), with the amount in solution increasing for Gradia and decreasing for Gradia Direct flo. Bisphenol A diglycidildimethacrylate (BIS GMA) was found in Filtek and Tetric materials. Triethylene glycol dimethacrylate (TEGDMA) was detected in all materials. On the other hand, there were no statistically significant differences in the amounts of TEGDMA detected across different time points in either of the tested materials. Monomers HEMA, TEGDMA, 4-dimethylaminobenzoic acid ethyl ester (DMA BEE) and BIS GMA in bulk-fill and conventional composites (high and low viscosity) can be eluted after polymerization. The good selection of composite material and proper handling, the following of the manufacturer's instructions for polymerization and the use of finishing and polishing procedures may reduce the elution of the unpolymerized monomers responsible for the possible allergic and genotoxic potential of dental resin composites.

Keywords: liquid chromatography coupled with triple quadrupole tandem mass spectrometry (LC-MS/MS); dental composites; bulk-fill composites; elution; residual monomer

1. Introduction

Light-cured dental composites are the materials of choice in restorative dentistry due to their esthetic properties, mechanical strength and applicability in minimally invasive procedures. The time-consuming incremental technique has recently been replaced by a bulk technique thanks to the discovery of bulk-filling composites. Recently, a brand-new class of resin-based composites known as bulk-fill composites has been introduced. Their main selling point is the ability to install and cure increments of up to 4 mm in a single step, reducing chairside time. These composites also have a fast activation time due to newly designed initiation mechanisms and higher translucency due to larger filler particles and lower filler loading. Bulk-fill composites simplify the clinical procedure because they can be used in thicker layers [1]. For this reason, manufacturers claim that the composite can control the polymerization process and ensure adequate depth of cure even when larger increments are used. The most important advantage offered by these materials is the time saved in placing the material and in polymerization, as well as the reduced sensitivity to the technique [2]. The molecular basis of these resin composites has been altered to allow greater incremental incorporation by reducing or replacing Bis-GMA, resulting in a lower viscosity monomer, and/or by replacing higher molecular weight monomers often based on Bis-EMA, TEGDMA, EBPDMA, and UDMA monomers. The incorporation of stress reducers and modification of filler content also contribute to the reduction of polymerization shrinkage. When bulk-fill composites are used, polymerization shrinkage should be reduced, which in turn allows for good marginal integrity and less cusp deformation in the final composite restoration [3].

The degree of conversion during polymerization refers to the ratio of monomer to polymer. In other words, the degree of conversion refers to the percentage of C=C bonds of the monomers present in the polymeric matrices that have undergone reaction. The internal standard refers to the percentage of C=C bonds determined from the ratio of cured to uncured monomers. The complete conversion of all monomers to polymers results in a conversion rate of 100 percent, but this is never achieved. The conversion rate is normally between 43 and 70%. Ten percent of the elution from resin composites is caused by free monomers [4]. Lower conversion rates result in more monomer eluting into the oral environment, which negatively affects the mechanical and physical properties of the material. Dental composites essentially consist of glass filler particles dispersed in methacrylate resin. The latter is photocurable and can be cured by radical polymerization when irradiated with visible light. The polymerization of multifunctional methacrylate monomers results in a densely crosslinked network and yields monomer conversions that rarely exceed 80% [5]. This suggests that residual monomers can elute from the restoration to the oral cavity. Dental composites consist of a few main components: organic matrix (monomers: 2 Hydroxyethyl metacrylate (HEMA), Bysphenil-glycidyl-methacrylate (Bis GMA) and/or Urethane-dimethacrylate (UDMA)), co-monomers (Ethylene glycol dimethylacrylate (EGDMA), Methyl ether methacrylate (DEGMA), Triethylene glycoldymethacrylate (TEGDMA)), inorganic fillers (quartz, borosilicate, lithium aluminum silicate glasses and amorphous silicas), photoinitiators (camphorquinone CQ, Lucirin TPO, PPD), co-initiator Ethyl 4-dymethyloamino benzoate (DMA BEE), inhibitors of polymerization (BHT) and photostabilizers (Benzophenone) [6–8]. Unreacted monomers might have an influence on the biocompatibility of the restorations and can cause local or systemic toxic effects [9–12]. The majority of the degradation products have probably not yet been identified. Lower conversion rates result in more monomer eluting into the oral environment, which negatively affects the mechanical and physical properties of the material. In water, 2–6 wt% in 70% ethanol, and 10% in methanol, the amount of eluting molecules varies [13]. The aging of composites can also lead to more porosity due to filler wear, water sorption, and chemical/enzymatic degradation, resulting in an increased release of unpolymerized monomers originally trapped in the polymer network. The ability of the residual monomers to penetrate the matrices and expand the space between the polymer chains, allowing the soluble chemicals to diffuse, was the reason for the use of ethanol as a solvent in the present investigation. It is claimed that the replacement of the composite in the oral cavity releases various components. Reportedly, these substances have estrogenic, genotoxic, mutagenic, and cytotoxic properties. Reportedly, the unpolymerized monomer can reach the pulp and cause negative pulpal reactions [14]. In order to reduce polymerization shrinkage, achieve adequate depth of cure and reduce the elution of components from conventional composites, it is necessary to apply the material in layers, whereas bulk-fill dental resin composites use the single-layer technique to achieve the same [15]. There are few literature data on the elution of monomers from bulk-fill composites. Polydorou et al. [16] investigated the elution of monomers from two conventional dental composites after different polymerization and storage times using LC-MS/MS. No significant difference was found between samples polymerized for 20 and 40 s, and only BisGMA and TEGDMA were detected. Manojlovic et al. [17] quantified the elution of the major monomers from four commercial composites using high-performance liquid chromatography and established a mathematical model of the elution kinetics. It was shown that TEGDMA was identified as the main compound released from dental composites analyzed by high-performance liquid chromatography (HPLC) [18]. Mass spectroscopy can basically be conceptualized as molecular scale. Tandem mass spectrometers (LC-MS/MS), sometimes referred to as MS/MS instruments, are devices used to chemically process molecules before weighing the results. Mass spectrometers use charged molecules (ions) in a vacuum to make these observations. HPLC, on the other hand, works with molecules in solution. The first step at MS is to convert the sample into a charged ion in the gas phase, which is followed by the measurement. While liquid chromatography separates mixtures with multiple components, mass spectrometry provides spectral information that can help identify (or confirm the suspected identity of) the individual separated components. It is a highly efficient chemical technique that combines the physical separation capabilities of liquid chromatography with the mass analysis capabilities of mass spectrometry. HPLC is less accurate and sensitive than LCMS, which was advantageous for this study to more accurately determine the elution of the monomers of the most commonly used high and low viscosity composites for dental fillings [19].

The objective of this study was to determine all possible residual monomers from conventional and bulk-fill composites (high and low viscosity) leaching from the materials after different time intervals (24 h, 7 days, and 28 days) using more accurate and sensitive liquid chromatography coupled with triple quadrupole tandem mass spectrometry (LC-MS/MS). For this study, the following two working hypotheses were made: (1) there is no elution of residual monomers after polymerization of conventional and filled resin composites; (2) there is no difference in elution between conventional and filled composites with high and low viscosity; (3) there is no difference in monomer elution measured after 24 h, 7 days and 28 days.

2. Materials and Methods

2.1. Sample Preparation

In this study, six commercially available bulk-fill dental composites were investigated and compared with six conventional composites (Table 1). Low- and high-viscosity composite samples were prepared by applying the resin material directly from the compule into a Teflon mold (diameter 5 mm and depth 2 mm). After placing the composite resin, the surface was covered with a transparent plastic Mylar strip, and the sample was light-cured according to the standard protocol (20 s of irradiation with 1200 mW/cm^2 in a wavelength range of 380–515 nm) using a LED-curing device, Bluephase G2 (Ivoclar Vivadent, Schaan, Liechtenstein), which was measured with the LED light-curing radiometer Bluephase Meter II (Ivoclar Vicadent, Schaan, Liechtenstein) and immersed in 5 mL of 75% ethanol solution at 37 °C. After a dark storage period of 24 h, 7 days or 28 days, the eluates were collected and analyzed (Figure 1). The study was approved by the Ethics Committee of the Faculty of Dentistry, College of Zagreb (number 05-PA-30-XXI-10/2020). For the measurement of the monomer elution and due to the high effectiveness of the LC-MS/MS method, a sample size of 5 samples was determined to be optimal for the study. The sample size was calculated using the G * power program based on the difference in numerical variables between measurements, setting a significance level of 0.05 and a power level of 0.8 (high power size of 0.8), and obtaining a minimum required sample size of 5 samples (number of replicates within each experimental group) per group.

Table 1. Composite materials used in study.

Name	Producer	Lot	Abbreviation	Matrix Composition Declared by Producer
Tetric Evo Ceram Bulk Fill	Ivoclar Vivadent, (Schaan, Lichtenstein)	82 O135539	TeCBf	Bis GMA, Bis-EMA, UDMA
Tetric EvoFlow Bulk Fill	Ivoclar Vivadent, (Schaan, Lichtenstein)	U34907	TefBf	Bis GMA, Bis EMA, UDMA
X-tra Fil	VOCO (Cuxhaven, Germany)	14385921	Xf	Bis GMA, TEGDMA, UDMA
Filtek Bulk Fill	3M ESPE (St. Paul, MN, USA)	N626709	Fbf	Bis GMA, Bis EMA, UDMA, Procrylat resin
Filtek Bulk Fill flow	3M ESPE (St. Paul, MN, USA)	N732765	Fbff	Bis GMA, Bis EMA, UDMA, Procrylat resin
SDR	DENTSPLY (Charlotte, NC, USA)	1610131	SDR	Modified UDMA, EBPADMA, TEGDMA
Gradia	GC (Tokyo, Japan)	1710312	G	UDMA, TEGDMA
Gradia Direct flo	GC (Tokyo, Japan)	1502041	GDf	UDMA, TEGDMA
Filtek Supreme	3M ESPE (St. Paul, MN, USA)	N763255	FS	Bis GMA, TEGDMA, UDMA
Filtek Supreme flow	3M ESPE (St. Paul, MN, USA)	6033A2	Fsf	Bis GMA, TEGDMA, UDMA
TetricEvo Ceram	Ivoclar Vivadent, (Schaan, Lichtenstein)	V16037	TeC	Bis GMA, Bis EMA, UDMA, TEGDMA
TetricEvo flow	Ivoclar Vivadent, (Schaan, Lichtenstein)	V02622	Tcf	Bis GMA, Bis EMA, UDMA, TEGDMA

Bis GMA (Bysphenil-glycidyl-methacrylate), Bis EMA (Ethoxylated bisphenol A glycol dimethacrylate), UDMA (Urethane-dimethacrylate), TEGDMA (Triethylene glycoldymethacrylate), EBPADMA (ethoxylated bisphenol A dimethacrylate).

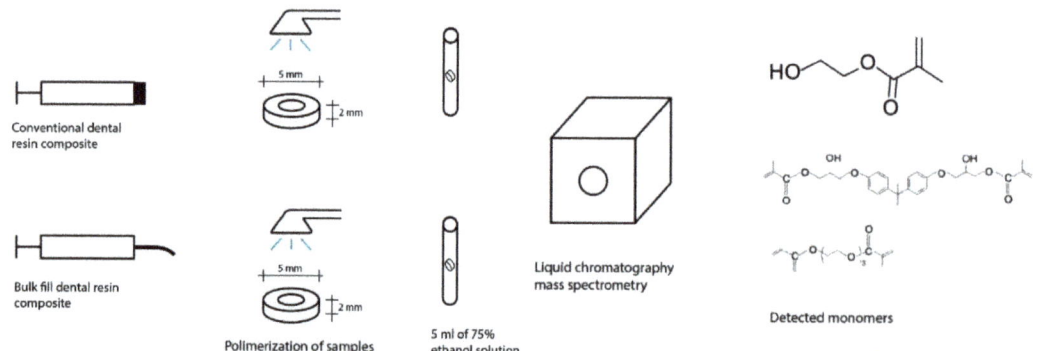

Figure 1. Sample preparation and monomer detection.

2.2. Analytical Technique

Liquid chromatography coupled with triple quadrupole tandem mass spectrometry (LC-MS/MS) was used to evaluate the presence of leachable compounds in the eluates. Sample preparation was performed as follows: 0.5 mL of ethanol-water extract was concentrated in vacuo (Martin Christ, Osterode am Harz, Germany) until dry, and the sample was reconstituted in 100 µL methanol. Quantitative and qualitative analysis was reconstituted using LC-MS/MS (Shimadzu LC system AC 20 coupled to ABSciex 3200 Qtrap tandem mass spectrometer system). All compounds were determined qualitatively by comparing their production mass spectra to the available internal production mass spectra library, and TEGD-MA, Bis GMA, HEMA, CQ and DMABEE were also measured quantitatively using calibration curves. Calibration curves were constructed with exactly 6 concentration levels (Eurachem Guide) ranging from 100 ng/mL–10,000 ng/mL (CQ and DMABEE), 1000 ng/mL–50,000 ng/mL (TEGDMA and Bis GMA), and 100 ng/mL–20,000 ng/mL (HEMA). Components were separated on a Poroshell 120 EC -C18HPLC column 2.1 × 100 mm 2.7 µm (Agilent Technologies), and 5 µL of sample was added via the autosampler. The column was maintained at 40 degrees Celsius. The mobile phases used were 63 mg of ammonium formate and 1 mL of formic acid to 1 L of deionized water (mobile phase A) and LiChrosolv methanol (mobile phase B). Separation was performed using a constant total flow rate of 400 µm/min, 30% of mobile phase B. A gradient flow was introduced after 0.5 min, reaching 100% of B after 3 min, held isocratic until 3.75 min and then decreased to 30% B at 3.76 min and held isocratic until the end of the run at 6.5 min. The mass spectrum was set to MRM mode for quantification using Solvent Blue 35 (SB -35) as the internal standard. The relationship between concentration and absorbance was plotted using the calculated areas under the peaks. The percentages of the different polymers in each study group were calculated. All measurements were performed once for each sample. The measurements were performed after 1, 7, and 28 days, respectively [20].

2.3. Statistical Test Methods

Data were recorded in an Excel spreadsheet (Microsoft, Seattle, WA, USA), and the mean of the technical replicates for each material and time point was calculated. Results were expressed as mean and standard deviation (SD). Statistical analysis and plots were generated using Medcalc (v11.4, MedCalc Software bv, Ostend, Belgium). Normality of the data was assessed with the Kolmogorov-Smirnov test. The amount of each monomer released at different time points was assessed using one-way analysis of variance or the Kruskal-Wallis test (one-way analysis of variance by ranks) and repeated-measures non-parametric Friedma analysis of variance. Comparisons between specific groups of materials at the same time point were made using ANOVA. In addition, individual comparisons between 2 groups were assessed with the t test. A p value less than 0.05 was considered statistically significant.

3. Results

The overall results showed that monomers leached from the polymerized samples into the ethanol were detected at most time points. The mean amounts (and standard deviation) of each monomer leached from each material at each time point are shown in Table 2. DMA-BEE was found in all samples analyzed, but the concentration detected in Gradia and Gradia Direct flo was well below the limit of quantitation (LOQ) of the method and was therefore considered to be zero. A statistically significant difference was found between the different time points in some of the materials analyzed (Fbf $p = 0.006$, FBFf $p = 0.002$, SDR $p = 0.004$ and TEC $p = 0.001$ samples, respectively). In each case, there was an increase in the concentration of the leachable component in the solution. *Bis GMA* was detected in nine different materials (Table 2). There was a statistically significant difference between time points in TeCBf ($p = 0.001$), FBFf ($p = 0.002$), TEC ($p = 0.033$), Fsf and FBf ($p < 0.001$) samples. The amount of leached compounds increased in samples TeCBf, FbFf and TEC and decreased in samples FSf and FBf with time. Detectable amounts of HEMA were found in only two of twelve sample materials. In both Gradia and Gradia Direct flo, the amount detected at different time points was statistically different, increasing in solution in Gradia (from 14.1 + 1.6 to 34.2 + 10; ANOVA $p = 0.022$) and decreasing in Gradia Direct flo (from 9.7 + 1.1 to 7.7 + 0–5; ANOVA $p = 0.036$). TEGDMA was detected in all materials. On the other hand, there were no statistically significant differences in the amounts detected in any of the tested materials at the different time points (Table 2). To evaluate the difference between conventional and bulk-fill materials or between low- and high-viscosity materials, the eluted values at the earliest time point for each analyte were compared between the different types. To reduce variability due to different commercial suppliers, results from the same supplier were grouped where possible (Table 3). For filtek materials, there was a statistically significant difference in the elution of DMA BEE between bulk (Filtek Bulk fill) and conventional (Filtek Supreme) high viscosity materials (BHV vs. CHV; t-test p value = 0.006). The difference between bulk fill (Filtek Bulk Fill flow) and conventional (Filtek Supreme flow) low viscosity composites approached but did not reach the significance threshold of 0.05 (BLV vs. CLV; t-test p = 0.056). There was no significant difference between high- or low-viscosity materials. Interestingly, the significant differences initially observed were not significant after the compounds were leached for 7 days or after 28 days (Table 4; Figure 2). The only comparison that reached statistical significance was the difference between DMA BEE, which was leached from conventional high- and low-viscosity filtek materials. The difference was barely significant and was mainly due to a lower variance in the measurements (t-test p-value p = 0.049; Table 4). There were statistically significant differences in all comparisons between TetricEvo Bulk and conventional materials and between high- and low-viscosity composite materials for DMA BEE (all p-values < 0.01). However, for the other composite materials (Gradia, Gradia Direct flo, X-tra fil and SDR), there was only a statistically significant difference between bulk-fill high and bulk-fill low viscosity (p = 0.006); other comparisons were not possible. After 28 days, most of the differences between the leached compounds diminished, and only the difference between DMA BEE, which was leached from conventional high- and low-viscosity composites, remained significant (t-test p = 0.022; Table 4; Figure 3). *Bis GMA* was not detected in the low-viscosity Filtek Bulk material, so some comparisons were not possible. In other cases, there was also no statistical significance of the leached BiS-GMA amounts in the other Filtek materials after 24 h (Table 3). At the final time point after 28 days, Bis GMA was not observed in the low-viscosity bulk, again preventing some comparisons. The remaining comparisons were not statistically significant. As with DMA BEE, the amounts of *BiS GMA* were significantly different in all Tetric materials (all $p < 0.01$) and were not detected at all in the other materials tested after 24 h. After 28 days, as with DMA BEE, the differences between materials were less significant and only the high-viscosity bulk materials and the conventional high-viscosity materials remained statistically significantly different (p = 0.001). *Bis GMA* was also not observed in the other materials after 28 days (Table 4). The differences in leached TEGDMA were significant only between high- and low-viscosity filtek materials (p = 0.037). On the other hand, the TEGDMA differences were significant only when comparing bulk-fill and conventional Tetric materials. The comparisons of the other materials were not statistically significant after 24 h, but after 28 days with TEGDMA the differences were more significant for the Filtek materials (Table 4). No significant differences were observed in Tetric and other materials. Since HEMA was not detected in any of the Filtek or Tetric materials, no comparisons could be made. The amounts observed in Gradia and Gradia Direct flo (high- and low-viscosity versions of the same material) were statistically significantly different at both 24 h (p = 0.018) and 28 days (p = 0.045). The large differences between the high- and low-viscosity materials are shown in Figure 3.

Table 2. Mean values (SD) of residual leachable compounds for each dental material at each time point.

Material	Category *	Compound	24 h Mean	SD	7 d Mean	SD	28 d Mean	SD	ANOVA **
Fbf	BHV	TEGDMA	25.9	(12.9)	43.7	(28.6)	33.2	(28.3)	0.685
		Bis GMA	2.2	(0.5)	0.0	(0.0)	0.0	(0.0)	<0.001
		DMA BEE	1.0	(0.4)	2.6	(0.9)	3.5	(0.3)	0.006
		HEMA	0.0	(0.0)	0.0	(0.0)	0.0	(0.0)	NA
TecBf	BHV	TEGDMA	9.9	(13.9)	31.0	(13.7)	93.3	(66.5)	0.097
		Bis GMA	3.8	(0.9)	6.8	(0.4)	7.3	(0.5)	0.001
		DMA BEE	3.8	(0.0)	3.8	(0.0)	3.8	(0.0)	0.124
		HEMA	0.0	(0.0)	0.0	(0.0)	0.0	(0.0)	NA
Xf	BHV	TEGDMA	127.4	(98.2)	251.4	(33.3)	121.7	(63.1)	0.109
		Bis GMA	5.0	(0.6)	6.8	(0.5)	5.7	(1.1)	0.082
		DMA BEE	3.8	(0.0)	3.8	(0.0)	3.8	(0.0)	0.159
		HEMA	0.0	(0.0)	0.0	(0.0)	0.0	(0.0)	NA
Fbff	BLV	TEGDMA	2.2	(3.4)	14.8	(13.4)	1.4	(2.4)	0.154
		Bis GMA	0.0	(0.0)	4.1	(0.7)	6.9	(2.1)	0.002
		DMA-BEE	1.5	(0.9)	2.7	(0.5)	4.6	(1.0)	0.009
		HEMA	0.0	(0.0)	0.0	(0.0)	0.0	(0.0)	NA
TefBf	BLV	TEGDMA	43.6	(41.1)	22.9	(23.5)	46.6	(25.3)	0.619
		Bis GMA	7.9	(1.1)	12.1	(4.3)	14.5	(11.5)	0.552
		DMA BEE	1.9	(0.3)	3.2	(1.2)	3.8	(3.2)	0.533
		HEMA	0.0	(0.0)	0.0	(0.0)	0.0	(0.0)	NA
SDR	BLV	TEGDMA	61.0	(3.7)	54.2	(16.2)	54.2	(15.3)	0.772
		Bis GMA	0.0	(0.0)	0.0	(0.0)	0.0	(0.0)	NA
		DMA BEE	0.8	(0.4)	1.9	(0.2)	2.0	(0.2)	0.004
		HEMA	0.0	(0.0)	0.0	(0.0)	0.0	(0.0)	NA
Fs	CHV	TEGDMA	191.1	(86.6)	239.6	(27.7)	128.3	(34.3)	0.127
		Bis GMA	3.1	(2.2)	2.1	(1.2)	6.5	(6.3)	0.403
		DMA BEE	3.8	(0.0)	3.8	(0.0)	3.8	(0.0)	0.234
		HEMA	0.0	(0.0)	0.0	(0.0)	0.0	(0.0)	NA
TeC	CHV	TEGDMA	65.9	(27.8)	29.7	(38.8)	4.5	(7.7)	0.091
		Bis GMA	7.9	(0.7)	13.4	(1.1)	16.1	(1.8)	0.001
		DMA BEE	1.3	(0.2)	2.5	(0.2)	3.4	(0.5)	0.001
		HEMA	0.0	(0.0)	0.0	(0.0)	0.0	(0.0)	NA
G	CSHV	TEGDMA	55.3	(48.9)	109.2	(24.1)	98.9	(35.9)	0.255
		Bis GMA	0.0	(0.0)	0.0	(0.0)	0.0	(0.0)	NA
		DMA BEE	0.0	(0.0)	0.0	(0.0)	0.0	(0.0)	0.43
		HEMA	14.1	(1.6)	25.1	(4.0)	34.2	(10.0)	0.022
Fsf	CLV	TEGDMA	93.7	(99.3)	64.9	(4.0)	51.2	(6.0)	0.671
		Bis GMA	6.6	(1.5)	4.2	(0.5)	4.0	(0.7)	0.033
		DMA BEE	3.2	(0.7)	2.7	(0.6)	3.1	(0.3)	0.558
		HEMA	0.0	(0.0)	0.0	(0.0)	0.0	(0.0)	NA
Tcf	CLV	TEGDMA	59.2	(26.6)	175.6	(114.9)	202.8	(99.2)	0.192
		Bis GMA	17.1	(2.7)	14.4	(2.7)	15.8	(1.6)	0.428
		DMA BEE	4.1	(0.5)	4.5	(0.9)	5.0	(0.5)	0.343
		HEMA	0.0	(0.0)	0.0	(0.0)	0.0	(0.0)	NA
GDf	CLV	TEGDMA	131.8	(44.1)	85.6	(46.5)	75.1	(12.2)	0.227
		Bis GMA	0.0	(0.0)	0.0	(0.0)	0.0	(0.0)	NA
		DMA BEE	0.0	(0.0)	0.0	(0.0)	0.0	(0.0)	0.245
		HEMA	9.7	(1.1)	8.0	(0.4)	7.7	(0.5)	0.036

* BHV—Bulk high viscosity material, BLV—Bulk low viscosity, CHV—Conventional high viscosity, CLV—Conventional low viscosity. ** Kruskal-Wallis test one way analysis of variance by ranks test result p-value. Significant results highlighted in bold. NA—Not applicable. TEGDMA (Triethylene glycoldymethacrylate), Bis GMA (Bysphenil-glycidyl-methacrylate), DMA BEE (4-dimethylaminobenzoic acid ethyl ester), HEMA (2-Hydroxyethyl methacrylate).

Table 3. Concentration of different compounds leached from different types of dental material preparations from the same manufacturer after 24 h.

Manufacturer and Analyte	BHV Mean	(SD)	BLV Mean	(SD)	CHV Mean	(SD)	CLV Mean	(SD)	t-Test p Value BHV vs. CHV	BLV vs. CLV	BHV vs. BLV	CHV vs. CLV
Filtek												
DMA BEE	1.0	(0.4)	1.5	(0.9)	3.8	(0.0)	3.2	(0.7)	**0.006**	0.056	0.449	0.294
Bis GMA	2.2	(0.5)	0.0	(0.0)	3.1	(2.2)	6.6	(1.5)	0.518	NA	NA	0.083
TEGDMA	25.9	(12.9)	2.2	(3.4)	191.1	(86.6)	93.7	(99.3)	0.082	0.252	**0.037**	0.269
HEMA	0.0	(0.0)	0.0	(0.0)	0.0	(0.0)	0.0	(0.0)	NA	NA	NA	NA
Tetric												
DMA BEE	3.8	(0.0)	1.9	(0.3)	1.3	(0.2)	4.1	(0.5)	**0.001**	**0.003**	**0.008**	**0.001**
Bis GMA	3.8	(0.9)	7.9	(1.1)	7.9	(0.7)	17.1	(2.7)	**0.003**	**0.005**	**0.007**	**0.004**
TEGDMA	9.9	(13.9)	43.6	(41.1)	65.9	(27.8)	59.2	(26.6)	**0.036**	0.611	0.249	0.779
HEMA	0.0	(0.0)	0.0	(0.0)	0.0	(0.0)	0.0	(0.0)	NA	NA	NA	NA
Other												
DMA BEE	3.8	(0.0)	0.8	(0.4)	0.0	(0.0)	0.0	(0.0)	NA	NA	**0.006**	NA
Bis GMA	5.0	(0.6)	0.0	(0.0)	0.0	(0.0)	0.0	(0.0)	NA	NA	NA	NA
TEGDMA	127.4	(98.2)	61.0	(3.7)	55.3	(48.9)	131.8	(44.1)	0.318	0.109	0.362	0.114
HEMA	0.0	(0.0)	0.0	(0.0)	14.1	(1.6)	9.7	(1.1)	NA	NA	NA	**0.018**

BHV—Bulk high viscosity material, BLV—Bulk low viscosity, CHV—Conventional high viscosity, CLV—Conventional low viscosity. Significant results highlighted in bold. NA—Not applicable. TEGDMA (Triethylene glycoldymethacrylate), Bis GMA (Bysphenil-glycidyl-methacrylate), DMA BEE (4-dimethylaminobenzoic acid ethyl ester), HEMA (2-Hydroxyethyl methacrylate).

Table 4. Concentration of different compounds leached from different types of dental material preparations from the same manufacturer after 28 days.

Manufacturer and Analyte	BHV Mean	(SD)	BLV Mean	(SD)	CHV Mean	(SD)	CLV Mean	(SD)	t-Test p Value BHV vs. CHV	BLV vs. CLV	BHV vs. BLV	CHV vs. CLV
Filtek												
DMA BEE	3.5	(0.3)	4.6	(1.0)	3.8	(0.0)	3.1	(0.3)	0.295	0.067	0.156	**0.049**
Bis GMA	0.0	(0.0)	6.9	(2.1)	6.5	(6.3)	4.0	(0.7)	NA	0.084	NA	0.567
TEGDMA	33.2	(28.3)	1.4	(2.4)	128.3	(34.3)	51.2	(6.0)	**0.021**	**<0.001**	0.192	**0.019**
HEMA	0.0	(0.0)	0.0	(0.0)	0.0	(0.0)	0.0	(0.0)	NA	NA	NA	NA
Tetric												
DMA BEE	3.8	(0.0)	3.8	(3.2)	3.4	(0.5)	5.0	(0.5)	0.323	0.547	0.991	**0.022**
Bis GMA	7.3	(0.5)	14.5	(11.5)	16.1	(1.8)	15.8	(1.6)	**0.001**	0.865	0.389	0.829
TEGDMA	93.3	(66.5)	46.6	(25.3)	4.5	(7.7)	202.8	(99.2)	0.149	0.057	0.319	0.075
HEMA	0.0	(0.0)	0.0	(0.0)	0.0	(0.0)	0.0	(0.0)	NA	NA	NA	NA
Other												
DMA BEE	3.8	(0.0)	2.0	(0.2)	0.0	(0.0)	0.0	(0.0)	NA	NA	**0.003**	NA
Bis GMA	5.7	(1.1)	0.0	(0.0)	0.0	(0.0)	0.0	(0.0)	NA	NA	NA	NA
TEGDMA	121.7	(63.1)	54.2	(15.3)	98.9	(35.9)	75.1	(12.2)	0.615	0.137	0.146	0.339
HEMA	0.0	(0.0)	0.0	(0.0)	34.2	(10.0)	7.7	(0.5)	NA	NA	NA	**0.045**

BHV—Bulk high viscosity material, BLV—Bulk low viscosity, CHV—Conventional high viscosity, CLV—Conventional low viscosity. Significant results highlighted in bold. NA—Not applicable. TEGDMA (Triethylene glycoldymethacrylate), Bis GMA (Bysphenil-glycidyl-methacrylate), DMA BEE (4-dimethylaminobenzoic acid ethyl ester), HEMA (2-Hydroxyethyl methacrylate).

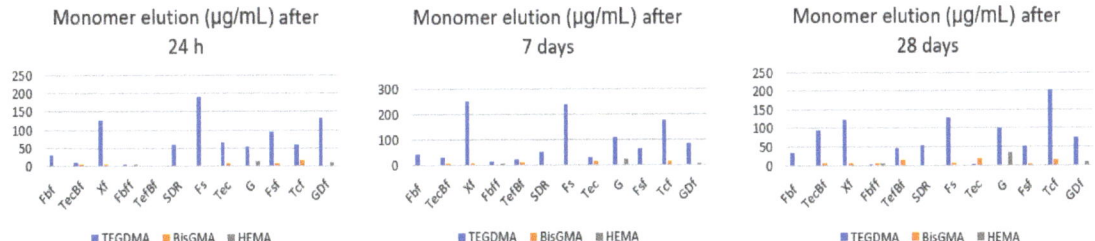

Figure 2. Mean amounts of leached TEGDMA, Bis GMA and HEMA at different time point. Fbf (Filtek Bulk Fill) TecBf (Tetric Evo Ceram Bulk Fill) Xf (X-tra Fil) Fbff (Filtek Bulk Fill flow) TefBf (Tetric EvoFlow Bulk Fill) SDR (SDR) Fs (Filtek Supreme) TeC (TetricEvo Ceram) G (Gradia) Fsf (Filtek Supreme flow) Tcf (TetricEvo flow) GDf (Gradia Direct flo). TEGDMA (Triethylene glycol dimethacrylate) Bis GMA (Bysphenil-glycidyl-methacrylate) HEMA (2-Hydroxyethyl methacrylate).

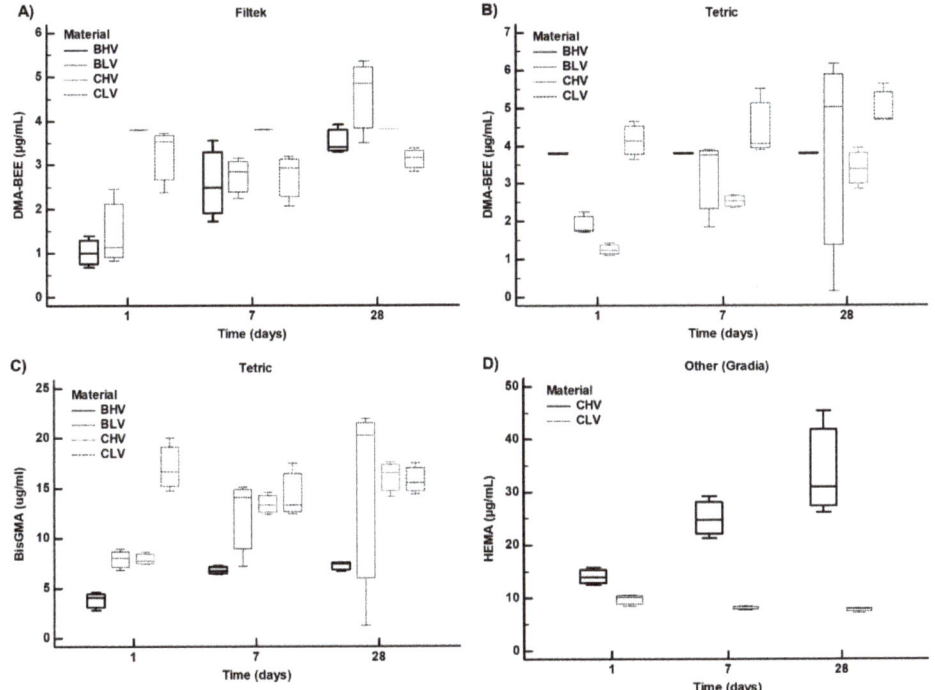

Figure 3. Amounts of leachable compounds detected by LC-MS/MS across different dental materials from the same provider (Filtek Panel (**A**); Tertic Panels (**B**,**C**); Gradia Panel (**D**)) in ethanol medium incubated at 37 Degrees Celsius for 24 h, 7 days or 28 days. BHV—Bulk high viscosity material, BLV—Bulk low viscosity material, CHV—Conventional high viscosity, CLV—Conventional low viscosity. Each material was sampled in three replicates and each replicate is represented by a mean of three LC-MS/MS measurements. Bis GMA (Bysphenil-glycidyl-methacrylate), DMA BEE (4-dimethylaminobenzoic acid ethyl ester), HEMA (2-Hydroxyethyl methacrylate).

4. Discussion

The aim of various previous studies was to determine the constituents extractable from polymerized resin composites. In most studies, only a few substances could be identified [21,22]. An important factor affecting the leaching of monomers is the type and molecular size of the monomers in the resin. Smaller molecules are leached faster than larger ones, and monomers with small molecular weight can be extracted in larger amounts than monomers with large molecular weight [23]. The various analytical methods used to determine leachable species from resin composites have been described by Ruyter and Oysaed [24]. In this study, we used liquid chromatography-tandem mass spectrometry (LC-MS/MS) to identify and quantify the elution of monomers. With the exception of Polydorou et al. [16], who studied the elution of monomers from two light-cured materials (nanohybrid and ormocer) after different curing times and different storage times, there is not much literature on the release of monomers from composites using this method. It is well known that eluted monomers can contribute to the cytotoxicity of composite resins. Geurtsen and Leyhausen [25] reported that cytotoxic aqueous resin eluates often contain high amounts of TEGDMA. In fact, the National Institute of Occupational Safety and Health has classified TEGDMA as an irritant to various tissues [26–28]. Spahl et al. [29] showed in their study that co-monomers and various additives, as well as impurities from the manufacturing process, were detected in all polymerized resin composites. Several in vitro studies have shown cytotoxic, genotoxic, mutagenic or estrogenic effects on the pulpal and gingival/oral mucosa due to the reactions of some monomers [30,31].

In the present study, the elution of TEGDMA, Bis GMA, HEMA, and DMA BEE from conventional and bulk-fill resin composites was investigated at three time intervals. The first and second hypotheses were rejected because residual monomers eluted after the polymerization of the materials, and there were also differences in elution between conventional and bulk-fill resin composites. The results of this study showed that TEGDMA was detected in all the materials studied, but there were no statistically significant differences in the amounts detected at different time points in any of the materials studied. This is consistent with other studies that have also found TEGDMA to be the main monomer eluting from composite resins [25]. However, differences in leached TEGDMA were significant only between high- and low-viscosity Filtek materials and bulk and conventional Tetric materials, and only after 28 days, with TEGDMA differences being more pronounced in Filtek materials. TEGDMA is a small monomer and elutes faster than larger molecules such as Bis GMA [32]. In this study, Bis GMA was detected in nine different materials. Only SDR, Gradia, and Gradia Direct flo did not contain Bis GMA at any of the time points. SDR is a flowable, single-component, fluoride-containing, visible-light-cured, radiopaque posterior composite restorative material designed for use as a base for Class I and II preparations. It has the typical handling characteristics of a "flowable" composite, but can be used in 4 mm increments with minimal polymerization stress. According to the manufacturer, eluted monomers from SDR can also irritate the skin, eyes, and oral mucosa [33]. HEMA is used in dental composites due to its hydrophilic application as a co-monomer of the organic resin matrix and was found in only a few tested materials. HEMA is known to cause cytotoxic and genotoxic effects [34]. In Gradia and Gradia Direct flo, the amount was statistically different, increasing in Gradia and decreasing in Gradia Direct flo. HEMA could be a degradation product of UDMA, which is a component of Gradia and Gradia Direct flo according to the MSDSs. Bis GMA was not detected in the low-viscosity material, so some comparisons were not possible and there was no statistical significance of the leached Bis GMA amounts in filtek materials after 24 h and at the last time point after 28 days. Bis GMA was not observed in the low viscosity bulk material, again preventing some comparisons. The amounts observed in Gradia and Gradia Direct Flo were not statistically significant at both 24 h and 28 days. In the study by Cebe et al., [2] the amount of eluted Bis GMA from Tetric Evo Ceram Bulk Fill and the amount of eluted TEGDMA and HEMA from X-tra Fill were higher than other composites, which was in contrast to our study where the TEGDM monomer was more eluted from all types of high- and low-viscosity bulk composites compared to Bis GMA, while HEMA was found only in Gradia composites.

DMA BEE is a co-initiator used in composites to accelerate the degradation of initiators into radicals and thus polymerization [35]. Various solvents such as distilled water, saliva, ethanol, methanol and acetonitrile have been used in studies to evaluate the leaching of monomers [25]. A 75% ethanol/water solution was the solution of choice in several studies to simulate and accelerate the aging of restorations [26]. The oral cavity represents an environment somewhere between water and more aggressive solvents (ethanol, methanol, acetonitrile) [30]. A 75 percent ethanol/water solution has a solubility parameter very close to that of oral fluid, resulting in maximum softening of the resin [36,37]. This solution is recommended by the United States Food and Drug Administration (FDA) guidelines (1976, 1988) as a clinically relevant mouth-simulating fluid and has been used in several studies [38–40]. Therefore, this solution was used in this study. The elution time of 24 h is based on previous findings [28] suggesting that almost all leachable substances are eluted within 24 h after polymerization. However, the elution of monomers is definitely not linear over time, and there are studies showing that the release of monomers lasts up to 30 days [16,37] or even up to one year after polymerization [16]. Therefore, additional time points of 7 and 28 days were also investigated. In this study, a statistically significant difference in the release of DMA BEE was found between Filtek Bulk and conventional high-viscosity materials and the difference between Filtek Bulk and conventional low-viscosity composites. The only comparison where a statistically significant difference was obtained was the amount of DMA BEE leached from conventional high viscosity and conventional low viscosity Filtek materials, but the difference was hardly significant and was mainly due to a smaller variation in the measurements. In all comparisons for Tetric Bulk and conventional and high- and low-viscosity preparations, there were statistically significant differences for DMA BEE, and there was also a statistical difference between X-tra fill and SDR, but other comparisons were not possible. After 28 days, the leaching DMA BEE of the conventional high- and low-viscosity material remained significant. There were also statistically significant differences between time points for samples TCbf, FBFf, TEC, FSf and FBf, and the amount of compound leached increased for samples TECbf, FBFF, TEC and decreased for samples FSf and FBf over time.

HEMA release showed a maximum increase on the 28th day for Gradia, which was in accordance with Altıntaş and Üşümez, [41], who investigated the residual monomer release from resin cements and reported the HEMA release amount from Nexus 2 (Kerr/Italy) cement to be lower in the 10th minute and much higher on the 21st day. Gradia Direct flo showed maximum increase in the first 24 h, with decreasing amounts of leached monomer after 28 days. This was similar to a study by Duruk et al. [42], who found that the amount of HEMA released from the resin cement of Ionolux (VOCO, Cuxhaven, Germany) was found to be very low for the 1st hour and higher on the first day in comparison to the 21st day. This situation may be due to the interaction of HEMA molecules with water, considering that HEMA is highly hydrophilic and the solution consists of 75% ethanol–25% water. For TEGDMA, the circumstances were different among materials because in some materials, TEGMA was higher after only 24 h compared to 28 days (X-tra Fil, SDR, Filtek Supreme, TetricEvo Ceram and Gradia Direct flo), and the highest amounts were found after 28 days for Filtek Bulk Fill, Tetric Evo Ceram Bulk Fill, Tetric EvoFlow Bulk Fill, Gradia and TetricEvo flow. The differences in filler particle type and monomer ratios specified by the manufacturer are assumed to be the cause of the residual monomer release seen between micro-hybrid and nano-hybrid composites. After one day and fourteen days, De Angelis et al. [43] used HPLC to evaluate the eluted monomer from the GrandioSO (VOCO) nanohybrid composite. According to their findings, TEGDMA levels became detectable after 24 h, while BIS-GMA levels were higher after 24 h than after 14 days. According to Duruk et al., after 24 h and 14 days, the amounts of TEGDMA released by the GrandioSO composite were much higher than on the 21st day when it was undetectable [42]. Additionally, after 24 h the amounts of BisGMA were higher than after 7 days or 28 days for Filtek Bulk Fill, Filtek Supreme flow, and TetricEvo flow, while for all others, the composites' amount of Bis GMA were higher after 28 days except for SDR, G, and GDf; no BisGMA were determined in either of the measured periods.

In the majority of studies, dilute ethanol, distilled water, and methanol have been used as solvents for testing the materials. In other protocols, elution was also studied in artificial saliva and various media commonly used for cell culture development. Artificial saliva and distilled water are both water-based solvents that can simulate intraoral conditions. Greater dissolution efficiency is a characteristic of organic solvents, likely due to better sorption, swelling, and penetration of the material. Since monomers are usually hydrophobic, similar differences between the main release in organic solvents and those based on water have been found in experiments [9]. In in vitro studies on dental materials and their properties, the environment of the oral cavity is usually mimicked to ensure the repeatability and stability of the applicable analytical procedures. Saliva is constantly produced in the oral cavity to clean the surfaces of teeth and dentures before being excreted by swallowing. Natural human saliva has a very complex and diverse composition that is influenced by numerous individual factors (including food intake, bacterial colonization, and others) that fundamentally affect intraoral pH. Because of these factors, it is difficult to produce a synthetic formula that exactly matches real saliva [44]. However, because real human saliva is unstable outside the oral canal, its use for this purpose is also unreliable. It appears to be very difficult to replicate the exact intraoral conditions, and this should be taken into account when evaluating the results of this or any other in vitro research that cannot fully correlate with the in vivo situation. Studies that looked at monomer solutions in artificial saliva also confirmed that the elution of bulk-fill composites was equivalent to that of conventional materials, despite their greater incremental thickness. The hydrophobicity of the base monomers and the final network properties of the resin matrix have a significant influence on monomer elution [45].

With higher monomer concentrations in the samples stored for 1 month compared to those stored for 24 h and 7 days, it was found in the present study that increasing the storage time resulted in higher amounts of Bis GMA and DMA BEE elution for all Tetric and Filtek composites, as noted in the study by Janani et al. [46]. Nazar at al. also used high-performance liquid chromatography analysis and reported that longer storage times resulted in statistically significant increases in BisGMA and UDMA amounts for both Tetric and Filtek materials [47]. There are few studies investigating the long-term elution of monomers over 1, 3, and 12 months using liquid chromatography tandem mass spectrometry and high-performance liquid chromatography [16,46]. However, the long-term effects of residual monomers on biocompatibility are still unclear. Due to the constant salivary flow in the oral environment, monomer concentrations are not expected to reach the cumulative levels determined in this study, while long-term chronic exposure and systemic adverse effects must also be considered when evaluating the potential toxicity of eluted compounds.

The monomer released from Gradia materials was HEMA (2-hydroxyethyl methacrylate). It is a tiny, low-molecular-weight monomer that is soluble in both types of solvents (130 g/mol). HEMA is a commonly used co-monomer in commercial resin-based products because its hydrophilic properties prevent the separation of water and hydrophobic co-monomers. However, some unfavorable physico-mechanical properties of HEMA have been documented, such as low conversion efficiency and water retention, which hinders effective polymerization [48]. In addition, HEMA showed some cytotoxicity that affected cell survival [49], which could be exacerbated by the water solubility of HEMA. The TEGDMA monomer was found in comparatively high amounts in all tested materials, especially in organic solvents. TEGDMA is a low-viscosity, low-molecular-weight molecule (286.32 g/mol) that is often added to composites to reduce the viscosity of the mixture and thus increase the degree of conversion (DC). Unfortunately, the larger DC of TEGDMA also leads to greater shrinkage of the material during polymerization. For this reason, TEGDMA is often replaced, at least in part, by another monomer that has a larger molecular mass and lower viscosity (e.g., Bis-EMA). There are reports of the cytotoxic effect of TEGDMA on human and gingival fibroblasts clinically associated with pulp infarction and necrosis [45]. As in other studies, our study confirmed that Bis-GMA has the lowest release, as it has the highest molecular mass (512.599 g/mol) and the lowest solubility in all types of solvents. Due to its high refractive index, low volatility, strong mechanical properties, low volumetric shrinkage after polymerization, diffusivity into tissue, and good adhesion to enamel, Bis GMA is a basic matrix compound that is generally useful [50]. However, the market for materials based on Bis GMA resins [51] such as composites based on Bis EFMA has begun to expand due to concerns about the viscosity of Bis GMA, which can negatively affect the mechanical properties of materials, and its potential cytotoxic effect in combination with BPA [52]. Bezgin et al. [53] measured the release of residual monomers with HPLC after 24, 48, and 72 h and also determine the effects of finishing and polishing procedures on the elution of Bis-GMA, TEGDMA, UDMA, and HEMA monomers from compomer and bulk-fill composite resins. The finishing and polishing procedures had a significant effect on reducing the quantity of UDMA release, so the Mylar strip also used in our study did not prevent the formation of the oxygen-inhibition layer, and final polishing was still essential to remove the resin-rich outer layer, which can be the source of unreacted monomers that elute into the oral cavity.

Chemicals are released in order of cytotoxic potential as determined by Reichl et al.: HEMA < TEGDMA < UDMA < Bis GMA. In their cytotoxicity study, they found that the EC50 values for HEMA and TEGDMA decreased from about 5 mmol/L (6 h) to about 0.6 mmol/L (48 h) and from about 3 mmol/L (6 h) to about 0.4 mmol/L (48 h), respectively. [54]. In this study, human gingival fibroblasts were exposed to Bis-GMA at a concentration of 0.087 mmol/L, UDMA at a concentration of 0.106 mmol/L, and HEMA at a concentration of 11.530 mmol/L. Such a decrease in the viability of TEGDMA was observed at 3.460 mmol/L. When dental resin materials with and without Bis-GMA were compared, those that released Bis-GMA and TEGDMA were found to have a higher potential for cytotoxicity and genotoxicity [48]. Numerous studies [55,56] have described the specific effects of monomers placed in direct contact with dental pulp cells, including inflammation and suppression of dentin mineralization. In our study, the TEGDMA concentrations of all tested materials were found to be below the hazardous concentrations for TEGDMA identified in some previous studies [25,43].

To determine the quantity of released compounds, most previously cited studies performed the analysis prevalently through the HPLC (high-performance liquid chromatography) or GC–MS (gas chromatography mass spectrometry) methods. The analytical methods of LC–MS (liquid chromatography mass spectrometry) and UPLC-MS/MS (ultraperformance liquid chromatography-tandem mass spectrometry) were used rarely and not so often, so we compared our results to other studies dealing with this method but also with HPLC, which is much popular in this type of study. LC–MS (liquid chromatography mass spectrometry) technique is based on the detection of the mass-over-charge ratio of a compound of interest and its daughter ions, leading to two extra parameters that are compound-specific. Susila et al. [57] measured the elution of the composites using Liquid Chromatography-Mass Spectrometry (LC-MS). They measured BisEMA, BisGMA, TEGDMA and UDMA elution in three different materials: polysiloxane-dimethacrylate (Ceram XTM), Silorane (Filtek P90TM) and dimethacrylate (RestofillTM). Dimethacrylate-based composites eluted more monomer and exerted strong cytotoxicity, which was similar to results found in our study where monomers from bulk-fill and conventional composites (high and low viscosity) were eluted from 2 mm thick samples after polymerization of 20 s with irradiance of 1200 mW/cm^2 with a LED curing unit.

Our results show that there are significant differences in the leachable components, depending in part on the type and consistency of the dental material studied. This was a pilot study with a smaller number of samples, and no correction was made for multiple testing. The number of samples tested for each material is a limiting factor for the study, but due to the large number of materials tested (which accounts for the uniqueness of this study) at multiple time points, it was not possible to increase the number of replicates for each material. The results should be confirmed with a larger number of samples, which is planned for future studies.

5. Conclusions

Within the limitations of the present quantitative study, it can be concluded that monomers (HEMA, TEGDMA, DMABEE, Bis GMA) can be eluted in bulk fill and conventional composites (high and low viscosity) after polymerization. The results indicate that the effect may be ambiguous, as apparently materials from different manufacturers release some monomers more than others. However, all but one material showed a high release of TEGDMA. The results of the present study show that the restorative materials investigated here are not chemically stable after polymerization, and the concentrations of eluted monomers can reach critical toxicity levels even after a single 2 mm thick restoration placement. Also, Mylar strips do not prevent the formation of the oxygen inhibition layer, and final polishing is still essential for the removal of the resin-rich outer layer, which may be the source of unreacted monomers eluting into the oral cavity. Thus, a good selection of composite material and proper handling, the following of the manufacturer's instructions for polymerization, and the use of finishing and polishing procedures can reduce the release of unpolymerized monomers from composite materials with possible genotoxic and cytotoxic potential to soft tissues and to the body in general.

Author Contributions: M.L.B. and E.K.; data curation, M.L.B. and E.K.; formal analysis, M.L.B. and E.K; funding acquisition M.L.B. and E.K.; investigation, H.S. and I.S.; methodology, M.L.B. and E.K.; project administration, M.L.B. and E.K.; resources, H.S. and I.S.; software, E.K., V.P. and A.K.; supervision, V.P. and A.K.; validation, M.L.B. and E.K.; visualization, M.L.B., E.K. and H.S.; writing—original draft, E.K.; writing—review and editing. All authors have read and agreed to the published version of the manuscript.

Funding: This research was funded by Croatian Science Foundation, grant number IP-2019-04-6183.

Institutional Review Board Statement: Not applicable.

Informed Consent Statement: Not applicable.

Data Availability Statement: The datasets generated and analyzed during the current study are available from the corresponding author on reasonable request.

Conflicts of Interest: The authors declare no conflict of interest. The funders had no role in the design of the study; in the collection, analyses, or interpretation of data; in the writing of the manuscript; or in the decision to publish the results.

References

1. Knezevic, A.; Ristic, M.; Demoli, N.; Tarle, Z.; Music, S.; Mandic, V.N. Composite Photopolymerization with Diode Laser. *Oper. Dent.* **2007**, *32*, 279–284. [CrossRef] [PubMed]
2. Cebe, M.A.; Cebe, F.; Cengiz, M.F.; Cetin, A.R.; Arpag, O.F.; Ozturk, B. Elution of monomer from different bulk fill dental composite resins. *Dent. Mater.* **2015**, *31*, e141–e149. [CrossRef] [PubMed]
3. Kim, R.J.-Y.; Kim, Y.-J.; Choi, N.-S.; Lee, I.-B. Polymerization shrinkage, modulus, and shrinkage stress related to tooth-restoration interfacial debonding in bulk-fill composites. *J. Dent.* **2015**, *43*, 430–439. [CrossRef]
4. Leprince, J.G.; Palin, W.M.; Hadis, M.A.; Devaux, J.; Leloup, G. Progress in dimethacrylate-based dental composite technology and curing efficiency. *Dent. Mater. Off. Publ. Acad. Dent. Mater.* **2013**, *29*, 139–156. [CrossRef]
5. Hosoda, H.; Yamada, T.; Inokoshi, S. SEM and elemental analysis of resin composites. *J. Prosthet. Dent.* **1990**, *64*, 669–676. [CrossRef]
6. Oysaed, H.; Ruyter, I. Water Sorption and Filler Characteristics of Composites for Use in Posterior Teeth. *J. Dent. Res.* **1986**, *65*, 1315–1318. [CrossRef] [PubMed]
7. Söderholm, K.-J.M. Filler leachability during water storage of six composite materials. *Scand J. Dent. Res.* **1990**, *98*, 82–88. [CrossRef]
8. Schweikl, H.; Spagnuolo, G.; Schmalz, G. Genetic and Cellular Toxicology of Dental Resin Monomers. *J. Dent. Res.* **2006**, *85*, 870–877. [CrossRef]
9. Van Landuyt, K.L.; Nawrot, T.; Geebelen, B.; De Munck, J.; Snauwaert, J.; Yoshihara, K.; Scheers, H.; Godderis, L.; Hoet, P.; Van Meerbeek, B.; et al. How much do resin-based dental materials release? A meta-analytical approach. *Dent. Mater.* **2011**, *27*, 723–747. [CrossRef]
10. Sideridou, I.; Tserki, V.; Papanastasiou, G. Effect of chemical structure on degree of conversion in light-cured dimethacrylate-based dental resins. *Biomaterials* **2002**, *23*, 1819–1829. [CrossRef]
11. Knezevic, A.; Zeljezic, D.; Kopjar, N.; Tarle, Z. Cytotoxicity of Composite Materials Polymerized with LED Curing Units. *Oper. Dent.* **2008**, *33*, 23–30. [CrossRef] [PubMed]

12. Knezevic, A.; Zeljezic, D.; Kopjar, N.; Tarle, Z. Influence of curing mode intensities on cell structure cytotoxicity/genotoxicity. *Am. J. Dent.* **2009**, *22*, 43–48. [PubMed]
13. Schulz, S.D.; Laquai, T.; Kümmerer, K.; Bolek, R.; Mersch-Sundermann, V.; Polydorou, O. Elution of Monomers from Provisional Composite Materials. *Int. J. Polym. Sci.* **2015**, *2015*, 617407. [CrossRef]
14. Pongprueksa, P.; De Munck, J.; Duca, R.C.; Poels, K.; Covaci, A.; Hoet, P.; Godderis, L.; Van Meerbeek, B.; Van Landuyt, K.L. Monomer elution in relation to degree of conversion for different types of composite. *J. Dent.* **2015**, *43*, 1448–1455. [CrossRef]
15. Atkinson, J.C.; Diamond, F.; Eichmiller, F.; Selwitz, R.; Jones, G. Stability of bisphenol A, triethylene-glycol dimethacrylate, and bisphenol A dimethacrylate in whole saliva. *Dent. Mater.* **2002**, *18*, 128–135. [CrossRef]
16. Polydorou, O.; König, A.; Hellwig, E.; Kümmerer, K. Long-term release of monomers from modern dental-composite materials. *Eur. J. Oral Sci.* **2009**, *117*, 68–75. [CrossRef]
17. Manojlovic, D.; Radisic, M.; Lausevic, M.; Zivkovic, S.; Miletic, V. Mathematical modeling of cross-linking monomer elution from resin-based dental composites. *J. Biomed. Mater. Res. Part B: Appl. Biomater.* **2013**, *101B*, 61–67. [CrossRef]
18. Stefova, M.; Ivanova, V.; Muratovska, I. Identification and quantification of Bis-GMA and Teg-DMA released from dental mate-rials by HPLC. *J. Liq. Chromatogr. Relat. Technol.* **2005**, *28*, 289–295. [CrossRef]
19. Zhang, Y.V.; Wei, B.; Zhu, Y.; Zhang, Y.; Bluth, M.H. Liquid Chromatography–Tandem Mass Spectrometry An Emerging Technology in the Toxicology Laboratory. *Clin. Lab. Med.* **2016**, *36*, 635–661. [CrossRef]
20. Demirel, M.G.; Gönder, H.Y.; Tunçdemir, M.T. Analysis of Monomer Release from Different Composite Resins after Bleaching by HPLC. *Life* **2022**, *12*, 1713. [CrossRef]
21. Wu, W.; McKinney, J. Influence of Chemicals on Wear of Dental Composites. *J. Dent. Res.* **1982**, *61*, 1180–1183. [CrossRef] [PubMed]
22. Cokic, S.M.; Duca, R.C.; De Munck, J.; Hoet, P.; Van Meerbeek, B.; Smet, M.; Godderis, L.; Van Landuyt, K.L. Saturation reduces in-vitro leakage of monomers from composites. *Dent. Mater.* **2018**, *34*, 579–586. [CrossRef] [PubMed]
23. Tuna, E.B.; Aktoren, O.; Oshida, Y.; Gencay, K. Elution of residual monomers from dental composite materials. *Eur J Paediatr Dent* **2010**, *11*, 110–114. [PubMed]
24. Ruyter, I.E.; Oysaed, H. Analysis and characterisation of dental polymers. *CRC Crit. Rev. Biocompat.* **1988**, *4*, 247–279.
25. Geurtsen, W.; Leyhausen, G. Chemical-Biological interactions of the resin monomer triethyleneglycol-dimethylacrilate (TEGDMA). *J. Dent. Res.* **2001**, *80*, 2046–2050. [CrossRef]
26. Forsten, L. Short- and long-term fluoride release from glass ionomers and other fluoride-containing filling materials In Vitro. *Scand J. Dent. Res.* **1990**, *98*, 179–185. [CrossRef]
27. Asmussen, E. Factors affecting the quantity of remaining double bonds in restorative resin polymers. *Eur. J. Oral Sci.* **1982**, *90*, 490–496. [CrossRef]
28. Ferracane, J. Elution of leachable components from composites. *J. Oral Rehabilitation* **1994**, *21*, 441–452. [CrossRef]
29. Spahl, W.; Budzikiewicz, H.; Geurtsen, W. Determination of leachable components from four commercial dental composites by gas and liquid chromatography/mass spectrometry. *J. Dent.* **1998**, *26*, 137–145. [CrossRef]
30. Schwengberg, S.; Bohlen, H.; Kleinsasser, N.; Kehe, K.; Seiss, M.; Walther, U.; Hickel, R.; Reichl, F. In vitro embryotoxicity assessment with dental restorative materials. *J. Dent.* **2005**, *33*, 49–55. [CrossRef]
31. Ruyter, I.E. Physical and Chemical Aspects Related to Substances Released from Polymer Materials in an Aqueous Environment. *Adv. Dent. Res.* **1995**, *9*, 344–347. [CrossRef]
32. Tanaka, K.; Taira, M.; Shintani, H.; Wakasa, K.; Yamaki, M. Residual monomers (TEGDMA and Bis-GMA) of a set visible-light-cured dental composite resin when immersed in water. *J. Oral Rehabilitation* **1991**, *18*, 353–362. [CrossRef] [PubMed]
33. Densply. SDR Scientific Compendium. 2011. Available online:. (accessed on 1 February 2014).
34. Gallorini, M.; Cataldi, A.; Di Giacomo, V. HEMA-induced cytotoxicity: Oxidative stress, genotoxicity and apoptosis. *Int. Endod. J.* **2014**, *47*, 813–818. [CrossRef]
35. Kullmann, W. Atlas of Parodontology with Glass Ionomer Cements and Composites. Carl Hanser Verlag: Munich, Germany; Vienna, Austria, 1990.
36. Chung, K.; Greener, E.H. Degree of conversion of seven visible light-cured posterior composites. *J. Oral Rehabilitation* **1988**, *15*, 555–560. [CrossRef]
37. Pearson, G.J.; Longman, C.M. Water sorption and solubility of resin-based materials following inadequate polymerization by a visible-light curing system. *J. Oral Rehabilitation* **1989**, *16*, 57–61. [CrossRef] [PubMed]
38. Rothmund, L.; Reichl, F.-X.; Hickel, R.; Styllou, P.; Styllou, M.; Kehe, K.; Yang, Y.; Högg, C. Effect of layer thickness on the elution of bulk-fill composite components. *Dent. Mater.* **2016**, *33*, 54–62. [CrossRef] [PubMed]
39. United States Food and Drug Administration (US FDA). *Reccomendations for Chemistry Data for Indirect Food Additives Petitions*; US FDA: Silver Spring, MD, USA, 1998.
40. Kim, J.-G.; Chung, C.-M. Elution from light-cured dental composites: Comparison of trimethacrylate and dimethacrylate as base monomers. *J. Biomed. Mater. Res. Part B: Appl. Biomater.* **2005**, *72*, 328–333. [CrossRef]
41. Altintas, S.H.; Usumez, A. Evaluation of monomer leaching from a dual cured resin cement. *J. Biomed. Mater. Res. B Appl. Biomater.* **2008**, *86B*, 523–529. [CrossRef]
42. Duruk, G.; Akküç, S.; Uğur, Y. Evaluation of residual monomer release after polymerization of different restorative materials used in pediatric dentistry. *BMC Oral Health* **2022**, *22*, 232. [CrossRef]

43. De Angelis, F.; Mandatori, D.; Schiavone, V.; Melito, F.P.; Valentinuzzi, S.; Vadini, M.; Di Tomo, P.; Vanini, L.; Pelusi, L.; Pipino, C.; et al. Cytotoxic and Genotoxic Effects of Composite Resins on Cultured Human Gingival Fibroblasts. *Materials* **2021**, *14*, 5225. [CrossRef]
44. Jakubik, A.; Przeklasa-Bierowiec, A.; Muszynska, B. Artifcial saliva and its use in biological experiments. *J. Physiol. Pharmacol.* **2017**, *68*, 807–813.
45. Alshali, R.Z.; Salim, N.A.; Sung, R.; Satterthwaite, J.D.; Silikas, N. Analysis of long-term monomer elution from bulk-fill and conventional resin-composites using high performance liquid chromatography. *Dent. Mater.* **2015**, *31*, 1587–1598. [CrossRef] [PubMed]
46. Janani, K.; Teja, K.V.; Sandhya, R.; Alam, M.K.; Al-Qaisi, R.K.; Shrivastava, D.; Alnusayri, M.O.; Alkhalaf, Z.A.; Sghaireen, M.G.; Srivastava, K.C. Monomer Elution from Three Resin Composites at Two Different Time Interval Using High Performance Liquid Chromatography—An In-Vitro Study. *Polymers* **2021**, *13*, 4395. [CrossRef]
47. Nazar, A.M.; George, L.; Mathew, J. Effect of layer thickness on the elution of monomers from two high viscosity bulk-fill com-posites: A high-performance liquid chromatography analysis. *J. Conserv. Dent.* **2021**, *23*, 497–504. [CrossRef] [PubMed]
48. Ahmed, M.; Yoshihara, K.; Yao, C.; Okazaki, Y.; Van Landuyt, K.; Peumans, M.; Van Meerbeek, B. Multiparameter evaluation of acrylamide HEMA alternative monomers in 2-step adhesives. *Dent. Mater.* **2020**, *37*, 30–47. [CrossRef]
49. Ginzkey, C.; Zinnitsch, S.; Steussloff, G.; Koehler, C.; Hackenberg, S.; Hagen, R.; Kleinsasser, N.H.; Froelich, K. Assessment of HEMA and TEGDMA induced DNA damage by multiple genotoxicological endpoints in human lymphocytes. *Dent. Mater.* **2015**, *31*, 865–876. [CrossRef]
50. Alrahlah, A.; Al-Odayni, A.-B.; Al-Mutairi, H.F.; Almousa, B.M.; Alsubaie, F.S.; Khan, R.; Saeed, W.S. A Low-Viscosity BisGMA Derivative for Resin Composites: Synthesis, Characterization, and Evaluation of Its Rheological Properties. *Materials* **2021**, *14*, 338. [CrossRef]
51. Becher, R.; Wellendorf, H.; Sakhi, A.K.; Samuelsen, J.T.; Thomsen, C.; Bølling, A.K.; Kopperud, H.M. Presence and leaching of bisphenol a (BPA) from dental materials. *Acta Biomater. Odontol. Scand.* **2018**, *4*, 56–62. [CrossRef]
52. Maserejian, N.N.; Trachtenberg, F.L.; Wheaton, O.B.; Calafat, A.M.; Ranganathan, G.; Kim, H.-Y.; Hauser, R. Changes in urinary bisphenol A concentrations associated with placement of dental composite restorations in children and adolescents. *J. Am. Dent. Assoc.* **2016**, *147*, 620–630. [CrossRef]
53. Bezgin, T.; Cimen, C.; Ozalp, N. Evaluation of Residual Monomers Eluted from Pediatric Dental Restorative Materials. *BioMed Res. Int.* **2021**, *2021*, 6316171. [CrossRef]
54. Reichl, F.X.; Esters, M.; Simon, S.; Seiss, M.; Kehe, K.; Kleinsasser, N.; Folwaczny, M.; Glas, J.; Hickel, R. Cell death effects of resin-based dental material compounds and mercurials in human gingival fbroblasts. *Arch Toxicol.* **2006**, *80*, 370–377. [CrossRef] [PubMed]
55. Murray, P.; Hafez, A.; Windsor, L.; Smith, A.; Cox, C. Comparison of pulp responses following restoration of exposed and non-exposed cavities. *J. Dent.* **2002**, *30*, 213–222. [CrossRef] [PubMed]
56. Modena, K.C.; Casas-Apayco, L.C.; Atta, M.T.; Costa, C.A.; Hebling, J.; Sipert, C.R.; Navarro, M.F.; Santos, C.F. Cytotoxicity and bio-compatibility of direct and indirect pulp capping materials. *J. Appl. Oral. Sci.* **2009**, *17*, 544–554. [CrossRef] [PubMed]
57. Susila, A.V.; Balasubramanian, V. Correlation of elution and sensitivity of cell lines to dental composites. *Dent. Mater.* **2016**, *32*, e63–e72. [CrossRef] [PubMed]

Disclaimer/Publisher's Note: The statements, opinions and data contained in all publications are solely those of the individual author(s) and contributor(s) and not of MDPI and/or the editor(s). MDPI and/or the editor(s) disclaim responsibility for any injury to people or property resulting from any ideas, methods, instructions or products referred to in the content.

Review

Functionalization of Carbon Nanotubes and Graphene Derivatives with Conducting Polymers and Their Applications in Dye-Sensitized Solar Cells and Supercapacitors

Mirela Văduva [1,*], Teodora Burlănescu [1,2] and Mihaela Baibarac [1]

[1] National Institute of Materials Physics, Atomistilor Street, No 405 A, 077125 Magurele, Romania; teodora.burlanescu@infim.ro (T.B.); barac@infim.ro (M.B.)
[2] Faculty of Physics, University of Bucharest, Atomistilor Street, No 405, 077125 Magurele, Romania
* Correspondence: mirela.ilie@infim.ro

Abstract: Recent progress concerning the development of counter electrode material (CE) from the dye-sensitized solar cells (DSSCs) and the electrode material (EM) within supercapacitors is reviewed. From composites based on carbon nanotubes (CNTs) and conducting polymers (CPs) to their biggest competitor, namely composites based on graphene or graphene derivate (GD) and CPs, there are many methods of synthesis that influence the morphology and the functionalization inside the composite, making them valuable candidates for EM both inside DSSCs and in supercapacitors devices. From the combination of CPs with carbon-based materials, such as CNT and graphene or GD, the perfect network is created, and so the charge transfer takes place faster and more easily. Inside composites, between the functional groups of the components, different functionalizations are formed, namely covalent or non-covalent, which further provide the so-called synergic effect. Inside CPs/CNTs, CNTs could play the role of template but could also be wrapped in a CP film due to π–π coupling enhancing the composite conductivity. Active in regenerating the redox couple I^-/I_3^-, the weakly bound electrons play a key role inside CPs/GD composites.

Keywords: solar cell; functionalization; graphene; carbon nanotube; conducting polymer; composites; power conversion efficiency; supercapacitors; capacitance

1. Introduction

In the context of a higher energy demand assigned to an increased population and thus an increased level of needs, together with the depletion of natural resources and environmental pollution, the focus on finding alternative sources of energy (such as green or renewable energy) has also increased. Of all the eligible sources, e.g., the energy of the sun, wind, water, and thermal waters, the first is currently attracting the most interest. If it were efficiently converted, the energy from 1 h of sunlight on the entire globe would be enough to cover the need for one year of electricity [1]. Therefore, much research has been conducted to fabricate devices for converting solar energy into electricity and, as far as possible, to store it using the same device. Currently, the conversion process is made using silicon-based solar cells, and the trend is to replace these classical devices with lower-cost materials whose high conversion efficiency is similar. One of the newly tested devices is the DSSCs. They are conventional devices built from a photo-anode and a counter electrode (CE), overlapped in a sandwich configuration, with a thin layer of electrolyte that fills the space between them. The photo-anode consists of a transparent conductive layer (TCL), which could be Indium-tin-oxide (ITO) or Fluorine-doped Tin Oxide (FTO), on which a thin layer of TiO_2 is deposited and dipped in a dye solution (usually N719). The outer part is represented by the CE, usually made of Pt, deposited on a transparent conductive layer (TCL). The electrolyte is represented by an I^-/I_3^- redox couple, mostly liquid.

DSSCs could also be bifacial, illuminated on both sides. They include a CE material with a double function, which works as a charge transfer agent and as a regenerator for the redox couple. The CE for this kind of DSSC is transparent, with illumination available from both the front and rear sides [2]. Of all DSSC components, the fundamental one used to convert the luminous energy into electric energy is represented by the photo-anode. The most common semiconductor used for this role inside DSSCs is TiO_2. The mechanism inside DSSCs consists of the path followed by the solar light from TCL until the separated charge (namely the electrons) is loaded into the external circuit. When the radiation enters the transparent conductive layer (for example, TiO_2) impregnated with dye, the radiation excites the dye molecules. Therefore, the dye molecules move to higher energy levels, namely the lowest unoccupied molecular orbital (LUMO), and from there, the electrons are promoted into the TiO_2 conduction band (CB) and move across to the external electric circuit. When the electrons arrive at the conductive transparent electrode (TCE), they are collected and transferred to the CE, which is usually made of platinum (Pt). The dye molecules remain in an oxidized state after light exposure and regenerate by accepting an electron from the electrolyte redox couple. After that, they return to the fundamental state. This mechanism of the charge transfer inside the DSSCs device is shown in Figure 1.

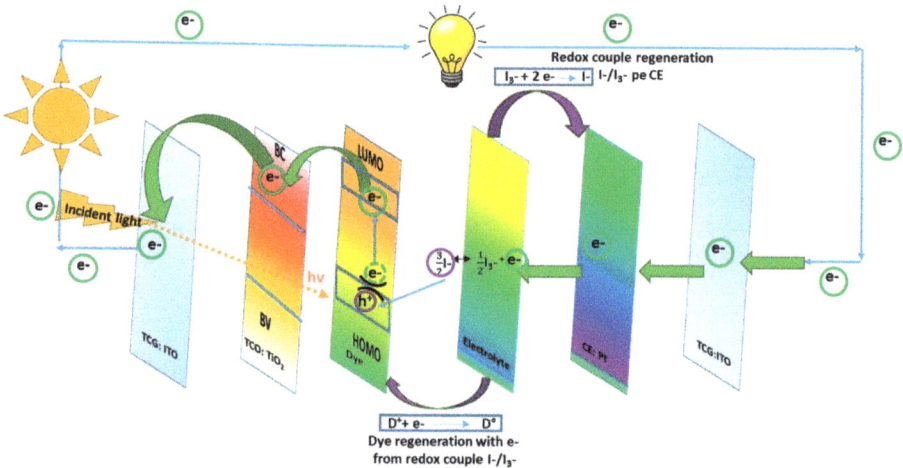

Figure 1. Mechanism inside a DSSC device.

Researchers who design and test DSSCs encounter many difficulties regarding sensitive issues about the way the components work inside the DSSCs. For example, the flexibility, long-term stability, active surface area (SA) and transparency of CE and TCL, the absorption efficiency of light, charge recombination, and so on. Amongst all of these, the main drawback of DSSCs remains the difficulty in controlling the charge recombination process, which is responsible for a major decrease in conversion efficiency. The last is the reason for being reported only on a few occasions: a conversion efficiency higher than 11% under diffuse daylight [3]. Excited dye molecules and other acceptor species from the electrolyte are involved in charge recombination processes, capturing electrons from the system and thus remaining unavailable to further interactions involved to complete the electric circuit. When using a flat surface of dye, less than 1% of incident monochromatic light is absorbed. One way of improving this performance is to increase the area of the active surface of TCL on which the dye is adsorbed, for example, by thermal sintering treatment of the TiO_2 before depositing onto ITO or FTO substrates [4]. Moreover, dye molecules play an important role in the main process of the DSSC mechanism. Attached to the TiO_2 surface, these absorb light, broadening the range of wavelengths to be absorbed. Then, the electrons are injected from the LUMO into the TiO_2 conduction band. From this point, the electrons enter the semiconductor layer and enter the external circuit. At the same time, the oxidized dye molecules regenerate to the neutral state by reduction of the redox species in the electrolyte solution. Because it is a complete circuit, it runs without material consumption by generating electricity from sunlight. A substantial amount of this energy is unfortunately lost through the recombination process of electrons inside TiO_2 with oxidized dye molecules or molecules in the electrolyte/redox medium. To increase the performance of the DSSCs, each component must be properly chosen. For example, the suitable sensitizer must fulfill several demands, such as: absorption in the full visible domain with the ability to use a higher percentage of light, affordable location of molecular orbitals (highest occupied molecular orbital (HOMO) and lowest unoccupied molecular orbital (LUMO)) to inject the electrons into the CB of the photo-anode and to help to regenerate the oxidized sensitizer from the redox electrolyte. In addition, the aggregation of the sensitizer molecules must be avoided by choosing a certain molecular structure and also the charge recombination from the TiO_2/electrolyte interface. To improve the charge injection, functional groups such as carboxyl and phosphonate are desirable and the sensitizer should be also photo- and thermic-resistant providing long-term stability for the DSSCs device. Considering all of the listed demands, several types of sensitizers have been reported into the literature, including porphyrins, phtalocyanines, and metal-free organic dyes [5]. From all of these sensitizers, the biggest efficiency was reported on ruthenium and porphyrin dye with the advantage of availability and ease of structural tuning, possessing high extinction molecular coefficients [6,7]. Only a few of them reported power conversion efficiency (PCE) values higher than 9% when combining with iodide.

An issue being increasingly studied is the relationship between the electrolyte couple and the dye. It seems that the preferred redox electrolyte couple is I^-/I_3^- due to several characteristics such as good solubility, low absorption of light, appropriate redox potential (0.35 V) providing dye rapid regeneration; this couple poses a very slow kinetic of recombination between the electrons from TiO_2 and the oxidized entity of the redox couple (I_3^-). They do not involve into the recombination reaction by contrast other sensitizers that bound I^- or I_3^-. Because the difference between the oxidation potential of the standard sensitizer (based on porphyrins and ruthenium, 1.1 V) and that of the redox couple (I^-/I_3^-, 0.35 V), which means that the reduction potential of the oxidized dye is 0.75 V, this process provides the biggest potential lost from the DSSCs devices [8]. This value must be reduced at least to half of the value in order to increase the PCE to 15%. In order to move towards this, some aspects should be considered. From the regeneration of the oxidized sensitizer with I^-, the reaction leads to the formation of the diiodide radical as a secondary product (I_2^-). Therefore, the redox potential of the I_2^-/I^- couple should be considered when determining the force of the sensitizer regeneration because I_2^- leads to I_3^- and I^- formation which is the main reason for decreasing the potential energy.

Another problem is the use of an expensive and rare CE material such as Pt, which implies the need to obtain Pt-free or low Pt content CE. Materials suitable for this position must be highly conductive, transparent, with a high rate of charge transfer, resistant to corrosion in electrolyte medium, low in cost, and available. Therefore, materials such as carbon and conducting polymers (CPs) structures or their combination thereof are eligible for this position [9]. Used alone, CPs do not perform very well as a CE inside DSSCs. In addition to their advantages, such as the conjugated structure, ease of preparation, availability and good stability, and possibility of depositing uniform thin film with a good adhesion on transparent conductive oxide (TCO), they have a relatively low conductivity. This drawback could be overcome by combining CPs with CNs for giving rise to composite materials with enhanced electrochemical and catalytic activity, able to ensure fast charge transfer from the external circuit to the electrolyte, supporting the regeneration process of the redox species. The reversible redox behavior of CPs is also a plus when it comes to contribution to a low charge transfer resistance (as in the case of polyaniline (PANI) and poly (3,4-ethylenedioxythiophene) (PEDOT)). As for the poor dispersibility of CNTs in common solvents and the tendency of GDs to overlap, leading to a graphite structure restoration, these problems are easily solved when CNTs and GDs are incorporated into the polymer matrix, so that charge transfer is accelerated and the specific active surface area is enlarged. Functionalized CNTs facilitate the formation of covalent bonds with the CPs through which the charge transfer occurs more easily and quickly; meanwhile, both GO and RGO, due to the functional groups on their surface, contribute to the uniform deposition of conductive polymers. Moreover, the charge transport is facilitated through π-π stacking between GD and CPs which creates multiple and shorter routes for both ion and electrons diffusion.

The first DSSC was made by depositing transparent layers of anatase on glass substrates, over which a thin layer of sensitizer, namely trimeric rhutenium complex, was dispersed. Using a sandwich configuration, an electrolyte thin layer containing the redox couple I^-/I_3^- filled the space between the Ru complex/TiO_2/glass and the CE made of glass covered with monolayers of Pt. At a filling factor of 0.76, the efficiency conversion was 7.9% and raised to 12% under diffuse light exposure [4].

From the entire mechanism of DSSCs, the materials used to design the CE are discussed here. The CE plays an important role, capturing the electrons from the external circuit and transporting them to the electrolyte where I_3^- accepts electrons and moves to I^-. To complete the electric circuit, following the dye regeneration pattern, I_3^- accepts two electrons and changes to I^-, which further regenerates through charge transfer from the CE during a reduction process (see Figure 2). The electron exchange involved in redox couple regeneration (electrons given up by I^-) is fast enough to ensure efficient dye regeneration, while I_3^- accepts electrons from the photoanode, and the process is slow enough to allow a high carrier collection efficiency.

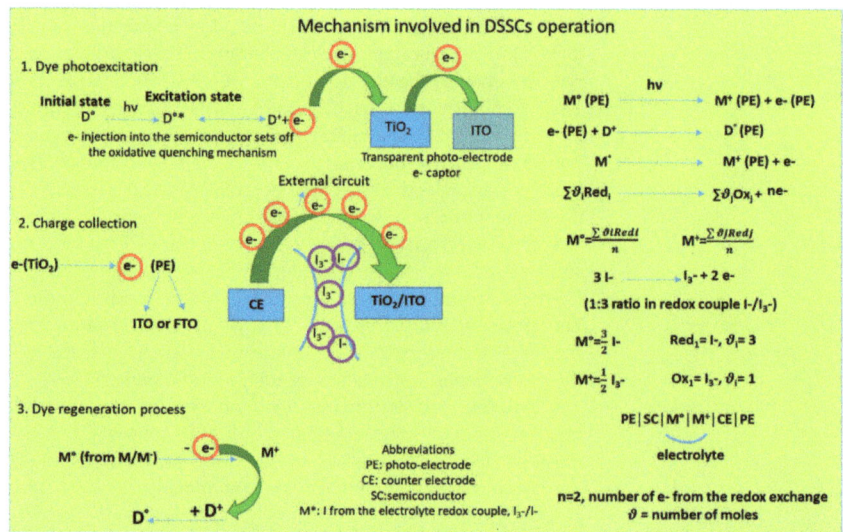

Figure 2. Mechanism of excitation and regeneration which underlines the DSSCs functionality.

The first materials tested as possible CEs were the carbon nanostructures such as graphene and CNTs, due to their high conductivity and low surface resistance, transparency, relatively fast charge transfer, availability, and affordability. Therefore, carbon nanostructures were used with this purpose from 1991 [10], when CNTs had been reported to improve the DSSCs performances, and this was followed by the use of graphene as a CE in DSSCs from 2013 [11–13]. In order to exceed the maxim PCE (7.88%) reached using carbon nanostructures [12], other materials were also tested, as for example their composites with CPs [13] or pristine CPs [14].

Combining CPs with CNTs, the perfect network is created and so the charge transfer takes place faster and more easily. Inside composites, different functionalizations, namely covalent or non-covalent, are formed between the functional groups attached on the surface of the CNT walls and the functional groups in the polymer backbone, which further provide the so-called synergistic effect.

2. Synthesis and Vibrational Properties of CPs-CNTs Composites

2.1. Synthesis of CPs-CNTs Composites

The main CPs which form the composites, used as CE, are poly(pyrrole) (PPy), PEDOT, and PANI. The CNTs, pristine [15] or trapped inside a gel [16], are the first carbon structures used as a CE material, replacing Pt. According to the literature, the composites based on CNTs and CPs were prepared by depositing CNTs on the substrates of the type FTO or ITO followed by: (i) chemical polymerization; (ii) electrochemical polymerization [16–21]; or other methods such as (iii) the precipitation of the already synthesized polymer [22] and (iv) the doctor blade method [23]. The one which provides good control of the thickness and uniformity of the grown layer is the electrochemical method that remains the most used of all. Before depositing the composite on a conductive glass substrate, certain preparations were necessary. Therefore, H. Li and colleagues treated an FTO substrate to improve its hydrophilicity by sonication into a mixture of ammonium hydroxide, water, and hydrogen peroxide (1:5:1 volume ratio) [24]. Then, on the already-prepared substrate, the CNTs were spin coated and then exposed to a heat treatment at 60 °C for 30 min. H. Li and co-workers have also reported preparation of PPy/CNTs composites using an in situ electro-polymerization technique. Prior to the preparation of the composite, the CNTs were treated to improve their solubility by functionalization with –COOH groups. This procedure was performed for improving the CNTs solubility in water by functionalization of the CNTs walls with –COOH functional groups [24]. Further, the composite based on CNTs and PPy was obtained from a mixture, containing 10 mM pyrrole, 20 mM sodium dodecyl sulphate (SDS), and 20 mM lithium perchlorate (LiClO$_4$), through cyclic voltammetry (CV) [19,24]. Using the same method, namely the electrochemical synthesis, a composite with a honeycomb morphology was obtained [25]. To prepare PANI- and single-wall carbon nanotubes (SWCNTs)-based composites by electrochemical synthesis, Bumika, M. and colleagues used sodium dodecylbenzene sulfonate (SDBS) to obtain a CNTs dispersion of carboxyl-functionalized SWCNTs (SWCNTs-COOH) with a weight ratio of SDBS: SWCNTs–COOH equal to 9:2, in 0.5 M H$_2$SO$_4$ solution [20]. The resultant dispersion was then mixed with a second mixture prepared from ZnO (6 wt.%) and 0.25 M aniline (ANI) solution and deposited on FTO through CV, in a three-electrode configuration cell, between −0.62 V and +1.2 V [20].

The second most used method to synthesize composites is the chemical polymerization. X. Liu and co-workers reported this method for the synthesis of CE composites materials based on three different CPs precursors and CNTs [16]. The chosen monomers were polymerized in the presence of CNTs embedded in the polyacrylic gel matrix (PAA). The process of incorporating CNTs into gel was conducted according to a protocol reported by Li and co-workers [26,27]. According to this protocol, 15 mL of CNTs aqueous homogenous dispersion was mixed with 1 g of hexadecyl trimethyl ammonium bromide and stirred at 80 °C for 10 min. Afterwards, 10 g of acrylic acid (AA) and 0.005 g of N, N-methylenebisacrylamide were added and stirred until homogenized. When all components were very well mixed, the polymerization reaction was started with potassium peroxydisulfate (KPS) (0.08 g). The reaction was carried out for 2 h, under vigorous stirring at 80 °C. The final product was freeze-dried for 72 h. After synthesizing the CNTs-PAA gel, pieces of it were dipped into solution of monomers, of ANI, 3,4-ethylenedioxythophene (EDOT), and pyrrole (Py), for 24 h, at room temperature so that the monomers could swell inside the gel. After dispersion of the swollen monomers inside CNT gels in KPS solution (0.03 M), an internal polymerization process took place with the formation of the corresponding conducting polymers. The final products were poly (AA-co-CNTs-Py), poly (AA-co-CNTs-co-ANI), and poly (AA-co-CNTs-co-EDOT) gel. As a part of the DSSCs device, the as-prepared composite gels were soaked into a liquid electrolyte containing tetrabutylammonium iodide, tetramethylammonium iodide, I_2, tetraethylammonium iodide, LiI, and tetrabuthylammonium iodide in N-methyl-2-pyrrolidone and acetonitrile (1:4 volume percentage) [16].

Other methods, rather physical methods, used to prepare a composite based on CPs and CNTs were reported by Abdul Almohsin, S.M. and colleagues [22] and respectively by Dowa, C. et al. [23], using the precipitation of CPs on top of CNTs/FTO and, respectively, the doctor blade method.

According to the last method, PANI and CNTs were dispersed in m-cresol at a 100 mg/mL concentration until a viscous paste was obtained. During mixing PANI and CNTs, a few drops of terpineol and ethyl cellulose (15 wt.% in ethanol) were added and the whole mixture was magnetically stirred for 3 h. The resultant paste was spread over an FTO-coated substrate by the doctor blade coating method. The coated layer was maintained at room temperature to dry and then thermally treated at 400 °C for half an hour. M-cresol interacts with the polymer chains and act as a dopant to PANI and, at the same time, it is a good dispersing agent for high CNT content. Mixing CNTs with m-cresol produces a thick, viscous solution that helps to deposit a more uniform and homogenous layer of CNTs on a substrate.

Composite materials used as CE after being deposited on FTO substrate show different features depending on the type of CPs, the carbon nanostructures inside the composite, the interaction between the two components of which is directly related to the type of synthesis. Moreover, when using carbon nanostructured inside the composites, there are some aspects regarding their functionalization to be eligible for further interaction with the CPs. Then, considering all these aspects, the vibrational properties will be further discussed, investigated using mainly Raman and FTIR spectroscopy and some parameters concerning the DSSCs electro-catalytic activity. Certain aspects will be followed in order to understand the interaction between the composite components and its influence on the light conversion efficiency, according to the information provided by the selected papers used in this work.

2.2. Vibrational Properties of CPs/CNTs

Prepared by in situ chemical polymerization of PAA–CNT gel soaked into monomer solution, poly (AA-co-CNTs-Py) was analyzed through IR spectroscopy and the recorded spectra revealed specific absorption bands assigned to Py, CNT, and PAA. According to Table 1 (inserted below) the specific bands were assigned to PPy ring vibrations (1547 and 1038 cm^{-1}), N-H in-plane deformation [28,29], N-C stretching vibrations, and C-H band stretching vibrations (1175 and 910 cm^{-1}), respectively. Meanwhile the band situated at 3421 cm^{-1} corresponds to the stretching vibration of the OH functional group of the PAA polymer, while the bands assigned to CNT are those located at 1450, 620, 1400, and 2945 cm^{-1} corresponding to wagging vibration, C-H out-of-plane vibration, the last two being characteristic of the methyl functional group and sp^3 hybridized carbon atom. Other bands are assigned to the external functional group attached to the CNTs walls, such as for example: for the carboxyl group, specifically the bending vibration of C=O bond in –COOH represented by the IR band at 1685 cm^{-1}, the C-O-C stretching vibration and the assigned band at 1140 cm^{-1} [30,31], the C=O stretching vibration and the bending vibration of the –OH bond in –COOH at 1719 and 1348 cm^{-1}. For a better understanding of all interactions between PPy and grafted CNTs with carboxyl groups, the IR vibrational structure of this composite is shown in Figure 3.

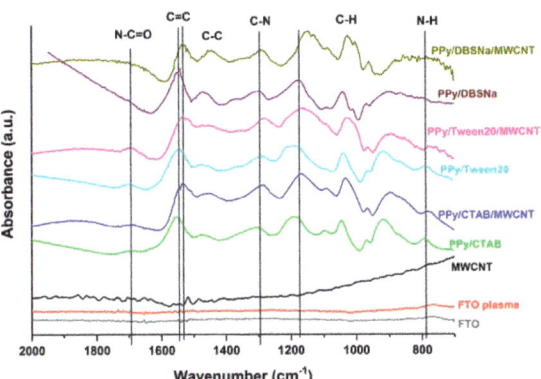

Figure 3. FTIR spectra of FTO, FTO plasma, MWCNT, PPy/CTAB, PPy/CTAB/MWCNT, PPy/Tween20, PPy/Tween20/MWCNT PPy/DBSNa, and PPy/DBSNa/MWCNT, where MWCNTs are non-covalently functionalized with PPy [17].

According to X. Liu et al., gel electrolytes with CNTs incorporated have a dual function inside DSSCs: to enlarge the SA, as well as increase conductivity, providing a high catalytic activity and thus contributing to short-circuit density (Jsc) enhancement [16].

Changes in the IR absorption bands of PPy have been reported in the case of the IR spectra of the composite based on PPy and CNTs, synthesized through the electropolymerization of Py in the presence of CNTs [17]. A shift to lower wavenumbers of the band located at 1530–1560 cm^{-1} and assigned to the C=C/C-C stretching vibration of the PPy chains indicates a higher delocalization length in the polyconjugated system [32], which means longer polymeric chains are formed in the presence of CNTs, because of a non-covalent interaction between the π-π bonds of PPy and CNT (see Figure 3).

The presence of functionalized multi-wall carbon nanotubes (MWCNTs) into the PPy matrix is revealed through an enhanced intensity of peaks and a little shift, as a consequence of the interaction between carboxyl group from MWCNTs and functional groups from PPy creating a network where electrons are transferred from one compound to the other. Further, inside the nanocomposite based on PPy and functionalized multi walled CNTs (FMWCNTs), the bond associated with C-H, C-C, and N-H vibration becomes weaker and instead the C-N bond becomes stronger. The fact that MWCNTs are wrapped in PPy is confirmed through the disappearance of IR bands, such as for example the IR bands located at 1198 and 2879 cm^{-1}, present in acid-treated MWCNTs [33,34]. As formed PPy-FMWCNTs have a better conductivity than pristine PPy, 250 S/cm vs. 35 S/cm due to a higher localization length of 10 nm vs. 1.55 nm for FMWCNTs, the value improved as a result of the large arrangement of the π conjugated structure [21].

The electrostatic interaction which takes place between the quinoid rings of both components, one donor (PPy) and the other e$^-$ acceptor (MWCNTs), determined a fast movement of charges inside the composite enhancing its conductivity. Morphologically, the PPy layer is porous and uniformly dispersed on the electrode surface, and embedded on FMWCNTs which creates a transport network for electrons that enhances the cathodic reaction of the redox couple I_3^-/I^- [35].

According to He and co-workers, using in situ chemical polymerization, CNTs are covalently bond to PPy, forming a new composite [36] (Figure 4) with a covalent bond between the nitrogen atom in the pyrrole ring and the sp^2-hybridized carbon atom in the CNT network. The best results have been recorded for an optimum of 2 wt.% SWCNTs on which values of 8.3 PCE have been reported. Compared to pristine PPy (PCE 6.3%), the improved results were assigned to the lower charge-transfer resistance (Rct) value.

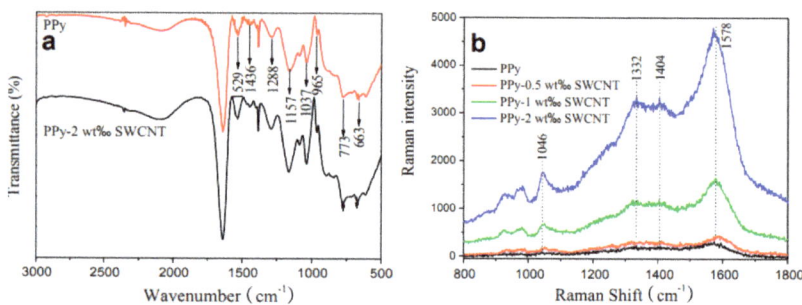

Figure 4. (**a**) FTIR and (**b**) Raman spectra of PPy and PPy-SWCNT composites, where SWCNTs are covalently functionalized with PPy [36].

In all Raman spectra of the PANI-CNTs composite, the CNT-specific lines were located at 1591 and 1334 cm^{-1} corresponding to the radial breathing mode (RBM) [37] and tangential mode [38] and the PANI lines located at 1581, 1052, 1083, 1330, 1370 cm^{-1} assigned to the vibrational modes of C=C stretching, C-H in plane deformation, and the aromatic ring stretching mode [39,40] (see Table 1 for more information). The most significant change has been recorded in the intensity of the Raman line located at 1334 cm^{-1} which increases with the amount of SWCNTs in the composite. The interaction between the components inside the composite was reported to be rather weak considering the lowest PL intensity band located at 520 nm [16]. The layer morphology, noted by PANI and CNT-based composites, varied from the honeycomb structure [24] to the axel sleeve structure obtained by co-polymerization [41] or a uniform film obtained by electrophoresis and CV [42].

It was already reported that the deposition of a PEDOT thin layer on the TCO substrate contributes significantly to a decrease in the surface resistance (SR) and as a direct consequence the conductivity increases. When PEDOT is prepared by the electropolymerization of EDOT, a layer of well-connected mesoporous composite is obtained on the CNTs film already cast on the TCO substrate. In this case, the CNTs play the role of template leading to porous nanostructured wires [43] but the CNTs can also be wrapped in PEDOT film by π-π coupling [44]. If another method is used, such as in situ chemical polymerization, covalent bonds are formed between PANI and CNTs, more precisely between –NH– in PANI and –C=, i.e., the sp^2 hybridized carbon atom of SWCNTs, and this bond contributes significantly to the acceleration of charge transfer between composite components [45]. The CE improves with increasing CNTs loading, regardless of the chosen synthesis method [46].

Composites based on PEDOT and CNT have been reported having different morphologies, such as porous wire nanostructures deposited on a CNT template [43], or core-shells structures, where CNT is the core and PEDOT is the shell [47], using the oxidative polymerization method. It was found that, when using an aligned structure or well-ordered CNTs, it could enhance photovoltaic performance, i.e., PCE, by decreasing resistance and increasing conductivity [15]. Different from the classical CNT and CP–based composite, CNTs could be also used as a filler or matrix with different CPs. The latter has been used to solve both the problem of the poor dispersion of CNTs in solvents and to improve their electronic conductivity. Another study reported as an electromaterial (EM) in DSSCs a poly(3,4-ethylenedioxythiophene)-poly(styrenesulfonate) (PEDOT: PSS)-based composite and CNTs, where PEDOT: PSS was used as a dispersing agent of CNTs [48]. The efficiency of the composites thus prepared was lower than those reported when Pt was used as a CE (8.5%).

Inside PANI and CNT-based composites, the resultant defects generate a higher surface area, and this is supported by an enhanced intensity of the D band peak in the Raman spectrum of the composite compared with the spectrum of the CNTs, which confirms the presence of sp^3-hybridized carbon atoms attached to the surface of CNTs. The value of the I_D/I_G ratio increases from 0.125 to 0.168 for the composite (see Figure 5) due to the increase in the number of defects at the edge of the CNTs [23].

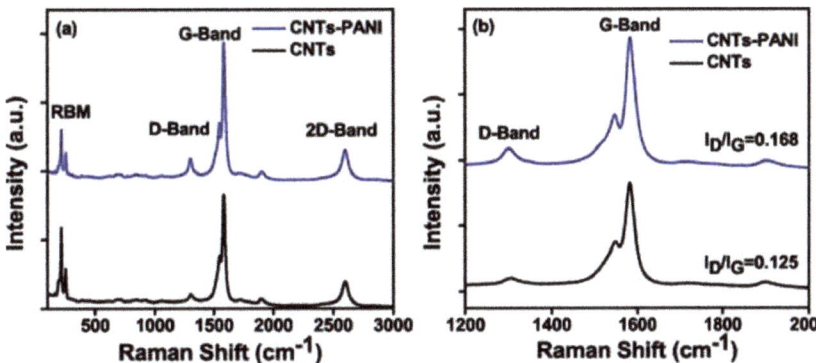

Figure 5. Raman spectra of (**a**) CNTs, CNTs-PANI films coated on FTO glass, and (**b**) a magnified version of the D and G band [23].

Table 1. Main specific IR and Raman bands assigned to the components used in the synthesis of CE composite materials (PPy, PANI, PEDOT, PAA, and CNTs).

CPs	Vibrational Modes Active in IR Spectroscopy	Wavenumber (cm^{-1})	Ref.	Vibrational Modes Active in Raman Spectroscopy	Wavenumber (cm^{-1})	Ref.
PPy	Vibration of pyrrole ring ν C–H ν N–C δs N–H (1038 and 1547 cm^{-1}) ν C=C and ν C–C, PPy ring vibrations	700–800 910 1175, 1210 1038 1547 1530–1560 1556	[16,28,29,32]	-	-	-
PANI	δ C–H of the quinoid ring ν C–N and δ C=C ν quinoid ring and δ benzoid ring	1133 1243 1301 1489 1564	[40,49–51]	Bipolaron and polaron bands, δ C-H, ν ring, ν C=C	940 990 1052 1083 1330 1334 1370 1581	[18,39]
PEDOT	ν C–C or C=C of the quinoide structure and ν thiophene ring	834 978 1187 1315 1356 1513	[52]	ν C-S-C bond in thiophene ring δ C-O-C bending vibration in ethylenedioxy group ν_{as} SO$_2$	834 978 1187 1315	[52]
PAA	ν OH	3421	[16]	-	-	-
CNTs	γ C–H ν C-O-C δ OH from –COOH ω C–H (1400 and 1450 cm^{-1}) δ C=O from –COOH (grafted to the CNTs wall) ν C=O ν CH$_3$	620 1140 1348 1400 1450 1685 1719 2945	[16]	RBM, E$_{2g}$ mode assigned to slightly disturbed graphite E$_{2g}$ mode of graphite wall	1334 1591	[16]

2.3. Performance of CPs/CNTs as CE in DSSCs

The efficiency of the presented composites varies from 1.67% [17], reported on PPy/CNTs, to 9.07% for PEDOT/CNTs [53], both composites synthesized through in-situ electropolymerization of monomers (see Table 2). Higher PCE values were obtained on composites when the dispersing agent (SDS) was used, the PCE value of 6.15% being close to that reported for Pt (6.36%) [20]. The good results were mainly attributed to the low Rct at the CE/electrolyte interface [18]. Also, very good results have been reported for structures obtained by chemical vapor deposition (CVD) of the polymer on the CNT film surface. Such an example is the subject of the study by W. Hou and co-workers, when the presence of PPy induced a PCE of 7.15 [54]. Comparing a covalent composite [36] with those in which the interaction between components is non-covalent [17], it would appear that those with a higher PCE are those in which covalent bonds are present (see Table 2).

Table 2. Synthesis and CE performance parameters (CE, FF, Jsc, and R_{CT}) of the CNTs/CPs composites.

Composite CPs/CNTs	Synthesis	PCE (%)	FF	J_{sc} (mA cm^{-2})	Rct ($\Omega \times$ cm^2)	Ref.
(a) CNTs/PPy (b) CNTs/PANI (c) CNTs/PEDOT	Electrochemical synthesis	6.82; 7.01; 7.2	0.69	13.73; 13.92; 14.11	1; 7.43; 7.5; 7.51	[16]
MWCNT-PEDOT: PSS	Physical mixing	6.1%	59.8	12.9	-	[18]
* h-PEDOT/MWCNTs	Electropolymerization	9.07	0.67	17.09	0.19	[53]
PPy/SDS/CNTs	Electrochemical polymerization	PPy-SDS-CNT 6.15	PPy-SDS-CNT:58.69	15.47	0.19	[19]
PPy/MWCNT/FTO	Electrochemical polymerization	1.67%	0.53	5.44	-	[17]
(A) Cu-PPy-CNT (B) PPy-CNT	Electrodeposition method	(a) 7.1% (b) 5.49	(a) 0.696 (b) 0.682	(a) 2.35 mA/cm^2 (b) 10.27	(a) 4.31 $\Omega \times$ cm^2 (b) 5.29	[21]
PPy-SWCNTs	Chemical polymerization	8.3%	0.71	15.68	8.15	[36]
PANI-SWCNTs	Electropolymerization	front ill **: 7.07%	0.53	17.5	0.18	[24]
PANI/SWCNT/ZnO nanorods	Polymer precipitation top of MWCNTs	-	-	-	-	[22]
PANI/SWCNT/ZnO	One-pot electrochemical synthesis	PS: 3.16 and PSZ: 3.81	PSZ (PANI-SWCNT-ZnO): 56	PSZ: 9.59	PSZ: 10.10	[20]

* honeycomb-like structure. ** illumination.

For describing the electro-catalytic activity of the CE, there are several important parameters and associated measurements which should be considered, namely CV, electrochemical impedance spectroscopy (EIS), Tafel polarization, chrono-amperometric studies, R_{CT}, surface layer resistance (Rs), PCE, Jsc, filler factor (FF), and so on. The information discussed in this section depends on the information provided by the papers selected for this study.

In addition, the photovoltaic performance of DSSCs is investigated though I-V measurements. EIS studies are performed to describe the charge transfer kinetic of the electrochemical system, more precisely the electro-catalytic activity of the CE versus the reduction process of the electrolyte, in a symmetric cell system. Using the Nyquist plot, the R_{CT} could be determined from the graphic which describes the charge mobility between the CE and the electrolyte. Specifically, the R_{CT} value depends on the diameter of the semicircle corresponding to the chemical capacitance (CPE) at the electrode/electrolyte interface. The semicircle on the right, located in the low frequencies area, represents the Nernst diffusion impedance (W), a measure of the electrolyte that controls the diffusion of the I_3^-/I^- redox species to the CE, in addition a lower value of R_{CT} indicating a faster charge transfer inside the symmetric cell. R_{CT} varies inversely with the Jsc recorded for I_3^- to I^- reduction process on CE. From the cyclic voltammograms recorded using electrodes of the type PEDOT/ITO, CNTs/ITO, ITO, and PEDOT-SWCNTs/ITO, the synergic effect of both components is presented. While on the pristine ITO electrode the redox peaks are missing, in the case of PEDOT/ITO, there are two peaks that are broad and far apart, and less reversible than those found on PEDOT-SWCNTs/ITO, revealing the idea that PEDOT alone could not be as efficient and that this could be extended to replacing the ITO substrate with SWCNTs as a better one for improved catalytic response. Another evidence that PEDOT and SWCNTS go well together is the efficiency value which is double in the case of PEDOT-SWCNTs than for pristine SWCNTs. A high filling factor and I-V measurement confirm the good catalytic performance of PEDOT-SWCNTs composites used as a CE for DSSCs devices [55].

According to Abdul Almohsin, S.M., the composites based on PANI and SWCNTs, prepared by one-pot electrochemical synthesis [22], present a much lower R_{CT} value than the pristine PANI (95 vs. 845 Ω), proving the contribution of the SWCNTs to the enhancement of charge transfer within the composite. The morphology of the CE films also contributes to the photoelectric performance of the DSSCs by influencing their transmittance and diffuse reflectance properties. The latter two are maintained at high values even when CNTs have been added [53]. The high conductivity of SWCNTs together with the layer morphology [24,53] doubled the photoelectric performances of the CE in DSSCs. This is the case of composites based on PPy and CNTs with a honeycomb-like structure [20]. The presence of a surfactant within the EM composite can influence its catalytic performance [17,19]. The surfactant induces an enhancement of the interactions between the polymer chains, increases the stability, and enhances the electropolymerization current, together with providing an easy charge exchange between the electrolytic medium and the polymers [56].

Benefiting from a morphology of the porous structure resulting from well-separated fibers formed from PANI-coated CNTs, as a result of the doctor blade synthesis method, the PCE value of CNTs-PANI (6.67%) almost reaches the value reported for Pt (7.70%) [23]. The low performance of this CE EM could be due to the thick catalytic layer with 6% PANI content, which provides a higher total internal resistance, but also the opaque nature that prevents the light reflection effect [43].

In terms of DSSCs performance, the external quantum efficiency (EQE), also called the incident monochromatic photon-to-electron conversion efficiency (IPCE), is an essential characterization method.

IPCE spectra showed a broad band in the 250–800 nm region, with a maximum value at 535 nm. From Figure 6b, it can be seen that the IPCE curves of DSSCs increase as a function of EM in the order CNTs < CNTs-PANI < Pt. The reported results obtained in the IPCE curves are well correlated with the JV curves (Figure 6a). Despite the fact that the performance of CNTs-PANI CE materials does not exceed that of the Pt electrode, CE composites based on CNTs and CPs could be a good alternative to replace Pt, at least economically.

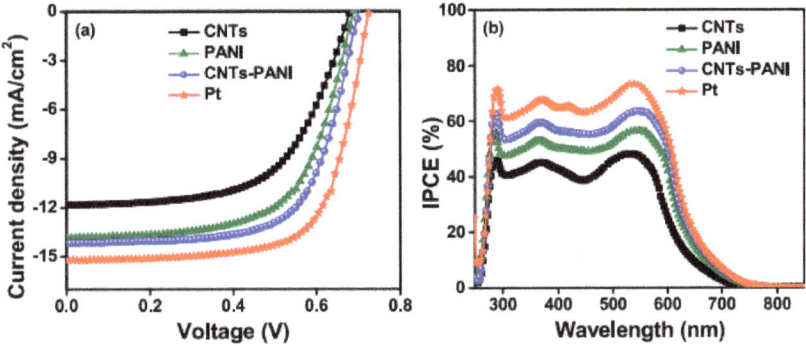

Figure 6. (a) JV characteristics of Ems, namely CNTs, PANI, CNTs-PANI, and Pt and (b) the incident photon to current conversion efficiency (IPCE) spectra of CNTs and CNTs-PANI-based DSSCs for 10 devices [23].

3. Synthesis and Vibrational Properties of CPs-GD Composites

3.1. Synthesis of CPs-GD Composites

After the incorporation of graphene into PEDOT, the electrochemical activity is reported as improved and the charge transport is much faster through the composite film [57]. The reduction potential of the redox couple is shifted towards negative potentials, compared to a Pt/ITO CE, which means it still has a higher resistance. When prepared by CVD, the characteristic parameters of the graphene layer are superior to those prepared using other techniques. For example, the I_{2D}/I_G ratio is about 2.5 while the D band has a low intensity and the transparency of the monolayer compared to four layers is of 97.4% vs. 90.6%, at λ = 550 nm.

Moving from mechanically mixing graphene [58,59] and CP to the deposition of a polymeric film on top of the graphene, ultimately by in situ polymerization [60,61], many methods of synthesizing composite that were designed as CE materials have been reported. Together with the organic sol-gel route for the synthesis of aerogel structures [62], all reported methods have in common the individual deposition of the polymer and the graphene/GD layer; this means that the resulting composite material is closer to a sandwich structure more than to a bulk one for which the polymerization process takes place simultaneously [63–65].

For the PANI/graphene composite, the reaction mixture was prepared by combining a solution of monomer, prepared from 1.2 mmoles of ANI dissolved in 50 mL acid solution (0.4 M HCl) with 50 mL of a second solution made by solubilizing 0.41 mmol ammonium persulphate (APS), as a polymerization initiator, in 0.4 M HCl. The mix of both solutions was poured into a previously cooled bath, into which the graphene-modified electrode was already been introduced, and left there for 30 min, and then immersed in a 1 M HCl solution until complete conversion occurred from polyaniline-emeraldine base (PANI-EB) to polyaniline-emeraldine salt (PANI-ES) [66]. Brought together, the pristine graphene, difficult to incorporate into a polymer matrix, and the PANI difficult to adhere to the FTO substrate also came with advantages such as providing appropriate support for nucleation and polymerization, due to the presence of GO, and the homogenous RGO dispersion provided by PANI, making them a winning combination for a CE material. At the interface of both PANI and RGO, interactions such as π-stacking, hydrogen bonding, electrostatic, and donor–acceptor take place [67]. In an acidic medium, between PANI, in the form of emeraldine salt (characterized by a polaron energetic state) and RGO with a negative charge storage capacity, a transfer of weakly bound electrons takes place, which is a good source of electrons for the regenerating of the electrolytic redox couple I^-/I_3^-.

A special method for preparing aerogels composite structure was reported recently by Mohan, K. et al. The conversion into a PANI/RGO aerogel (PANI/RGOA) composite was made by preparing two dispersions of GO and, respectively, PANI nanotube (PANI NT), in deionized water, of 10 mg/mL in each suspension. In the preparation of RGOA, GO aqueous suspension was mixed with resorcinol, formaldehyde, and sodium carbonate (0.337, 0.362 g and, respectively, 1.6 mg) and stirred for 30 min. The two suspensions were then mixed in 100 mL flasks, sealed, and maintained at 85 °C for 3 days to obtain PANI NT/RGO aerogels, washed with deionized water, and freeze-dried for 24 h. At the end of the process, the aerogel pastes were deposited through the doctor blade method onto FTO glass and heated to 80 °C to obtain the CE. The aerogel pastes contained 10 mg of PANI/RGO aerogel dispersed in 0.5 mL Nafion solution [62]. The PANI/RGO aerogel prepared via the organic sol-gel route presented a high surface area and excellent catalytic activity towards the reduction of I^-/I_3^- in electrolyte [62]. PANI/GD composites have been also synthesized using economical methods such as the spray method [68]. First, the RGO layer was deposited on the FTO surface from 0.015 g of RGO mixed with 2.5 mL of acetic acid and 0.02 Triton X-100, and finally 100 mL of ethanol was added, and the entire mixture was ultrasonicated for 1 h. The resulted mixture was sprayed on the FTO surface, thermally annealed at 100 °C, and finally sintered at 250 °C. Then, 0.015 g RGO powder was mixed with an appropriate amount of PANI, 2.5 mL acetic acid, and binding agent Triton X-100, for 10 min; after that 1 mL dispersion of SnO_2 in ethanol was added, followed by pouring 100 mL of ethanol. The entire mixture was then sonicated for 1 h and at the end, PANI/RGO/SnO_2 was obtained. The suspension was sprayed onto a heated FTO support and sintered at 250 °C. The resulting composite has a porous structure, with many interconnections and pathways between components that provide an enhanced electrocatalytic effect in the triiodide ion (I_3^-) reduction reaction at the CE. The porosity of the composite increase when SnO_2 was introduced into the RGO structure provided a better adhesion for the interaction with CP. Comparing the PANI/graphene composite with the PPy/graphene composite, the way of synthesizing is very similar, so the composites materials were obtained by mixing the already-prepared polymer with the graphene precursor. Therefore, PPy was first obtained from mixing the aqueous solution of polyvinyl alcohol (PVOH) with ferric chloride, followed by the addition of pyrrole monomer, and the mixture was stirred in an ice bath for 4 h to obtain powder in suspension. The resulting powder was then mixed with graphene oxide (GO) powder in 10 mL of water. To this mixture, 0.088 g ascorbic acid was added, with the task of playing two roles, binder and reducing agent. The suspension was then deposited onto the FTO substrate using the doctor blade method [61]. In addition, the mechanical mixing method is not to be neglected in the preparation of PEDOT: PSS and graphene-based composites [58,59]. Using ultrasonication to mix PEDOT: PSS with nanoporous RGO helped to exfoliate RGO and insert a thin layer of graphene into the PEDOT: PSS matrix [59].

3.2. Vibrational Properties of CPs-GD

Regarding the vibrational properties of CP and graphene-based composites, leaving aside the characteristics of the IR and Raman spectra signature of each PANI, PPy, PEDOT, PEDOT: PSS, graphene, GO, and RGO, which will be also mentioned here in Table 3 inserted below, some changes were observed in their composite spectra that are relevant to be discussed here in terms of the type of functionalization between the two components inside the composites.

For further clarification, each individual polymer was discussed as follows: in the case of PANI, the active conductive form of PANI is ES, and after preparation and conversion it is important to check the presence of PANI-ES. Therefore, the conversion of PANI-EB to PANI-ES is confirmed by the presence of N in a protonated state. The distribution of N states shows the existence of pyridinic and pyrolic nitrogen [69].

Further evidence for the conversion from PANI-EB to PANI-ES is represented by the presence of the absorption band located at 330 nm, assigned to the π-π* electronic transition of the benzoid ring, which is blue-shifted from 330 nm (for bulk PANI) to 313 nm, a minor peak at 347 nm corresponding to π-π* electronic transition of PANI-EB, and the split peak from 360 nm assigned to the localized polaron-π* transition from the conductive form of PANI, the emeraldine salt. Another two bands located at 366 nm and 434 nm, corresponding to localized polarons and π-polaron transition, also confirm the formation of PANI-ES [70]. In addition, the main band characteristic of the insulating form of PANI-EB, located at 630 nm, assigned to local charge transfer between the quinoid ring and the adjacent imine-phenyl-amine is absent [71]. Further, for the PANI/RGO aerogel composite, both the Raman and FTIR spectra reveal the presence of both PANI and RGO [62] (see Figure 7), the bands from the spectra being overlapped, and the main change in their profile when analyzing the PANI/RGO spectrum seems to be the change in the main bands profile, described by an enhancement in their intensity rather than the appearance of new bands. The aerogel composite prepared by introducing PANI in GO suspension also presents new bands in the FTIR absorption spectrum, located at 3421, 1304, 1113, and 812 cm^{-1}, assigned to stretching vibration of N-H bond, stretching vibration of C-N bond, stretching vibration of N-Q-N, and asymmetric stretching vibration of 1,4 double replaced benzoid ring.

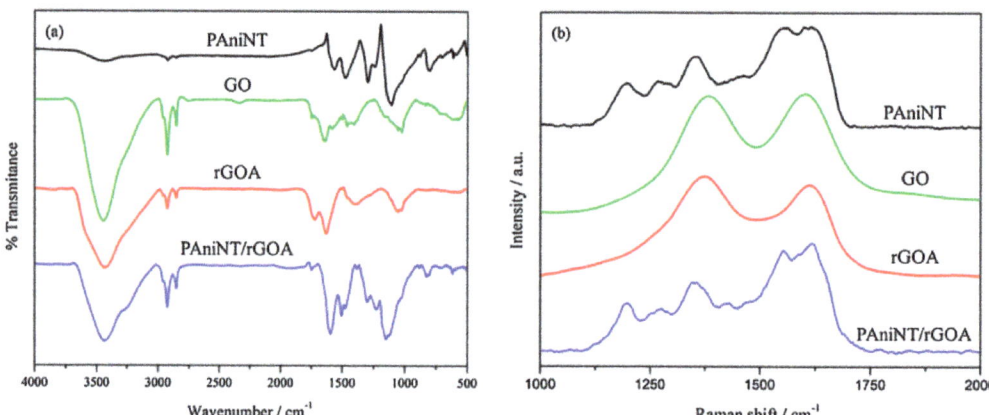

Figure 7. FTIR (**a**) and Raman (**b**) spectra of PANI NT, GO, RGOA, and PANI NT/RGOA [62].

The conversion of GO into RGO during synthesis was confirmed by the disappearance of the bands associated with the groups with oxygen and a decrease in intensity of the bands located at 3440, 1736, and 1402 cm^{-1}, as revealed from the PANI/RGO FTIR spectrum. Complementary to the FTIR spectrum, the Raman spectra reveal the bands specific to PANI, located at 1620, 1550, 1350, and 1200 cm^{-1}, associated with C-C vibration, imine vibration C=N, semi-quinoid polaronic vibration C-N$^+$, and the in-plane bending vibration of the C-H bond corresponding to the quinoid ring, respectively [72]. The bands specific to GO and RGO, the G and D bands, are located at 1608 and 1360–1385 cm^{-1}, respectively. H. Mohan and co-workers reported an enhancement of the I_G/I_D ratio during aerogel formation, namely from 0.98 to 1.5, due to the conversion of GO to RGO associated with a decrease in the number of the oxygen groups from the surface, which also leads to a retrained sp^2 domain surface. Nevertheless, the vibrational spectrum of PANI/RGOA does not differ much from the PANI spectrum, the main change consists of a decrease in intensity of the band located at 1620 cm^{-1}, assigned to the presence of RGOA inside the composite structure. From the morphology point of view, the tubular fiber structure of PANI is weaving with the wrinkled paper appearance of the graphene compound, where there in not graphene or PANI. PANI is attached at the graphene foil surface without aggregation. Inside the aerogel structure, PANI interacts with RGO through π-π stacking at the basal plane level.

The non-covalent interaction between PANI and RGO, exemplified according to Figure 8, is described with the help of the red shift of the IR band, related to the PANI matrix, located at 1560, 1480, 1295, 1241, 1125, and 797 cm^{-1}, as a consequence of a π-π interaction and hydrogen bonding between the basal planes of RGO and the PANI backbone [73]. Due to the strong interaction of RGO hydrophobic planes and PANI backbone [73,74], the Raman lines of RGO/PANI are shifted towards low wavenumbers. The peaks corresponding to the PANI signature have been red-shifted due to π-π interaction and hydrogen bonding between the basal planes of RGO and the PANI backbone [73], describing the non-covalent interaction between the components (PANI-RGO); a shift towards low wavenumbers has been also observed in the Raman spectra of RGO/SnO$_2$/PANI as a consequence of the strong interaction of RGO hydrophobic planes and the PANI backbone [73,74].

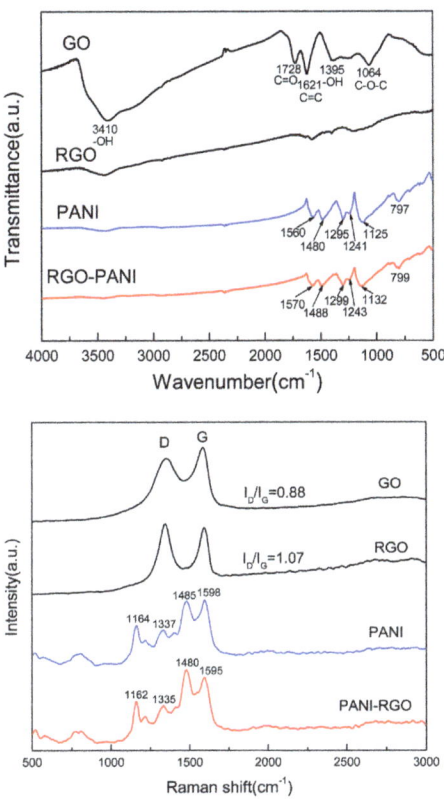

Figure 8. FT-IR and Raman spectra of GO, RGO, PANI, and PANI-RGO nanocomposites [73].

When analyzing the composite based on PPy and GD, it seems that the CP is the prevalent component of the composite as in the cases of the PANI-based composite described above. Thus, the signal of PPy prevails on the spectrum of the composite [61]: in the Raman spectra of the composite with RGO, two main bands were revealed, located at 1341 and 1559 cm^{-1} and assigned to the stretching vibrations of the pyrrole ring and C=C bond from Py, respectively. When added to RGO, the main PPy bands shifts to 1350 and 1590 cm^{-1} upon increasing the amount of RGO as a consequence of the overlaps with the D and G bands of RGO [75]. Comparing the aerogel carbonic structures, the polymer gel with the classic shape of composite used as a CE material and electrolytes for DSSCs, the first attract more interest. For example, from the carbonic aerogel category, graphene aerogels have a mesoporous three-dimensional structure due to the interconnected graphene sheets, a structure which provide special properties such as a high electric conductivity [76] and large volume of the pores. These features facilitate the charge transport and the mass transfer of the redox species. Therefore, their special structure recommends them as appropriate substitutes for Pt CEs. Very good results have been reported when aerogels based on carbon structures have replaced CEs in DSSCs [77]. An increase in PCE from 7.07% [77] to 8.83% was reported [78].

Porosity could be successfully enhanced through other ways as for example by metallic oxide particles inside the CE material (carbonic structure, GD, RGO). The increased porosity of the resultant composite provides better adhesion for the interaction with CP and as a result, efficiency improved considerably (from 4.7% to 6.25%) when introducing oxide metallic NPs inside the RGO matrix. For example, SnO_2 NPs increase the electrical conductivity [79] through contribution to the relaxation process of charge carriers [80]. All of these NPs are additional catalytic sites at the CE surface which are conducting and enhance the active surface area improving the DSSCs performances. Another aspect which could improve the PCE is the treatment of the photo-anode, containing two layers of TiO_2 on the FTO, with $TiCl_4$ (8.68%) [68]. The treatment was performed by dipping the photo-anode in a 0.04 M $TiCl_4$ solution, at 70 °C, for 30 min, and sintering at 450 °C for 45 min [68].

Table 3. Synthesis and CE performance parameters (CE, FF, J_{sc}, and R_{CT}) of CPs/GD composites.

CPs/Graphene Composite	Synthesis	CE (%)	FF	J_{sc} (mA/cm^2)	Rct (Ω)	Ref.
PANI/graphene	In situ chemical polymerization	3.58	0.473	10.683	0.346	[81]
PANI/graphene	In situ chemical polymerization	7.45	62.23	15.504	-	[60]
PANI/RGO aerogel	Organic sol-gel route	5.47	0.59	11.5	14.36	[62]
RGO/SnO_2 NPs/PANI	Spray method	8.68	63	18.6	23.5	[68]
PPy/RGO	Chemical polymerization	0.05%	0.28	0.4	-	[61]
Graphene-Si_3N_4/PEDOT: PSS	Mechanically mixture	5.24%	0.71	10.16	49.13	[58]
PEDOT: PSS-PG	Ultrasonication	9.57%	16	76	0.92	[59]

3.3. Performance of CPs/GD as CE in DSSCs

The electro-catalytic properties of the CE materials inside the DSSCs are evaluated through EIS, CV, I-V, and Taffel polarization curves. From all these measurements, parameters such as R_{CT}, RS, FF, J_{sc}, and J_0 are obtained, which are key factors that fully describe the CE and DSSCs performances, respectively. For all the parameters mentioned above, there is a short description at the CNTs/PCs section. Additional factors, e.g., the thickness of the CE layer, are also very important when considering the catalytic activity of a CE composite. The catalytic activity is directly related to the reduction process rate at the CE surface (I_3^-/I^-) and the number of the catalytic active sites. The lowest efficiency was reported for DSSCs fabricated with the RGOA counter electrode (3.29%) with an open circuit-voltage (VOC) of 755.81 mV, short-circuit current density (J_{SC}) of 7.59 mA cm^{-2}, and a fill factor (FF) of 57.66% compared to a DSSC with a PAniNT counter electrode which exhibits a slightly higher efficiency of 4.13% with a VOC of 775.11 mV, J_{SC} of 9.09 mA cm^{-2}, and FF of 58.61% [62]. The improved efficiency, reported when PAniNT is added to the RGOA matrix, is due to the higher catalytic activity of the polymer. Once again, the efficiency of the PANI/RGO composite at an optimum PANI loading (1:1 in this case) (5.47%) almost reaches the PCE reported when the CE was Pt (5.54%) asserting its capability to replace the costly Pt counter electrode in DSSCs (Figure 9).

Inside CPs/graphene or GD composites, the amount of the carbonaceous material within the composite determines the photoelectric performances. Thicker layers provide more catalytic sites and therefore enhance the rate of the reduction processes at the CE. Another parameter important to be followed is the filler factor (FF). When the FF value is higher than the value it has in conventional photovoltaic cells (50%), this parameter describes a successful limitation of the recombination processes at the CP/carbonaceous material interface. The chemical capacitance (Cμ) is used to estimate the electro-catalytic performance of the composite, evaluating its stability and durability. Cμ depends on the active surface area and the pseudo-capacitive charging effect of PEDOT [82,83]. High capacitance value is correlated with lower R_{CT}, higher Jsc, and FF, respectively (see Table 3). Related to Cμ is the double layer capacitance (Cdl), used to indicate the catalytic activity and the porosity of the composite structure. In terms of capacitance, the diffusion of I_3^- and I^-, and the rate of the reduction process, DSSCs with CEs made of composites based on PEDOT: PSS and RGO have very good PCE values, almost reaching the value reported for Pt.

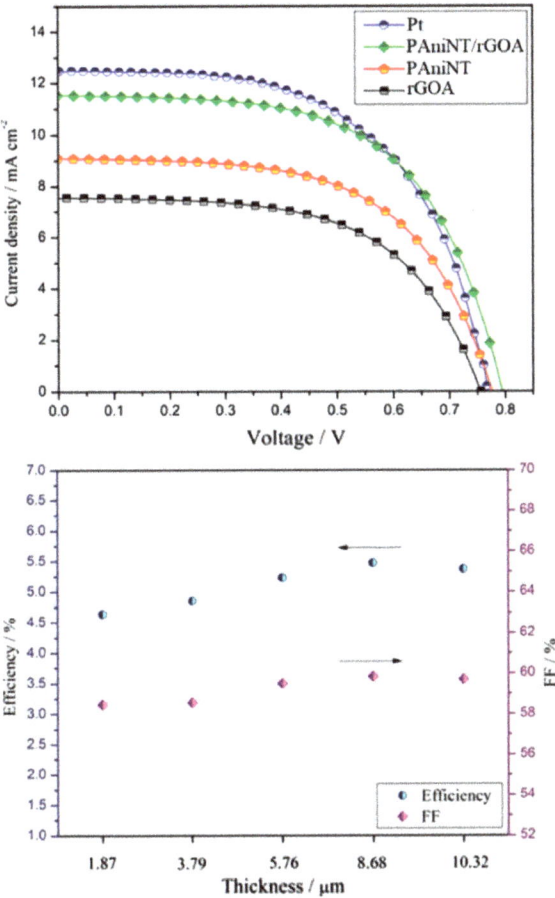

Figure 9. J-V characteristics of DSSCs fabricated with PAniNT, RGOA, PAniNT/rGOA, and Pt counter electrode (**top figure**) and photovoltaic parameters of DSSCs fabricated with PAniNT/rGOA CEs of different thicknesses under irradiation of 100 mW cm^{-2} light (**bottom figure**) [62].

There are only a few occasions when the R_{CT} is not correlated with a high catalytic activity. One of these cases was reported for composites based on PANI and graphene where, although the R_{CT} of PANI/Gr is higher compared to the Pt R_{CT}, the PCE values are very close (3.589% for Gr/PANI and 3.976 for Pt) [66]. This could be due to the fact that the Nyquist plot model used for Pt is not appropriate to be used for carbonaceous compounds.

As a general observation based on the analysis of different combinations of graphene and CPs, Chawarambwa, F.L. and co-workers concluded that the addition of a carbonaceous compound reduces the internal resistance of the composite [58]. The reduction process that occurs at CE involve two steps:

$$I_3^- + 2\,e^- \rightarrow 3\,I^- \quad (1)$$

$$3\,I_2 + 2\,e^- \rightarrow 2\,I_3^- \quad (2)$$

and the two potential peaks, positive and negative, correspond to the catalytic activity at the $CE/I_3^-/I^-$ interface and to the activity of the I_2/I_3^- from the electrolyte/dye interface [84]. With these two parameters, positive and negative potential peaks the list of parameters describing the catalytic performance of CE inside DSSCs is extended.

The list of parameters used to describe the catalytic activity of CE continues with the charge density peaks (Jox and Jred) together with the peak separation (ΔEp) of the reduction and oxidation peak potentials from the redox couple ($2I_3^-/I^-$). An excellent electro-catalytic behavior is revealed through a high value of current density (Jsc) and low peak separation. According to Dissanayake, M.A.K.L. and co-workers, for the RGO/PANI composite decorated with SnO_2 nanoparticles a higher value of Jox (1.1 mA/cm^2) was recorded, compared to RGO (0.53 mA/cm^2) and RGO/SnO_2 (0.98 mA/cm^2 Jox), revealing an improved catalytic activity of the first, followed by a peak separation value for RGO/SnO_2/PANI close to the one recorded for Pt (0.21 V). A lower ΔEp corresponds to higher electro-catalytic activity in the reduction process of the triiodide ion (I_3^-).

There are also inconveniences regarding the quality of composite components. For example, the incomplete reduction of RGO leads to low conductivity involving low charge transfer in the DSSCs [61] and another problem in determining the optimum amount of graphene compound used inside the composite, above the set limit, with the graphene layer overlap causing slow charge transport.

4. Synthesis and Vibrational Properties of CPs-CNs Composites as EM for Supercapacitors

4.1. Introduction in the Supercapacitors Cells

Recently, numerous studies have been reported on the use of CP-based composite materials, especially PANI and CNs of the type CNTs, graphene, and their derivatives or hybrids, resulted from the combination of the two, in a wide range of applications, in particular, as EMs in DSSCs [16,20,22,24,60,62,68,81] and supercapacitors [85–101]. In the composite configuration, the two components, i.e., CP and the carbon-based material, make their main contribution, the former through chemical stability, mechanical resistance, and redox behavior (PEDOT, PANI) and the latter through conductivity character and a high specific surface area. While in the previous chapter the discussion was mostly focused on DSSCs, this chapter highlights a brief description of the role of PANI/CNs composites as an EM in supercapacitors, where CNs are represented by CNTs and RGO. The description of the two types of electrode composite materials aims to obtain an overview of the aspects that control the synthesis of composites, significantly influencing the electrochemical capacitive performance of the material, the fundamental properties of these EMs being evaluated in terms of specific capacitance and charge/discharge rate.

The explanation for the fact that PANI is the most common CP used in supercapacitor EM lies in its excellent specific capacitance [102], plus the advantages of obtaining relatively simple and economical synthesis by chemical polymerization in an aqueous medium, as well as high stability in air [103]. Its main drawback, correctable by combining with CNs is poor conduction and low stability when used in repeated cycles. The latter is due to the significant changes taking place in the volume of the polymer matrix, being closely related to the processes of doping and de-doping [103]. The combination of PANI and CNs of the type CNTs or graphene or its derivatives is not accidental. Carbon-based materials exhibit very good capacitive behavior (often used as a current collector), high porosity that comes with a large specific surface area, and high chemical and mechanical stability. Thus, by introducing CNs into the polymer matrix, capacitance retention increases during repeated charge/discharge cycles, thus increasing cyclic stability and the charge/ion transfer rate between electrode and electrolyte, as well as the specific capacitance of the device, even at high current densities [104].

The capacitive behavior of the two components, CP and CNs, is different but contributes equally to the value of the final capacitance. PANI and other CPs with redox behavior exhibit pseudo-capacitive or Faradaic behavior where charge storage at the electrode surface occurs through the oxidation–reduction reactions that take place between the EM and the ions in the electrolyte, according to the mechanism reported by Jain D. et al. [86]:

$$CP \to CP_n^+ + ne^- \text{ (p doping)}; \quad (3)$$

$$CP + ne^- \to CP_n^- \text{ (n doping)}. \quad (4)$$

In the case of PANI, since it can store charges both in the electric double layer (EDLC) and through the Faradaic charge mechanism, it has been successfully used as an EM for supercapacitors [105]. On the other hand, carbonaceous compounds exhibit both pseudo-capacitive and non-faradaic behavior specific to double-layer capacitors, so called because in non-Faradaic behavior charge storage occurs in the electric double layer at the electrode-electrolyte interface. For storage capacity evaluation, often are recorded (a) cyclic voltammograms at different potential scanning speeds; (b) charge/discharge galvanostatic curves, when a dependence of voltage as a function of time, and (c) for a more detailed study, the Nyquist diagram is made based on electrochemical impedance spectroscopy (EIS) measurements used for the study of charge transfer within the composite, where the main traced parameter is the charge transfer resistance constant which is calculated from the diameter of the semicircle in the Nyquist graph [106]. A low value of charge transfer resistance corresponds to a high value of specific capacitance [107]. Due to the different nature of the two or even three components, in the case of composites, Faradaic and non-Faradaic behaviors occur simultaneously, leading to a greater storage capacity.

For carbon-based compounds, such as graphene, pseudocapacitive behavior is evidenced by a rectangular profile of cyclic voltammograms. This type of behavior is also visible in PANI and other CPs with reversible redox behavior, such as polypyrrole [108,109]. Comparing the cyclic voltammograms profile of the individual components and the composite, we notice a number of differences such as the pseudo-capacitive contribution of PANI [110,111], evidenced by the deviation from the symmetrical triangular shape visible in the case of the composite with CNs. Analysis of charge–discharge galvanostatic (CDG) curves helps to understand the behavior of EM. The profile of the voltage (E)–time (t) curves is triangularly symmetrical in the case of carbon-based materials with ideal capacitive behavior such as CNTs.

The most important parameters of a supercapacitor are the specific energy and power density, which are calculated according to the relationships:

$$E = 1/2M \times Cd \times V^2 \text{ and} \quad (5)$$

$$P = 1/4M \times (ESR) \times V^2, \quad (6)$$

where Cd is the discharge capacitance (F/cm^2), V is the voltage on the initial range of the discharge curve (excluding IR voltage drop), ESR is the series-equivalent resistance, M is the mass of the EM. ESR includes the interface contact resistance and diffusion resistance of electrolyte ions, as well as particles between electrodes that determine the power density of the supercapacitor [112]. Another indicator of the performance of the EM is the discharge capacitance that is evaluated from the linear portion of the discharge curves, using the relationship:

$$Cd = I \times \Delta t \, \Delta V, \quad (7)$$

where Cd is the discharge capacitance, I is the discharge current, Δt is the time interval over which discharge occurs and ΔV is the voltage variation. The specific capacitance (Cs) is calculated for both the individual EM and the assembly of two electrodes analogous to a supercapacitor and is calculated according to the relationship:

$$Cs = I \times \Delta t \, \Delta V \times M = Cd \, M, \text{ for an electrode}; \quad (8)$$

$$Cs = 2 \times Cd\ M \text{ for an assembly analogous to a supercapacitor.} \qquad (9)$$

Electrochemical impedance measurements evaluate the transport ability of materials. By interpreting the Nyquist graph, important information about the bulk material resistance (Rb) and charge transfer resistance (R_{CT}) is obtained. The graph is divided into three zones, the area of low, medium and high frequencies. In the first portion, in the area of low frequencies, the diagram has a step aspect and is an indicator of the capacitive nature of the material [113], while in the middle area, the profile of the graph is usually linear, and at high frequencies it is in the form of a semicircle. In the high-frequency region, the impedance spectrum depends on the load transport process, while in the low-frequency region mass transport processes dominate. For example, in the case of the PANI-Fe_3O_4/RGO composite [113], analyzing the spectra through EIS, it is observed that in the case of both RGO/Fe_3O_4 and the ternary composite, the charge transfer resistance is low in the high-frequency region and according to the appearance of the graph portion at low frequencies, namely the linear profile, with a slope of about 70°, indicating an ideal capacitive behavior of PANI nanorods [113]. PANI's nanorod structure has recently been shown to significantly improve the capacitive performance of composites used as EMs in supercapacitors, where the energy storage capacity of a redox material is described by the equation: q t = q electrolyte + q dl + q electrode, where q t is the total charge stored in the electrode, q electrolyte is the charge stored due to the electrolyte, q dl represents the storage capacity of the double layer, and q electrode the charge stored in the active material of the redox electrode [87].

4.2. Synthesis and Vibrational Properties of PANI-RGO Composites as Well as Their Performance in Supercapacitors

A very clear application for highlighting the performance of a supercapacitor was made by S. Mondal and his collaborators, who, using the template method, synthesized a ternary composite EM based on PANI and RGO decorated with Fe_3O_4 nanoparticles. Mondal S. and colleagues made electrodes from this ternary composite for a supercapacitor that supported the operation of an LED bulb for 30 min [87]. The material synthesized by them had a very good stability, with a capacitance retention of 78% after 5000 cycles. Thus, the transition metal oxide, in the structure of the composite, contributes to increasing its stability and storage capacity. Among the transition metal oxides used in the manufacture of supercapacitors, Fe_3O_4 stands out by the large potential window in which it is active, between −1.2 and +0.25 V, benefiting from a theoretical Cs of 2299 F/g plus increased availability, being from natural sources, low cost, and low toxicity, which recommends it for use as an electrode material with pseudo-capacitive properties.

The presence of PANI and RGO in the nanocomposite structure results in the formation of a conductive network due to the π-π conjugation between PANI and RGO-Co_3S_4. This phenomenon reduces the electron path length and the diffusion length of the ions in the active material, thus increasing the inner active space, and the possibility of storing more charge weight.

PANI has been used with good results as an EM for supercapacitors, as it can store charges in the EDLC as well as through the Faradaic charge mechanism. In combination with graphene, results on capacitive efficiency of PANI/RGO composite EMs have been reported both in binary combination [88–91,114] and in ternary composites [87,92,93] and in combination with ionic liquids [94] whose pseudo-capacitive effect is already known [94,115]. Thus, Meriga, V. et al. reported the EM PANI chlorosulfonate/RGO composite with a specific capacitance of 120 F/g in 2015 [91], a value that gradually increased with the addition of transition metal oxide, Fe_3O_4, to 283.4 F/g at a current density of 1 A/g [114] and 797.5 F/g at 0.5 A/g with a very good capacitance retention of 92.43% of baseline, after 1000 cyclic voltammograms for the ternary composite with Co_3S_4, in which covalent functionalization of RGO-Co_3S_4 with PANI takes place [92]. The Cs of the RGO composite also increases when doping RGO with N takes place (282 F/g at 1 A/g) [90], an important aspect in obtaining superior capacitive performance being the porous 3D structure of the synthesized EM, which increases its interaction with the electrolyte facilitating ion transfer at the electrode/electrolyte interface. It has also been shown that the use of 3D structures made of PANI/RGO composite gel leads to greatly improved Cs values, as is the case of the study reported by Wang, Z. et al. [89] on the PANI/RGO composite when a Cs value was recorded of 423 F/g at 0.8 A/g, with a retention of 75% after 1000 cyclic voltammograms. The synthesis morphology of PANI and RGO is also a very important parameter that significantly influences the specific capacitance value of the tested EM; thus, the composite formed by PANI nanowires in combination with 3D type structure of the N-doped RGO leads to a Cs of 385 F/g at 0.5 A/g [116]. Of all of the capacitive values reported, the largest are still maintained on ternary composites with single or double oxides of transition metals, with an additional pseudo-capacitive effect, such as MnO_2 [117], $NiCo_2O_4$ [93], and others (see Table 4), with a capacitance of 1090.2 F/g at 0.5 A/g and 1235 F/g at 60 A/g, respectively.

Table 4. PANI-RGO composites, synthesis methods, morphology, and capacitive performance.

Composite	Morphology	Synthesis Method	C_s (F/g)	E (Wh/kg)	P (KW/kg)	Ref.
Ternary composite rGO/Fe$_3$O$_4$/PANI	3D Nanorods of PANI doped with RGO decorated with Fe$_3$O$_4$	Template method	283.4	47.7	550	[87]
RGO/PPy/Cu$_2$O-Cu (OH)$_2$		Electrochemical polymerization	997 la 10 A/g,	20	8000 19,998.5	[118]
PANI-RGO	Globular or nano rods PANI on the surface of the RGO	In situ oxidative polymerization Pe RGO	797.5 F/g la 0.5 A/g, 92.43% after 1000 cycles			[92]
PANI-RGO	3D Porous composite PANI/RGO, with a specific surface of 228 m^2/g	Oxidative polymerization	420 F/g la 0.2 A/g, 80% after 6000 cycles at 2 A/g	9.3 for symmetric supercapacitor	0.1	[96]
PANI-RGO	RGO sheets randomly aggregated and closely linked together, uniformly coated by PANI nanofibers	Polymerization method surfactant-assisted	444 F/G la 0.6 A/g	13.36 W × h/kg	1.03 kW/kg	[86]
3D composite of the type RGO doped with N-PANI	PANI nanowires	In situ chemical polymerization	282 F/g la 1 A/g, 64.5% after 1000 cycles	-	-	[90]
PANI-RGO	Planar sheets of RGO, granular matrix of chlorosulfonated PANI	Chemical oxidation	120 F/g for PANI-RGO, 94% RGO	-	-	[91]
PANI/RGO		3D structure printing	1329 mF/cm^2 423 F/g at 0.8 A/g	-	-	[89]
PANI- tannic acid -RGO	Micro-fibrillary network of PANI	In situ oxidative chemical polymerization	268.5 F/g t 10 mV/s	1.68 la 0.5 A/g in symmetric supercapacitors	115	[88]
PANI-RGO- carbon fiber, ternary composite	Aggregate sheets with fine layers of PANI	Electrochemical method	430 F/g at 10^{-3} Hz	-	-	[95]
RGO/CNT-PANI	Fiber-shaped electrodes, skeleton/skin structure	GO reduction, PANI electrodeposition	193.1 F/cm^3 at 1 A/cm^3, 80.6% after 2000 cycles	0.98 (mW × H/cm^3)	16.25 (mW/cm^3)	[97]
NiCo$_2$O$_4$/PANI/rGO	Granular shape of PANI emeraldine base	Chemical polymerization	1235 F/g t 60 A/g, 78% after 3500 cycles	45.6 W × h/kg	610.1 kW/kg	[93]
PANI-RGO	Composite gel with 3D structure, porous	Self-assembly followed by a reduction process	808 F/g at 53.33 A/g	-	-	[98]
RGO-PANI	Nano-rods	Chemical polymerization	524.4 F/g at 0.5 A/g, 81.1% after 2000 cycles at 100 mV/s	-	-	[99]
PANI-RGO	Fibrillary morphology	In situ chemical polymerization	250 F/g	-	-	[100]
RGO-ion liquid/PANI (RGO-IL/PANI)	Excellent flexibility	In situ chemical polymerization	RGO-IL: 193 F/g at 1 A/g with 87% after 2000 cycles at 5 A/g	24.1	501	[94]
RGO-PANI	Dendritic nanofibers of PANI	Chemical polymerization	1337 F/g at 15 A/g, 81.25% after 5000 cycles	-	-	[114]
PANI-RGO- nanocellulose	Fibers	Chemical synthesis	79.71 F/g	Power density from 110.45 to 50.65 W/kg	-	[101]

In the case of composites based on CPs and CNs, as is the case with PANI/RGO/carbon fibers (CFs), some changes associated with the pseudo-capacitive contribution of PANI show the E-t-profile. Thus, the E-t profile of the composite deposited on CF has two areas: the first area corresponds to a shoulder due to faradaic processes (PANI) and the second area is linear, representing the capacitive process [95].

In the case of the covalent functionalization of RGO with PANI, the reported specific capacitance was 797.5 F/g at 0.5 A/g at a retention of 92.43% after 1000 cycles [92].

A three-dimensional structure of N-doped and PANI-coated RGO has also been reported by Liu, Z. et al. [90] resulting from the in situ chemical polymerization synthesis of

PANI in the presence of N-doped RGO using β-MnO_2 oxidant and polystyrene sacrificial microspheres to create the 3D structure of the RGO network.

Another type of morphology, namely well-separated planar sheets of RGO, without aggregation or association, embedded in the granular matrix of chlorosulfonated PANI, was reported by Meriga, V. and co-workers [91]; this composite material was synthesized by the chemical oxidation polymerization method of ANI in the presence of APS. EMs can also be printed as 3D structures using PANI/GO gel, obtained by the self-assembly method of PANI and GO in a mixture of NMP and water [89], with a capacitance of 423 F/g at 0.8 A/g for the PANI-RGO composite. In order to decrease the toxic compounds used in the reduction process of GO to RGO, Zhao, X. et al. tested tannic acid, a non-toxic compound successfully used for GO to RGO conversion, which influenced the morphology of the obtained composite, namely the PANI microfibrillar network [88]. After 24 h of reaction with tannic acid, the resulting RGO shows a higher specific surface area, a better distribution of PANI fibers on the surface, and a higher specific capacitance compared to the hydrazine-reduced PANI and RGO-based composite. The composite presented above recorded an energy density of 1.68 at 0.5 A/g, and respectively, a power density of 115 KW/kg in the symmetric supercapacitor.

A high capacitance (808 F/g at 53.33 A/g) [98] has been reported for binary PANI/RGO composites, synthesized by two successive processes of self-assembly in a mixture of water and NMP followed by the three-dimensional reduction of the assembly. The highest capacitance value of the PANI-RGO binary composite was reported by Nguyen, Van H. and co-workers (1337 F/g at 15 A/g) [114]. The composite material was prepared using a two-step synthesis method, this being in the form of thin distinct GO sheets with differentiated, wafer-like edges, coated with dendritic PANI nanofibers of 100 nm diameter and micron length. The very low size morphology of PANI is attributed to the chain structure of PANI molecules formed on the surface of RGO sheets [114]. On the other hand, ternary PANI-RGO-oxide composites have been synthesized by a number of methods that significantly influence the morphology of the resulting material. Thus, 3D nano rod-like structures of RGO-doped PANI decorated with Fe_3O_4 particles have been obtained by in situ polymerization [87]; 3D structures of the type RGO/PPy/Cu_2O-Cu$(OH)_2$ with very high power densities 8000 KW/kg were obtained by in situ electrochemical polymerization of RGO/PPY on Nickel foam followed by Cu_2O-Cu$(OH)_2$ deposition by chronoamperometry [118]. The granular morphology of PANI-EB in the $NiCo_2O_4$/PANI/rGO composite was reported by Rashti, A. et al., the $NiCo_2O_4$ particles were distributed on the PANI functionalized RGO lattice [93]. The ternary composite was tested in a three-electrode configuration, showing a capacitance value of 1235 F/g at 60 A/g, and in the solid-state asymmetric supercapacitor, the composite acted as cathode and the activated carbon acted as anode, the cell showing a specific capacitance of 262.5 F/g at 1 A/g, with a retention of 78% after 3500 cycles at a working potential of +1.5 V.

Charge storage properties of the composites were tested by GCD. In the case of ternary composite made of RGO, Fe_3O_4, and PANI [87] the electrochemical study together with the charge–discharge curves, recorded at 1 A/g, revealed the capacitance behavior of all stages of the composite from binary to ternary, namely for GO, RGO/Fe_3O_4 (RGF), and RGO/Fe_3O_4/PANI. Therefore, the quasi-triangular shape of the GCD spectrum corresponding to RGO/Fe_3O_4 and RGO/Fe_3O_4/PANI compared to the ideal triangular GO spectrum shape indicates the presence of two types of capacitances, namely the double-layer capacitance associated with GO and the pseudocapacitance of both PANI and Fe_3O_4. The longer charge–discharge time is directly related to an improved charge storage mechanism.

Electrochemical impedance spectroscopic (EIS) studies of GO, rGF, and rGFP composite have been measured in the range 0.1 to 100 kHz. They were performed to evaluate the transport processes within the composites. The impedance spectrum with the two main regions, namely the high-frequency region and low-frequency region is governed by the charge transport process, respectively, by the mass transport process. In the case of RGO/Fe_3O_4 and RGO/Fe_3O_4/PANI in the higher frequency region (Figure 10c), the semicircle is negligible, revealing the significantly low interfacial charge transfer resistance. In this region, both resistance (8.3 Ω) and charge transfer are low (2.03 Ω).

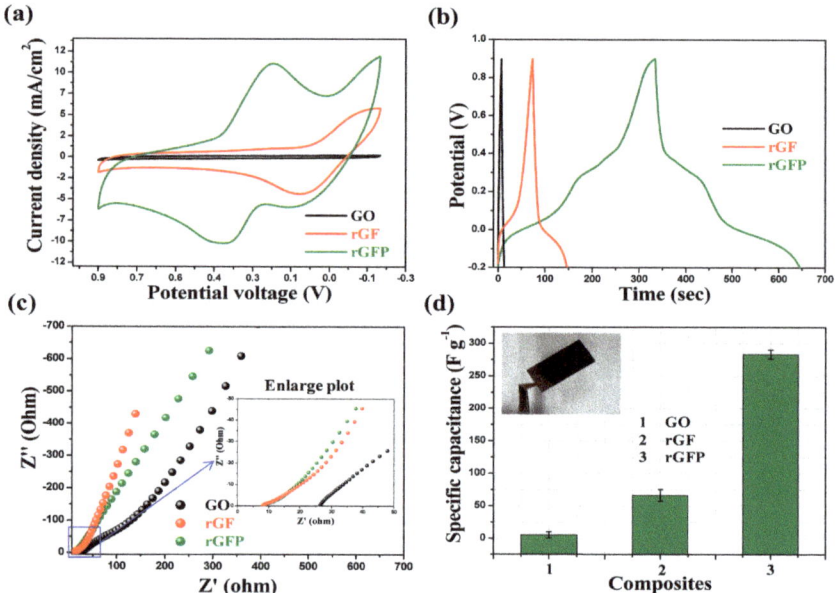

Figure 10. Electrochemical measurement (using the three-electrode system) of synthesized composites: (a) CV study (scan rate of 50 mV/s), (b) GCD study (at 1 A/g current density), (c) EIS study (Nyquist plot, in the 0.1–100,000 Hz range), and (d) a bar plot for specific capacitance. All measurements were made in a 0.5 M H_3PO_4 aqueous solution [87].

The slope of the vertical line, at the low-frequency region, is around ~70° indicating the nearly ideal capacitive behavior of the nanorods composite. The specific capacitance (Cs) value for synthesized rGFP ternary composites is high (283.4 F/g) compared to GO (5.5 F/g), respectively, to the binary RGF (66.4 F/g) (Figure 10).

When the EM was represented by RGO-ionic liquid/PANI (RGO-IL/PANI) [94], the reported capacitance was 193 F/g at 1 A/g due to RGO modification with ionic liquid, and a retention capacity of 87% after 2000 cycles at 5 A/g.

Composite materials based on CNTs and CPs, in particular PANI, with applications in the field of supercapacitors have also been reported. These composites generally obtained by oxidative polymerization of ANI in the presence of CNTs [119], respectively, by electrochemical polymerization of ANI on CNTs [120], exhibit three-dimensional structures with superior mechanical and electrical properties, a porous structure, and the ion diffusion property [104] (see Table 5). According to the study conducted by Malik, R. [104], a combination of vertically aligned N-doped CNTs in the form of the CNTs sheets were uniformly coated with a PANI layer, the final material having a core-shell morphology. Thus, the values of Cs reported were 359 F/g and 128 F/g at current densities of 4.95 A/g and 2.42 A/g, respectively, within the solid symmetric supercapacitor with PVA/H_2SO_4 type gel electrolyte composite layers between [104]. Vertically aligned N-doped CNTs have three times the specific surface area compared to classical N-doped CNTs and a pore size between 5 and 10 nm which favors PANI deposition and the rapid diffusion of electrolyte ions. According to Haq and co-workers [121], the N-doped centers in CNTs are the ANI polymerization initiation centers. Vertically aligned CNTs were grown on Ni-coated N-doped CNTs sheets by the CVD method from acetylene in an NH_3 plasma atmosphere [104]. While N-doped CNT substrate sheets (NCNTs) behave as a current collector, vertically aligned CNTs grown on NCNT sheets mediate and accelerate electron transfer from PANI to NCNT layers. The structure retains the core-shell morphology even after 30 cyclic voltammograms and the capacitance retention is 80% after 5000 cyclic voltammograms plus the open porous diode structure which allows ion diffusion while maintaining high ionic conductivity [122–124]. Upon covalent interaction between PANI and CNTs, the composite obtained by depositing porous PANI over CNTs with an interconnected pore structure exhibits a Cs of 1266 F/g at 1 A/g, i.e., a retention of 83% after 10,000 cyclic voltammograms [107]. The porous morphology results in more efficient ion transport, faster charge transport due to electron delocalization, and better PANI stability in redox processes. PANI grafting to CNTs leads to the extension of PANI conjugation and thus to increased conductivity up to 3009 S/cm [125]. Charge accumulation and Faradaic redox-reducing properties are improved due to the pore structure and specific surface of the activated MWCNT and PANI-based EM (A-MWCNTs/PANI). Activation of MWCNTs is performed by treatment with HNO_3 for better dispersion of MWCNTs in the polymer matrix. The Cs value is significantly improved from 42 F/g for A-MWCNTs to 201 F/g after introducing PANI into the EM. This is reflected in the higher current density increase in activated MWCNTs compared to the original ones, the activation positively influencing their porosity. The cyclic voltammograms show an almost rectangular and symmetrical profile, especially at low scan rates, indicating the high capacitive performance of the electrode–electrolyte system.

A frequently encountered problem in the behavior of EMs is maintaining a constant value of Cs at different values of current density. One material that meets this requirement is that reported by Ramana. G.V. and co-workers [119], namely the PANI/CNT composite obtained by oxidative chemical polymerization of ANI, with core-shell morphology and a Cs of 368.4 F/g. According to this study, the functionalization of CNTs prior to composite synthesis allows control of the oxidation state of PANI and can significantly improve the capacitive performance of the material. Another group of researchers, led by Zhou, H. [126], attempted to increase the capacitive performance of a PANI-CNT composite by using electrochemically expanded graphite (ExGP) as a current collector. By reducing the resistance at the EM/current collector interface they demonstrated increased electrochemical capacitance. The EM/current collector couple, mentioned above, has a capacitance of 826.7 F/g, much higher compared to other PANI/CNTs electrodes. The device made from the superposition of two PANI/CNT/ExGP type electrodes with a PVA/H_2SO_4 electrolyte layer interleaved is a flexible, solid-state supercapacitor with a power of 7.1 kW/kg at an energy density of 12 Wh/kg with a good cycle stability.

The CNTs used in the composite synthesis were functionalized with -COOH groups by exposure to concentrated acidic medium ($H_2SO_4:HNO_3$ = 3:1 in volume percent). Due to the expanded graphite substrate, the PANI/CNT morphology is different, thus CNTs are evenly distributed along the PANI fibers, the close contact between the two suggesting the dual role of CNT, as a binding agent and conductive additive, which shortens the distance between the relatively dispersed PANI fibers, facilitating electrical contact between them. The overall appearance is of a core-shell structure, and the length of the CNTs is longer due to the coating with the polymer layer—which increases the specific surface area and thus the number of active centers available in pseudo-capacitive reactions. The formation of the core-shell structure occurs due to the electrostatic interaction between the negatively charged functionalized nanotubes ($CNT-COO^-$) and the positive charges along the polymer chain of PANI. The higher specific capacitance of PANI/CNTs/ExGPs compared to PANI/ExGPs is evidenced by a larger cyclic voltammograms area and more intense redox maxima, respectively (Figure 11).

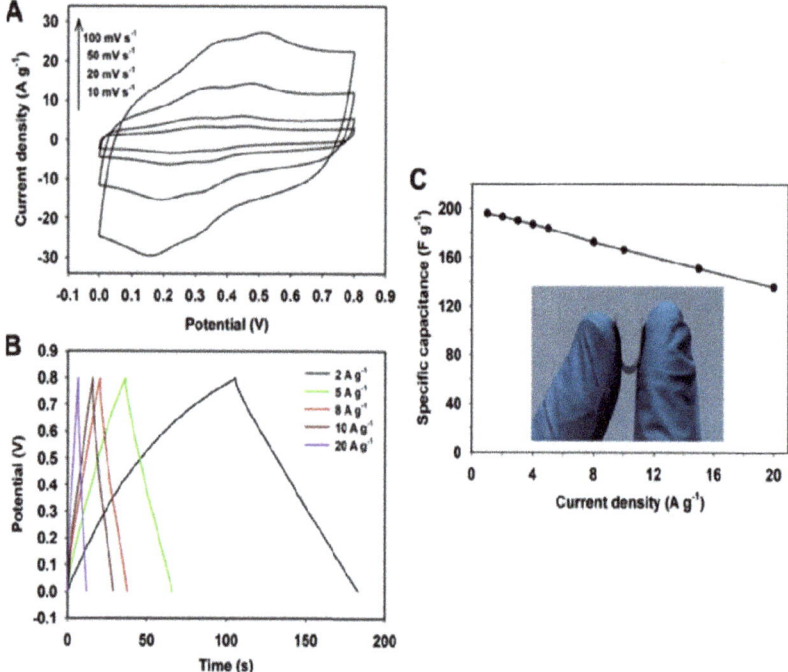

Figure 11. (**A**) Cyclic voltammograms profiles at the scan rates ranging from 10 to 100 mV s^{-1}, (**B**) GCD curves at the current densities from 2 to 20 A g^{-1}, and (**C**) plot of specific capacitance vs. GCD current density for the symmetric PANI-CNT/ExGP-based supercapacitor. The inset in (**C**) is an image showing the mechanical deformation of the PANI-CNT/ExGP-based supercapacitor [126].

Figure 12 shows the FTIR spectra of CNT, carboxylated CNT, PANI, and PANI-CNT. The FTIR spectrum of CNTs-COOH compared to CNT exhibits an additional peak (1715 cm^{-1}), due to C=O stretching vibration in carboxyl groups [127], which is a proof that carboxyl groups were introduced into CNT during the functionalization process. The -COO$^-$ groups on the surface of CNTs neutralize the positive charges in the oxidized PANI chain, during electrochemically oxidative polymerization, thereby forming the CNT@PANI core-shell structure mentioned above. The PANI-CNT spectrum is almost similar with PANI spectrum the only difference being that the typical C=C bonds in PANI located at 1565 and 1487 cm^{-1} are red-shifted to 1555 and 1477 cm^{-1} in the FTIR spectrum of the composite, due to the interaction between CNT and p-electrons of PANI chains [128].

CNTs could be prepared before interaction with PANI by the ball-milling process and subsequent exposure to acid solution. The resultant composite, MWCNT-PANI, exhibited a capacitance of 837.6 F/g at 1 mV/s [129]. The increase in anodic and cathodic current densities with increasing scan rate indicates a good material operating rate, which is accompanied by the large cyclic voltammograms area relative to that of the CNT indicating a large pseudo-capacitance of the electrode. Due to the electrostatic interactions of the positive charges in the PANI chain and the anions in the solution, floccules are formed that allow the ions to access their surface, reducing the distance between them and the PANI, favoring the charge/discharge processes. A special type of PANI/CNT composites are those in which the CNTs are open at one end (so-called partially unzipped CNTs), and play a role in favoring the access of ions from the electrolyte to the EM during charge–discharge processes, contributing decisively to increasing the mechanical strength of the EM. CNTs can be opened by several methods, such as chemical interaction with acids [130], intercalation and exfoliation of lithium ions [131], catalytic processes [132], electrical [133], and physical–

chemical methods [134]. The CNT unbonding process consists of the longitudinal splitting of the nanotube, a procedure that results in one or more layers of graphene or a combination of inner tube and graphene nanosheets, depending on the number of CNT walls. While the first mentioned methods used to unzip the CNTs involve the use of chemical compounds, the last two are cleaner methods to shape the final product, namely graphene nanoribbons (GNR). While the electric method as it says uses the electrical current to tailor a graphene sheet in high vacuum conditions, the physical–chemical method assumes the coverage of CNTs casted on substrate (Silicon wafer) with a thick film of poly methyl methacrylate (PMMA), followed by heat treatment to strengthen the CNTs-PMMA structure. After that, it follows the exfoliation of the resulted CNTs-PMMA film from the substrate using the KOH solution. The final step represents the etching process which occurs by exposing the sample to 10 W Argon plasma, for various periods of time, resulting in one or multi-layered GNRs or a combination between GNR and CNTs depending of the CNTs number of walls.

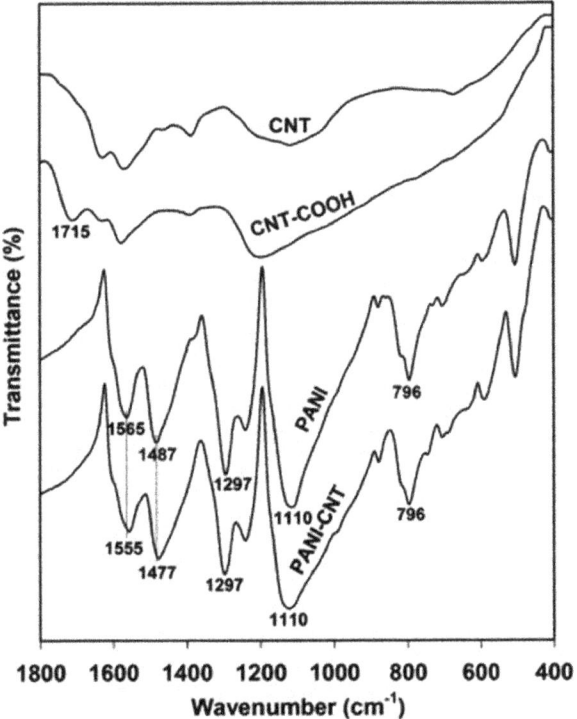

Figure 12. FT-IR spectra of CNT, CNT-COOH, PANI, and PANI-CNT.

Table 5. PANI-CNT composites, synthesis methods, morphology, and capacitive performance.

Composite	Morphology	Synthesis Method	Cs (F/g)	Ref
PANI/CNTs	Homogeneously co-dispersed open tubes together with graphene nanostructures resulting from the process of opening tubes coated with a uniform polymer layer *	In situ oxidative polymerization	762 with 81% retention after 1000 cyclic voltammograms	[85]
A-MWCNTs/PANI	Thick bundles of CNTs	In situ polymerization	248 at 0.25 A/g, and 99.2 F/g at 5 A/g	[103]
PANI/NCNT (CNT dopat cu N)	Vertically aligned nanotubes grown perpendicular to horizontally aligned CNTs coated with a uniform layer of polymer, observed by increasing the diameter of the CNTs	Electrochemical deposition	359 at 4.95 A/g with 82% retention at 46.87 mA/cm^2	[104]
Porous PANI/CNTs	Compact morphology	Chemical grafting by Ani interaction with -NH$_2$-functionalized CNTs	1266 F/g at 1 A/g, 83% after 10000 cyclic voltammograms	[107]
CNT-PANI	Shell-core structure	Oxidative chemical polymerization	368.4 F/g at various current densities	[119]
Composite expanded graphite (ExGP)/PANI-CNT	Interconnected fiber microstructures interleaved with CNTs	Electrochemical co-deposition	826.7 F/g	[127]
PANI/CNT	Polymer deposition on the surface of CNTs	In situ chemical polymerization	837.6 F/g at 1 mV/s, 68% after 3000 cyclic voltammograms	[129]
PANI/MWCNTs	Nanofibrous structure	In situ chemical polymerization	554 F/g at 1 A/g	[135]

* Outer tube of partially destroyed MWCNTs transformed into graphene nano-ribbons and core-like inner tube in the PANI composite; a decreased diameter of PANI fibers was observed suggesting the coating of hybrid MWCNTs with a uniform polymer layer.

5. Conclusions

Composite materials used as CEs after being deposited on FTO substrate present different features depending on the CPs type, the carbon nanostructures inside the composite, the interaction between both components which is directly related to the type of synthesis, and so on. Composites based on CNTs and CPs were prepared through different methods, starting with CNTs deposited on FTO or ITO substrates, followed by polymerization of monomers, precipitation of the polymer on top of CNTs, or by one-pot electrochemical synthesis from a CNTs dispersion mixed with a monomer solution. The only one which provides good control of the thickness and uniformity of the grown layer is the electrochemical method. Excellent results were also reported when the CVD technique was used to synthesize polymers on top of CNTs. Through CPs-CNTs composites, a special morphology is the honeycomb-like structure obtained on the CNTs substrate using a PMMA template (in the case of PPy and PEDOT composites with MWCNTs), when the CPs is electropolymerized on top.

Inside CPs/CNTs composites, CNTs play the role of template leading to porous nanostructured wires but the CNTs were wrapped in CPs film through π-π coupling (as in the case of CNTs-PEDOT composite). The use of aligned structures or well-ordered CNTs have been reported to enhance the PCE value of DSSC by decreasing the resistance and increasing the conductivity.

An innovation in the CE materials field was also the incorporation of CNTs inside a gel electrolyte, to improve the DSSC PCE performance. The purpose was achieved through the significant contribution of CNTs to enlarge the SA of the composite, increasing the conductivity and the Jsc and thus providing a high catalytic activity. Depending on the synthesis method used for their preparation, CNTs were reported as covalently (PPy-CNTs) and non-covalently functionalized with CPs. The first were obtained through the reflux technique followed by in situ polymerization and the second through all the other methods mentioned above.

Non-covalent functionalization of CNTs with CPs was revealed in the changes observed in the IR and Raman spectra of the individual component and on the composite. Due to the interaction between π-π bonds from the polymer structure and MWCNTs, the bands assigned to the C=C/C-C stretching vibration shift (case of PPy chains) to lower frequencies. The shift to lower frequencies is due to short polymer chains formation and could be accompanied by a decrease in the IR main band intensity (bands that could be associated to C-H, C-C, and N-H or C-N, depending on the CPs backbone structure).

Part from the interactions specific for non-covalent functionalization, namely the electrostatic interaction, which takes place between the donor (usually aromatic or quinoid ring from the CPs) and the acceptor (usually the CNTs), determines a fast movement of charges inside the composite enhancing its conductivity. CNTs wrapped within the CPs could create a transport network for electrons improving the cathodic reaction of the redox couple I_3^-/I^- (in the case of FMWCNTs-PPy composite).

When CNTs are covalently functionalized with CPs, new bonds are formed, namely covalent bonds, between CPs and CNTs, which are formed between –NH- from CPs and –C= from SWCNTs (inside PANI-SWCNTs). Through these covalent bonds, the charge transfer between the components of the composite is accelerated.

In the case of CPs/graphene or GD composites, all of the synthesis methods, namely mechanically mixing graphene and CP, the deposition of a polymeric film on top of graphene through in situ chemical polymerization, and the organic sol-gel route, have in common the individual deposition of the polymer and the graphene/GD layer. This is the reason why the morphology of the resulted composites is close to a sandwich structure. Regarding CPs/GD composites, GD (for example GO) provides appropriate support for the nucleation and polymerization of the monomers, precursors of CPs; meanwhile, the CPs ensures the homogenous dispersion of GD. Therefore, the combination of GD with CPs lead to the formation of composite materials as a promising candidate for a CE in the DSSC device. Special structures of CPs/GD composites play an important role as CEs in DSSC. The aerogel structure obtained by the organic sol-gel route, followed by freeze-drying, presents a high surface area and excellent catalytic activity towards the reduction of I^-/I_3^- in the electrolyte. The three-dimensional structure of graphene aerogels provides special properties such as a high electric conductivity and large volume of pores, facilitating the charge transport and the mass transfer of redox species.

At the interface of both CPs and GD the non-covalent functionalization is revealed through interactions such as π-stacking, hydrogen bonding, electrostatic, and donor–acceptor take place. The exchange of charge carriers between CPs and GD contributes actively to the electrolyte regeneration process providing an appropriate number of electrons. The weak bounded electrons which travel from CPs to GD (RGO) also represent a significant source of electrons for regenerating the electrolyte redox couple I^-/I_3^-. Regarding the ternary composites, namely metallic NPs/CPs/CNT or GD, the presence of metallic catalytic nanoparticles on the composite surface has been proved to contribute significantly to an improved PCE (Cu/PPy/MWCNTs, RGO/SnO$_2$/PANI). Not only at the surface but also inside the GD, oxide metallic NPs lead to increased porosity by creating interconnections and pathways between the components of the composite. This fact leads to enhancement in the reduction rate of I_3^- at the CE (redox process) and therefore efficiency improves considerably.

The best performance was reported for CPs/GD composites obtained through a mechanical mixture of graphene and PEDOT: PSS with a reported PCE value of 9.57%, followed closely by a modified GD with SnO$_2$-PANI composite with 8.68% PCE. In the case of CPs/CNTs composites, the best results were reported on PEDOT-MWCNTs, 9.07% followed by Cu-PPy-CNT, 7.1% PCE.

Composites based on PANI and CNs shaped as CNTs and RGO can also be used as EM in supercapacitor devices due to the combination of the superior mechanical and electrical properties of CNTs and RGO, ion diffusion properties, as well as both the conductive and pseudo-capacitor character of CP. Capacitive performance requires that the EM has a high capacitance value and a cyclic retention as close as possible to the initial capacitance value. Within the composite, both components contribute to the increase in Cs, on the one hand the CNs acts as a current collector, with a high specific surface area and superior mechanical strength that determines the cyclic stability of the EM into whose structure it enters, and on the other hand the CP is the active component, responsible for ion diffusion and electron transfer due to the extensive delocalization that takes place between the CP and the CNs, either RGO or CNTs. From the series of the CPs/RGO composites, the best capacitive performance has been reported for ternary composites with oxides of transition metals, which exhibit an additional pseudo-capacitive effect, e.g., MnO_2, $NiCo_2O_4$, Co_3S_4. As important as a high Cs value is, the cyclic retention should be as close as possible to the initial value after exposing the EM to several charge/discharge cycles. The different composite morphologies obtained from combinations of CNs, respectively, e.g., RGO with CP, are also an important indicator in predicting capacitive performance.

In the case of the PANI-CNTs composites, the morphologies range from compact to porous structures, from core-shell structures to nanofibers, with a capacitance value of 1266 F/g at a current density of 1 A/g and a retention of 83% after 10,000 cycles reported for the composite with CNTs covalently functionalized with PANI, demonstrating a higher cyclic stability in the case of CNTs composites as opposed to those with RGO but also a significant contribution of the type of functionalization within the composite. Very good results were also obtained by reducing the resistance at the electrode–electrolyte interface by introducing an exfoliated graphite substrate acting as a current collector.

6. Outlook

In terms of conversion efficiency, many leaps have been made, starting from around 4.04% PCE [136] to 9.07% reported in the case of composites based on CPs and CNTs [54].

The biggest issue remains to limit the recombination processes from the TiO_2/electrolyte interface to obtain a more efficient conversion. Energy is lost when the photons, with higher energy than the threshold value, dissipate the excess of energy releasing heat instead of generating more electrons of low energy, the so-called thermal/thermalized electrons originating from the energy of a single absorbed photon.

To limit the recombination, many ways could be tried, such as, for example, to recover the lost energy, which remains unabsorbed and unconverted, to capture and reabsorb it through the same system, or use an additional system for a more efficient conversion. This could be achieved further by using radiation-absorbent materials. These kind of materials have a band structure similar to quantum dots (for example, PbSe, known to produce, at high yield, multiple excitons from a single absorbed photon) or quantum wells and to find a way to extract and transfer photogenerated carriers from the quantum dot structure to produce electricity in the external circuit.

Another issue regards the distance between the photoanode and the electrode. Inside DSSCs, light conversion is assured by the TiO_2 particles network which collects the charge carriers (e^-). The absorption of light is performed by the molecules of dye deposited on the nano-porous surface of TiO_2 and by the interfacial contact distance, between the photoanode and the CE, kept low by the help of the electrolyte, sandwiched between them.

Eligible materials that could play the role of the electrolyte could be materials with mini-bands or intermediate bands such as the organic CPs or composites. Inside these kinds of materials, different energies of incident photons could promote the absorption at different isolated energetic levels, providing different voltages. A liquid CP should penetrate the porous structure of the solid photoanode and collect other types of carriers to complete the circuit from inside the cell.

With regard to supercapacitors, the main problem which is now in progress is the stability during cyclic exposure which could be diminished by using current collectors, not necessarily metallic foil but compounds with fibrous morphology eventually covered with thin metallic film. Another distinct drawback addresses the leaching of the electrolyte from the devices, a drawback which could be managed by encapsulation of the entire storage device in order to improve the Cs and to avoid the degradation of the EM through delamination. Apart from nano-carbonic materials, which represent the best candidate for EM in supercapacitors due to their high compatibility, cyclic retention, and resistance to corrosion, nano-cellulose may be a future good option due to its property of uniformity when shaped in thin films together with mechanical resistance.

There are also other types of polymers involved in energy storage as a part of the supercapacitor devices, namely the biopolymers such as alginate or chitosan which in combination with inorganic materials, such as molybdenum acid salts or manganese dioxide (electrolytically synthesized) lead to excellent charge storage properties [137,138]. The hybrid materials based on molybdenum acid salts have the general formula $MMoO_4$ (where M is a divalent cationic metal such as, for example, Co, Mn, Ni, Ca, Fe, Cu, or Pb). They have the advantage of easy synthesis, excellent efficiency, cycling stability, and they are environmentally friendly. For instance, cobalt molybdate exhibits a Cs of 1558 F/g together with an excellent cyclic retention, while chitosan-modified $CoMoO_4$ exhibits a capacitance of 135 F/g at the current density of 0.6 A/g and an energy density of 31 W h/kg, with a cyclic retention of 60% over 2000 cycles [139,140]. There are also other combinations of molybdenum salts within composites used as supercapacitors materials, such as $NiMoO_4$ and $NiCo_2O_4$ with a Cs equal to 2474 F/g [141], while the $MnMoO_4$/graphene hybrid composite shows a Cs of 364 F/g in the three-electrodes configuration [142]. All of these materials are the future trend in charge storage devices. The combination between biopolymers [137,138] and molybdenum salt hybrids is a perfect match due to the strong adhesion of the molybdate to the polymer network [143]. Similar to chitosan, another widely used biopolymer is alginate [137]. In combination with electrolytic manganese dioxide (namely EMD), this delivered five-times-higher capacitance than pristine EMD (487 vs. 94 F/g) at 1 mA/cm^{-2} in highly alkaline aqueous electrolyte. Coupled with activated carbon, the EMD composite exhibited a capacitance of 52 F/g and a cyclic retention of 94% over 5000 cycles [137]. Therefore, considering the trend towards more natural materials as EMs in supercapacitors, the composites based on biopolymers and hybrid materials based on molybdenum acid salts represent a future alternative to be exploited in this field due to their high capacitance and long cyclic stability.

Author Contributions: M.V.: writing—original draft preparation, M.V. and M.B.: conceptualization, methodology, writing—review and editing, visualization, T.B. writing-review and editing, M.B.: funding acquisition. All authors have read and agreed to the published version of the manuscript.

Funding: This work is funded by the Core Program of the National Institute of Materials Physics, granted by the Romanian Ministry of Research, Innovation, and Digitization through the project PC3-PN23080303.

Institutional Review Board Statement: Not applicable.

Data Availability Statement: Not applicable.

Conflicts of Interest: The authors declare no conflict of interest. The funders had no role in the design of the study; in the collection, analyses, or interpretation of data; in the writing of the manuscript; or in the decision to publish the results.

References

1. Goldemberg, J.; Johansson, T.B. (Eds.) *World Energy Assessment Overview: 2004 Update*; Prepared for United Nations Development Programme, United Nations Department of Economic and Social Affairs, and the World Energy Council; United Nations Development Programme: New York, NY, USA, 2004; p. 88.
2. Gao, J.; Yang, Y.; Zhang, Z.; Yan, J.; Lin, Z.; Guo, X. Bifacial quasi-solid-state dye-sensitized solar cells with Poly (vinyl pyrrolidone)/polyaniline transparent counter electrode. *Nano Energy* **2016**, *26*, 123–130. [CrossRef]
3. Li, Z.-Q.; Chen, W.-C.; Guo, F.-L.; Mo, L.E.; Hu, L.-H.; Hu, L.-H.; Dai, S.-Y. Mesoporous TiO_2 Yolk-Shell Microspheres for Dye-sensitized Solar Cells with a High Efficiency Exceeding 11%. *Sci. Rep.* **2015**, *5*, 14178. [CrossRef] [PubMed]

4. O'Regan, B.; Grätzel, M. A low-cost, high-efficiency solar cell based on dye-sensitized colloidal TiO$_2$ films. *Nature* **1991**, *353*, 737–740. [CrossRef]
5. Zhang, S.; Yang, X.; Numata, Y.; Han, L. Highly efficient dye-sensitized solar cells: Progress and future challenges. *Energy Environ. Sci.* **2013**, *6*, 1443–1464. [CrossRef]
6. Mishra, A.; Fischer, M.K.R.; Bauerle, P. Metal-free organic dyes for dye-sensitized solar cells: From structure: Property relationships to design rules. *Angew. Chem. Int. Ed.* **2009**, *48*, 2474–2499. [CrossRef] [PubMed]
7. Liang, M.; Chen, J. Arylamine organic dyes for dye-sensitized solar cells. *Chem. Soc. Rev.* **2013**, *42*, 3453–3488. [CrossRef] [PubMed]
8. Boschloo, G.; Hagfeldt, A. Characteristics of the iodide/triiodide redox mediator in dye-sensitized solar cells. *Acc. Chem. Res.* **2009**, *42*, 1819–1826. [CrossRef] [PubMed]
9. Susmitha, K.; Kumari, M.M.; Berkmans, A.J.; Kumar, M.N.; Giribabu, L.; Manorama, S.V.; Raghavender, M. Carbon nanohorns based counter electrodes developed by spray method for dye sensitized solar cells. *Sol. Energy* **2016**, *133*, 524–532. [CrossRef]
10. Iijima, S. Helical microtubules of graphitic carbon. *Nature* **1991**, *354*, 56–58. [CrossRef]
11. Wang, G.; Xing, W.; Zhuo, S. Nitrogen-doped graphene as low-cost counter electrode for high-efficiency dye-sensitized solar cells. *Electrochim. Acta* **2013**, *92*, 269–275. [CrossRef]
12. Yue, G.; Wu, J.; Xiao, Y.; Huang, M.; Lin, J.; Fan, L.; Lan, Z. Platinum/graphene hybrid film as a counter electrode for dye-sensitized solar cells. *Electrochim. Acta* **2013**, *92*, 64–70. [CrossRef]
13. Peng, Y.; Zhong, J.; Wang, K.; Xue, B.; Cheng, Y.-B. A printable graphene enhanced composite counter electrode for flexible dye-sensitized solar cells. *Nano Energy* **2013**, *2*, 235–240. [CrossRef]
14. Tai, Q.; Chen, B.; Guo, F.; Xu, S.; Hu, H.; Sebo, B.; Zhao, X.-Z. In situ prepared transparent polyaniline electrode and its application in bifacial dye-sensitized solar cells. *ACS Nano* **2011**, *5*, 3795–3799. [CrossRef] [PubMed]
15. Suzuki, K.; Yamaguchi, M.; Kumagai, M.; Yanagida, S. Application of carbon nanotubes to counter electrodes of dye-sensitized solar cells. *Chem. Lett.* **2003**, *32*, 28–29. [CrossRef]
16. Liu, X.; Liang, Y.; Yue, G.; Tu, Y.; Zheng, H. A dual function of high efficiency quasi-solid-state flexible dye-sensitized solar cell based on conductive polymer integrated into poly (acrylic acid-co-carbon nanotubes) gel electrolyte. *Sol. Energy* **2017**, *148*, 63–69. [CrossRef]
17. Kulicek, J.; Gemeiner, P.; Omastová, M.; Mičušík, M. Preparation of polypyrrole/multi-walled carbon nanotube hybrids by electropolymerization combined with a coating method for counter electrodes in dye-sensitized solar cells. *Chem. Pap.* **2018**, *72*, 1651–1667. [CrossRef]
18. Yun, D.J.; JinJeong, Y.; Ra, H.; Tae, J.-M.; An, K.; Rhee, S.-W.; Jang, J. Systematic optimization of MWCNT-PEDOT:PSS composite electrodes for organic transistors and dye-sensitized solar cells: Effects of MWCNT diameter and purity. *Org. Electron.* **2018**, *52*, 7–16. [CrossRef]
19. Lee, J.H.; Jang, Y.J.; Kim, D.W.; Cheruku, R.; Thogiti, S.; Ahn, K.-S.; Kim, J.H. Application of polypyrrole/sodium dodecyl sulfate/carbon nanotube counter electrode for solid-state dye-sensitized solar cells and dye-sensitized solar cells. *Chem. Pap.* **2019**, *73*, 2749–2755. [CrossRef]
20. Bumika, M.; Mallick, M.K.; Mohanty, S.; Nayak, S.K.; Palai, A.K. One-pot electrodeposition of polyaniline/SWCNT/ZnO film and its positive influence on photovoltaic performance as counter electrode material. *Mat. Lett.* **2020**, *279*, 128473. [CrossRef]
21. Rafique, S.; Rashid, I.; Sharif, R. Cost effective dye sensitized solar cell based on novel Cu polypyrrole multiwall carbon nanotubes nanocomposites counter electrode. *Sci. Rep.* **2021**, *11*, 14830. [CrossRef]
22. AbdulAlmohsin, S.M.; Khedhair, A.A.; Ajeel, S.K. Facile synthesis of polyaniline-single wall carbon nanotube nanocomposite as hole transport material and zinc oxide nanorodes as metal oxide to integration solid-state dye-sensitized solar cells. *Univ. Thi-Qar J.* **2019**, *14*, 1–13. [CrossRef]
23. Dawo, C.; Iyer, P.K.; Chaturvedi, H. Carbon nanotubes/PANI composite as an efficient counter electrode material for dye sensitized solar cell. *Mater. Sci. Eng. B* **2023**, *297*, 116722. [CrossRef]
24. Li, H.; Xiao, Y.; Han, G.; Li, M. Honeycomb-like polypyrrole/multi-wall carbon nanotube films as an effective counter electrode in bifacial dye-sensitized solar cells. *J. Mater. Sci.* **2017**, *52*, 8421–8431. [CrossRef]
25. Li, H.; Xiao, Y.; Han, G.; Hou, W. Honeycomb-like poly(3,4-ethylenedioxythiophene) as an effective and transparent counter electrode in bifacial dye-sensitized solar cells. *J. Power Sources* **2017**, *342*, 709–716. [CrossRef]
26. Li, Q.H.; Tang, Q.W.; Lin, L.; Chen, X.X.; Chen, H.Y.; Chu, L.; Xu, H.T.; Li, M.J.; Qin, Y.C.; He, B.L. A simple approach of enhancing photovoltaic performances of quasi-solid-state dye-sensitized solar cells by integrating conducting polyaniline into electrical insulating gel electrolyte. *J. Power Sources* **2014**, *245*, 468–474. [CrossRef]
27. Li, Q.H.; Chen, X.X.; Tang, Q.W.; Cai, H.Y.; Qin, Y.C.; He, B.L.; Li, M.J.; Jin, S.Y.; Liu, Z.C. Enhanced photovoltaic performances of quasi-solid-state dye sensitized solar cells using a novel conducting gel electrolyte. *J. Power Sources* **2014**, *248*, 923–930. [CrossRef]
28. Guo, H.F.; Zhu, H.; Lin, H.Y.; Zhang, J. Polypyrrole–multi-walled carbon nanotube nanocomposites synthesized in oil–water microemulsion. *Colloid Polym. Sci.* **2008**, *286*, 587–591. [CrossRef]
29. Ham, H.; Choi, Y.; Jeong, N.; Chung, I. Single wall carbon nanotubes covered with polypyrrole nanoparticles by the miniemulsion polymerization. *Polymer* **2005**, *46*, 6308–6315. [CrossRef]
30. Tang, Z.; Wu, J.; Li, Q.; Lan, Z.; Fan, L.; Lin, J.; Huang, M. The preparation of poly (glycidyl acrylate)–polypyrrole gel-electrolyte and its application in dye sensitized solar cells. *Electrochim. Acta* **2010**, *55*, 4883–4888. [CrossRef]

31. Mudiyanselage, T.; Neckers, D. Photochromic superabsorbent polymers. *Soft Matter* **2008**, *4*, 768–774. [CrossRef]
32. Zerbi, G.; Gussoni, M.; Castiglioni, C. Vibrational spectroscopy of polyconjugated aromatic materials with electrical and nonlinear optical properties. In *Conjugated Polymers*; Springer Netherlands: Dordrecht, The Netherlands, 1991; Volume 259, pp. 435–507. [CrossRef]
33. Peng, S.; Wu, Y.; Zhu, P.; Thavasi, V.; Mhaisalkar, S.G.; Ramakrishna, S. Facile fabrication of polypyrrole/functionalized multiwalled carbon nanotubes composite as counter electrodes in low-cost dye-sensitized solar cells. *J. Photochem. Photobiol. A Chem.* **2011**, *223*, 97–102. [CrossRef]
34. Chakraborty, G.; Gupta, K.; Meikap, A.K.; Babu, R.; Blau, W.J. Synthesis, electrical and magnetotransport properties of polypyrrole-MWCNT nanocomposite. *Solid State Commun.* **2012**, *152*, 13–18. [CrossRef]
35. Ghani, S.; Sharif, R.; Bashir, S.; Zaidi, A.A.; Rafique, M.S.; Ashraf, A.; Shahzadi, S.; Shaista, R.; Kamboh, A.H. Polypyrrole thin films decorated with copper nanostructures as counter electrode for dye-sensitized solar cells. *J. Power Sources* **2015**, *282*, 416–420. [CrossRef]
36. He, B.; Tang, Q.; Luo, J.; Li, Q.; Chen, X.; Cai, H. Rapid charge transfer I polypyrrole single wall carbon nanotube complex counter electrodes: Improved photovoltaic performances of dye-sensitized solar cells. *J. Power Sources* **2014**, *256*, 170–177. [CrossRef]
37. Hlura, H.; Ebbesen, T.W.; Tanigaki, T.; Takahashi, H. Raman studies of carbon nanotubes. *Chem. Phys. Lett.* **1993**, *202*, 509–512. [CrossRef]
38. de Heer, W.A.; Bacsa, W.S.; Chatelain, A.; Gerfin, T.; Humphrey-Baker, R.; Forro, L.; Ugarte, D. Aligned carbon nanotube films: Production and optical and electronic properties. *Science* **1995**, *268*, 845–847. [CrossRef] [PubMed]
39. Liu, J.S.; Tanaka, T.; Sivula, K.; Alivisatos, A.P.; Fréchet, J.M.J. Employing End-Functional Polythiophene to Control the Morphology of Nanocrystal–Polymer Composites in Hybrid Solar Cells. *J. Am. Chem. Soc.* **2004**, *126*, 6550–6551. [CrossRef]
40. Liu, Y.-C.; Hwang, B.-J.; Jian, W.-J.; Santhanam, R. In situ cyclic voltammetry-surface-enhanced Raman spectroscopy: Studies on the doping–undoping of polypyrrole film. *Thin Solid Film.* **2000**, *374*, 85–91. [CrossRef]
41. Zhang, D.W.; Li, X.D.; Chen, S.; Tao, F.; Sun, Z.; Yin, X.J.; Huang, S.M. Fabrication of double-walled carbon nanotube counter electrodes for dye-sensitized solar cells. *J. Solid State Electrochem.* **2010**, *14*, 1541–1546. [CrossRef]
42. Siuzdak, K.; Klein, M.; Sawczak, M.; Wroblewski, G.; Sloma, M.; Jakubowska, M.; Cenian, A. Spray-deposited carbon-nanotube counter electrodes for dye-sensitized solar cells. *Phys. Status Solidi A* **2016**, *213*, 1157–1164. [CrossRef]
43. Tison, Y.; Giusca, C.E.; Stolojan, V.; Hayashi, Y.; Silva, S.R.P. The inner shell influence on the electronic structure of double-walled carbon nanotubes. *Adv. Mater.* **2008**, *20*, 189–194. [CrossRef]
44. Yang, S.; Huo, J.; Song, H.; Chen, X. A comparative study of electrochemical properties of two kinds of carbon nanotubes as anode materials for lithium ion batteries. *Electrochim. Acta* **2008**, *53*, 2238–2244. [CrossRef]
45. Li, W.Z.; Wen, J.G.; Sennett, M.; Ren, Z.F. Clean double-walled carbon nanotubes synthesized by CVD. *Chem. Phys. Lett.* **2003**, *368*, 299–306. [CrossRef]
46. Belin, T.; Epron, F. Characterization methods of carbon nanotubes: A review. *Mat. Sci. Eng. B* **2005**, *119*, 105–118. [CrossRef]
47. Peigney, A.; Laurent, C.; Flahaut, E.; Bacsa, R.R.; Rousset, A. Specific surface area of carbon nanotubes and bundles of carbon nanotubes. *Carbon* **2001**, *39*, 507–514. [CrossRef]
48. Fan, B.; Mei, X.; Sun, K.; Ouyang, J. Conducting polymer/carbon nanotube composite as counter electrode of dye sensitized solar cells. *Appl. Phys. Lett.* **2008**, *93*, 143103. [CrossRef]
49. Duchet, J.; Legras, R. Demoustier-Champagne, S. Chemical synthesis of polypyrrole: Structure–properties relationship. *Synth. Met.* **1998**, *98*, 113–122. [CrossRef]
50. Gonçalves, A.B.; Mangrich, A.S.; Zarbin, A.J.G. Polymerization of pyrrole between the layers of α-Tin (IV) Bis (hydrogenphosphate). *Synth. Met.* **2000**, *114*, 119–124. [CrossRef]
51. Xiao, Y.M.; Wu, J.H.; Lin, J.Y.; Yue, G.T.; Lin, J.M.; Huang, M.L.; Lan, Z.; Fan, L.Q. A dual function of high performance counter-electrode for stable quasi-solidstate dye-sensitized solar cells. *J. Power Sources* **2013**, *241*, 373–378. [CrossRef]
52. Yue, G.T.; Wu, J.H.; Xiao, Y.M.; Lin, J.M.; Huang, M.L.; Lan, Z. Application of poly (3,4-ethylenedioxythiophene): Polystyrenesulfonate/polypyrrole counter electrode for dye-sensitized solar cells. *J. Phys. Chem. C* **2012**, *116*, 18057–18063. [CrossRef]
53. Hou, W.; Xiao, Y.; Han, G.; Zhou, H. Electro-polymerization of polypyrrole/multi wall carbon nanotube counter electrodes for use in platinum-free dye-sensitized solar cells. *J. Mater. Sci.* **2017**, *190*, 720–728. [CrossRef]
54. Li, H.; Xiao, Y.; Han, G.; Zhang, Y. A transparent honeycomb-like poly(3,4-ethylenedioxythiophene)/multi-wall carbon nanotube counter electrode for bifacial dye sensitized solar cells. *Org. Electron.* **2017**, *50*, 161–169. [CrossRef]
55. Aitola, K.; Borghei, M.; Kaskela, A.; Kemppainen, E.; Nasibulin, A.G.; Kauppinen, E.I.; Lund, P.D.; Ruiz, V.; Halme, J. Flexible metal-free counter electrode for dye solar cells based on conductive polymer and carbon nanotubes. *J. Electroanal. Chem.* **2012**, *683*, 70–74. [CrossRef]
56. Hernández-Ferrer, J.; Ansón-Casaos, A.; Martínez, M.T. Electrochemical synthesis and characterization of single-walled carbon nanotubes/polypyrrole films on transparent substrates. *Electrochim. Acta* **2012**, *64*, 1–9. [CrossRef]
57. Lee, K.S.; Lee, Y.; Lee, J.Y.; Ahn, J.-H.; Park, J.H. Flexible and Platinum-Free dye-sensitized solar cells with conducting-polymer-coated graphene counter electrodes. *ChemSusChem* **2012**, *5*, 379–382. [CrossRef] [PubMed]
58. Chawarambwa, F.L.; Putri, T.E.; Son, M.-K.; Kamataki, K.; Itagaki, N.; Koga, K.; Shiratani, M. Graphene-S3N4 nanocomposite blended polymer counter electrode for low-cost dye-sensitized solar cells. *Chem. Phys. Lett.* **2020**, *758*, 137920. [CrossRef]

59. Sudhakar, V.; Singh, A.K.; Chini, M.K. Nanoporous reduced graphene oxide and polymer composites as efficient counter electrodes in dye-sensitized solar cells. *ACS Appl. Electron. Mater.* **2020**, *2*, 626–634. [CrossRef]
60. Mehmood, U.; Karim, N.A.; Zahid, H.F.; Asif, T.; Younas, M. Polyaniline/graphene nanocomposites as counter electrode materials for platinum free dye-sensitized solar cells (DSSCSs). *Mater. Lett.* **2019**, *256*, 126651. [CrossRef]
61. Loryuenyong, V.; Khadthiphong, A.; Phinkratok, J.; Watwittayakul, J.; Supawattanakul, W.; Buasri, A. The fabrication of graphene-polypyrrole composite for application with dye-sensitized solar cells. *Mater. Today Proc.* **2019**, *17*, 1675–1681. [CrossRef]
62. Mohan, K.; Bora, A.; Roy, R.S.; Nath, B.C.; Dolui, S.K. Polyaniline nanotube/reduced graphene oxide aerogel as efficient counter electrode for quasi-solid state dye sensitized solar cell. *Sol. Energy* **2019**, *186*, 360–369. [CrossRef]
63. Saranya, K.; Rameez, M.; Subrania, A. Developments in conducting polymer based counter electrodes for dye-sensitized solar cells–an overview. *Eur. Polym. J.* **2015**, *66*, 207–227. [CrossRef]
64. Gunasekera, S.S.B.; Perera, I.R.; Gunathilaka, S.S. Conducting Polymers as Cost Effective Counter Electrode Material in Dye-Sensitized Solar Cells. In *Solar Energy*; Springer: Berlin/Heidelberg, Germany, 2020; pp. 345–371. [CrossRef]
65. Hou, W.; Xiao, Y.; Han, G.; Lin, J.-Y. The applications of polymers in solar cells: A review. *Polymers* **2019**, *11*, 143. [CrossRef] [PubMed]
66. Shahid, M.U.; Mohamed, N.M.; Muhsan, A.S.; Bashiri, R.; Shamsudin, A.E.; Zaine, S.N.A. Few-layer graphene supported polyaniline (PANI) film as a transparent counter electrode for dye-sensitized solar cells. *Diam. Relat. Mater.* **2019**, *94*, 242–251. [CrossRef]
67. Rana, U.; Malik, S. Graphene oxide/polyaniline nanostructures: Transformation of 2 D sheet to 1 D nanotube and in situ reduction. *Chem. Commun.* **2012**, *48*, 10862–10864. [CrossRef] [PubMed]
68. Dissanayake, M.A.K.L.; Kumari, J.M.K.W.; Senadeera, G.K.R.; Anwar, H. Low cost, platinum free counter electrode with reduced graphene oxide and polyaniline embedded SnO2 for efficient dye sensitized solar cells. *Sol. Energy* **2021**, *230*, 151–165. [CrossRef]
69. Nechiyil, D.; Vinayan, B.; Ramaprabhu, S. Tri-iodide reduction activity of ultra-small size PtFe nanoparticles supported nitrogen-doped graphene as counter electrode for dye-sensitized solar cell. *J. Colloid Interface Sci.* **2017**, *488*, 309–316. [CrossRef]
70. Rehim, M.H.A.; Youssef, A.M.; Al-Said, H.; Turky, G.; Aboaly, M. Polyaniline and modified titanate nanowires layer-by-layer plastic electrode for flexible electronic device applications. *RSC Adv.* **2016**, *6*, 94556–94563. [CrossRef]
71. Borah, R.; Banerjee, S.; Kumar, A. Surface functionalization effects on structural, conformational, and optical properties of polyaniline nanofibers. *Synth. Met.* **2014**, *197*, 225–232. [CrossRef]
72. Xu, C.; Shi, X.; Ji, A.; Shi, L.; Zhou, C.; Cui, Y. Fabrication and characteristics of reduced graphene oxide produced with different green reductants. *PLoS ONE* **2015**, *10*, e0144842. [CrossRef]
73. Zhang, Y.; Liu, J.; Zhang, Y.; Liu, J.; Duan, Y. Facile synthesis of hierarchical nanocomposites of aligned polyaniline nanorods on reduced graphene oxide nanosheets for microwave absorbing materials. *RSC Adv.* **2017**, *7*, 54031–54038. [CrossRef]
74. Mitra, M.; Kulsi, C.; Chatterjee, K.; Kargupta, K.; Ganguly, S.; Banerjee, D.; Goswami, S. Reduced graphene oxide-polyaniline composites-Synthesis, characterization and optimization for thermoelectric applications. *RSC Adv.* **2015**, *5*, 31039–31048. [CrossRef]
75. Lim, S.P.; Pandikumar, A.; Lim, Y.S.; Muang, N.; Lim, H.N. In-situ electrochemically deposited polypyrrole nanoparticles incorporated reduced graphene oxide as an efficient counter electrode for platinum-free dye-sensitized solar cells. *Sci. Rep.* **2014**, *4*, 5305. [CrossRef] [PubMed]
76. Xu, Y.; Bai, H.; Lu, G.; Li, C.; Shi, G. Flexible graphene films via the filtration of water-soluble noncovalent functionalized graphene sheets. *J. Am. Chem. Soc.* **2008**, *130*, 5856–5857. [CrossRef] [PubMed]
77. Xue, Y.; Liu, J.; Chen, H.; Wang, R.; Li, D.; Qu, J.; Dai, L. Nitrogen-doped graphene foams as metal-free counter electrodes in high-performance dye-sensitized solar cells. *Angew. Chem.-Int. Ed.* **2012**, *51*, 12124–12127. [CrossRef] [PubMed]
78. Yang, Y.; Zhao, B.; Tang, P.; Cao, Z.; Huang, M.; Tan, S. Flexible counter electrodes based on nitrogen-doped carbon aerogels with tunable pore structure for high-performance dye-sensitized solar cells. *Carbon N. Y.* **2014**, *77*, 113–121. [CrossRef]
79. Diantoro, M.; Kholid, M.; Yudiyanto, A.A. The Influence of SnO2 Nanoparticles on Electrical Conductivity, and Transmittance of PANI-SnO2 Films. *IOP Conf. Ser. Mater. Sci. Eng.* **2018**, *367*, 012034. [CrossRef]
80. Biswas, S.; Bhattacharya, S. Influence of SnO2 nanoparticles on the relaxation dynamics of the conductive processes in polyaniline. *Phys. Lett. Sect. A Gen. At. Solid State Phys.* **2017**, *381*, 3424–3430. [CrossRef]
81. Shahid, M.U.; Mohamed, N.M.; Muhsan, A.S.; Khatani, M.; Bashiri, R.; Zaine, S.N.A.; Shamsudin, A.E. Dual function passivating layer of graphene/TiO2 for improved performance of dye-sensitized solar cells. *Appl. Nanosci.* **2018**, *8*, 1001–1013. [CrossRef]
82. Sudhagar, P.; Nagarajan, S.; Lee, Y.-G.; Song, D.; Son, T.; Cho, W.; Heo, M.; Lee, K.; Won, J.; Kang, Y.S. Synergistic Catalytic Effect of a Composite (CoS/PEDOT: PSS) Counter Electrode on Triiodide Reduction in Dye-Sensitized Solar Cells. *ACS Appl. Mater. Interfaces* **2011**, *3*, 1838–1843. [CrossRef]
83. Li, Q.; Wu, J.; Tang, Q.; Lan, Z.; Li, P.; Lin, J.; Fan, L. Application of Microporous Polyaniline Counter Electrode for Dye-Sensitized Solar Cells. *Electrochem. Commun.* **2008**, *10*, 1299–1302. [CrossRef]
84. Peng, S.; Tian, L.; Liang, J.; Mhaisalkar, S.G.; Ramakrishna, S. Polypyrrole Nanorod Networks/Carbon Nanoparticles Composite Counter Electrodes for High-Efficiency Dye-Sensitized Solar Cells. *ACS Appl. Mater. Interfaces* **2012**, *4*, 397–404. [CrossRef] [PubMed]
85. Fathi, M.; Saghafi, M.; Mahboubi, F.; Mohajerzadeh, S. Synthesis and electrochemical investigation of polyaniline/unzipped carbon nanotube composites as electrode material in supercapacitors. *Synth. Met.* **2014**, *198*, 345–356. [CrossRef]

86. Jain, D.; Hashmi, S.A.; Kaur, A. Surfactant assisted polyaniline nanofibers-reduced graphene oxide (SPG) composite as electrode material for supercapacitors with high rate performance. *Electrochim. Acta* **2016**, *222*, 570–579. [CrossRef]
87. Mondal, S.; Rana, U.; Malik, S. Reduced Graphene Oxide/Fe_3O_4/Polyaniline Nanostructures as Electrode Materials for All-Solid-State Hybrid Supercapacitor. *J. Phys. Chem. C* **2017**, *121*, 7573–7583. [CrossRef]
88. Zhao, X.; Gnanaseelan, M.; Jehnichen, D.; Simon, F.; Jurgen, P. Green and facile synthesis of polyaniline/tannic acid/rGO composites for supercapacitor purpose. *J. Mater. Sci.* **2019**, *54*, 10809–10824. [CrossRef]
89. Wang, Z.; Zhang, Q.; Long, S.; Luo, Y.; Yu, P.; Tan, Z.; Bai, J.; Qu, B.; Yang, Y.; Shi, J.; et al. Three-Dimensional Printing of Polyaniline/Reduced Graphene Oxide Composite for High-Performance Planar Supercapacitor. *ACS Appl. Mater. Interfaces* **2018**, *10*, 10437–10444. [CrossRef]
90. Liu, Z.; Li, D.; Li, Z.; Liu, Z.; Zhang, Z. Nitrogen-doped 3D reduced graphene oxide/polyaniline composite as active material for supercapacitor electrodes. *Appl. Surf. Sci.* **2017**, *422*, 339–347. [CrossRef]
91. Meriga, V.; Valligatla, S.; Sundaresan, S.; Cahill, C.; Dhanak, V.R.; Chakraborty, A.K. Optical, electrical, and electrochemical properties of graphene based water soluble polyaniline composites. *J. Appl. Polym. Sci.* **2015**, *132*, 42766. [CrossRef]
92. Li, Y.; Zheng, Y. Preparation and electrochemical properties of polyaniline/reduced graphene oxide composites. *J. Appl. Polym. Sci.* **2018**, *135*, 46103. [CrossRef]
93. Rashti, A.; Wang, B.; Hassani, E.; Feyzbar-Khalkhali-Nejad, F.; Zhang, X.; Tae-Sik, O. Electrophoretic Deposition of Nickel Cobaltite/Polyaniline/rGO Composite Electrode for High-Performance All-Solid-State Asymmetric Supercapacitors. *Energy Fuels* **2020**, *34*, 6448–6461. [CrossRef]
94. Dong, C.; Zhang, X.; Yu, Y.; Huang, L.; Li, J.; Wu, Y.; Liu, Z. An ionic liquid-modified RGO/polyaniline composite for high-performance flexible all-solid-state supercapacitors. *Chem. Comm.* **2020**, *56*, 11993–11996. [CrossRef] [PubMed]
95. Almeida, D.A.L.; Couto, A.B.; Ferreira, N.G. Flexible polyaniline/reduced graphene oxide/carbon fiber composites applied as electrodes for supercapacitors. *J. Alloys Compd.* **2019**, *788*, 453–460. [CrossRef]
96. Moyseowicz, A.; Gryglewicz, G. Hydrothermal-assisted synthesis of a porous polyaniline/reduced graphene oxide composite as a high-performance electrode material for supercapacitors. *Compos. Part B* **2019**, *159*, 4–12. [CrossRef]
97. Liu, D.; Du, P.; Wei, W.; Wang, H.; Wang, Q.; Liu, P. Skeleton/skin structured (RGO/CNTs)@PANI composite fiber electrodes with excellent mechanical and electrochemical performance for all-solid-state symmetric supercapacitors. *J. Colloid Interface Sci.* **2018**, *513*, 295–303. [CrossRef] [PubMed]
98. Wu, J.; Zhang, Q.; Wang, J.; Huang, X.; Bai, H. A self-assembly route to porous polyaniline/reduced graphene oxide composite materials with molecular-level uniformity for high-performance supercapacitors. *Energy Environ. Sci.* **2018**, *11*, 1280–1286. [CrossRef]
99. Chen, N.; Ren, Y.; Kong, P.; Tan, L.; Feng, H.; Luo, Y. In situ one-pot preparation of reduced graphene oxide/polyaniline composite for high-performance electrochemical capacitors. *Appl. Surf. Sci.* **2017**, *392*, 71–79. [CrossRef]
100. Kumar, N.A.; Choi, H.-J.; Shin, Y.R.; Chang, D.W.; Dai, L.; Baek, J.-B. Polyaniline-Grafted Reduced Graphene Oxide for Efficient Electrochemical Supercapacitors. *ACS Nano* **2012**, *6*, 1715–1723. [CrossRef]
101. Hsu, H.H.; Khosrozadeh, A.; Li, B.; Luo, G.; Xing, M.; Zhong, W. An Eco-Friendly, Nanocellulose/RGO/in Situ Formed Polyaniline for Flexible and Free-Standing Supercapacitors. *ACS Sustain. Chem. Eng.* **2019**, *7*, 4766–4776. [CrossRef]
102. Wang, K.; Huang, J.; Wei, Z. Conducting Polyaniline Nanowire Arrays for High Performance Supercapacitors. *J. Phys. Chem. C* **2010**, *114*, 8062–8067. [CrossRef]
103. Lee, S.-Y.; Kim, J.-I.; Park, S.-J. Activated carbon nanotubes/polyaniline composites as supercapacitor electrodes. *Energy* **2014**, *78*, 298–303. [CrossRef]
104. Malik, R.; Zhang, L.; McConnell, C.; Schott, M.; Hsieh, Y.-Y.; Noga, R.; Alvarez, N.T.; Shanov, V. Three-dimensional, free-standing polyaniline/carbon nanotube composite-based electrode for high-performance supercapacitors. *Carbon* **2017**, *116*, 579–590. [CrossRef]
105. Ryu, K.S.; Kim, K.M.; Park, N.-G.; Park, Y.J.; Chang, S.H. Symmetric redox supercapacitor with conducting polyaniline electrodes. *J. Power Sources* **2002**, *103*, 305–309. [CrossRef]
106. Xiong, S.; Wei, J.; Jia, P.; Yang, L.; Ma, J.; Lu, X. Water-processable polyaniline with covalently bonded single-walled carbon nanotubes: Enhanced electrochromic properties and impedance analysis. *ACS Appl. Mater. Interfaces* **2011**, *3*, 782–788. [CrossRef] [PubMed]
107. Che, B.; Li, H.; Zhou, D.; Zhang, Y.; Zeng, Z.; Zhao, C.; He, C.; Liu, E.; Lu, X. Porous polyaniline/carbon nanotube composite electrode for supercapacitors with outstanding rate capability and cyclic stability. *Compos. Part B* **2019**, *165*, 671–678. [CrossRef]
108. Li, Y.; Louarn, G.; Aubert, P.H.; Alain-Rizzo, V.; Galmiche, L.; Audebert, P.; Miomandre, F. Polypyrrole-Modified Graphene Sheet Nanocomposites as New Efficient Materials for Supercapacitors. *Carbon* **2016**, *105*, 510–520. [CrossRef]
109. Jiang, L.I.; Lu, X.; Xie, C.M.; Wan, G.J.; Zhang, H.P.; Youhong, T. Flexible, Free-Standing TiO2–Graphene–Polypyrrole Composite Films as Electrodes for Supercapacitors. *J. Phys. Chem. C* **2015**, *119*, 3903–3910. [CrossRef]
110. Mao, L.; Zhang, K.; Chan, H.S.O.; Wu, J. Surfactant—stabilized graphene/polyaniline nanofiber composites for high performance supercapacitors. *J. Mater. Chem.* **2012**, *22*, 80. [CrossRef]
111. Yu, A.; Chabot, V.; Zhang, J. *Electrochemical Supercapacitors for Energy Storage and Delivery, Fundamentals and Applications*; CRC Press: London, UK; New York, NY, USA, 2013. [CrossRef]

112. Wang, L.; Ye, Y.; Lu, X.; Wen, Z.; Li, Z.; Hou, H.; Song, Y. Hierarchical Nanocomposites of Polyaniline Nanowire Arrays on Reduced Graphene Oxide Sheets for Supercapacitors. *Sci. Rep.* **2013**, *3*, 3568. [CrossRef]
113. Anothumakkool, B.; Torris, A.T.A.; Bhange, S.N.; Unni, S.M.; Badiger, M.V.; Kurungot, S. Design of a High Performance Thin All-Solid-State Supercapacitor Mimicking the Active Interface of Its Liquid-State Counterpart. *ACS Appl. Mater. Interfaces* **2013**, *5*, 13397–13404. [CrossRef]
114. Van Nguyen, H.; Lamiel, C.; Kharismadewi, D.; Van Tran, C.; Shim, J.J. Covalently bonded reduced graphene oxide/polyaniline composite for electrochemical sensors and capacitors. *J. Electroanal. Chem.* **2015**, *758*, 148–155. [CrossRef]
115. Du, X.; Wang, C.; Chen, M.; Jiao, Y.; Wang, J. Electrochemical Performances of Nanoparticle Fe_3O_4/Activated Carbon Supercapacitor Using KOH Electrolyte Solution. *J. Phys. Chem. C* **2009**, *113*, 2643–2646. [CrossRef]
116. Meng, Y.; Wang, K.; Zhang, Y.; Wei, Z. Hierarchical Porous Graphene/Polyaniline Composite Film with Superior Rate Performance for Flexible Supercapacitors. *Adv. Mater.* **2013**, *25*, 6985–6990. [CrossRef] [PubMed]
117. Wu, Q.; Chen, M.; Wang, S.; Zhang, X.; Huan, L.; Diao, G. Preparation of sandwich-like ternary hierarchical nanosheets manganese dioxide/polyaniline/reduced graphene oxide as electrode material for supercapacitor. *Chem. Eng. J.* **2016**, *304*, 29–38. [CrossRef]
118. Asen, P.; Shahrokhian, S. A High Performance Supercapacitor Based on Graphene/Polypyrrole/Cu_2O–$Cu(OH)_2$ Ternary Nanocomposite Coated on Nickel Foam. *J. Phys. Chem C* **2017**, *121*, 6508–6519. [CrossRef]
119. Ramana, G.V.; Srikanth, V.V.S.S.; Padya, B.; Jain, P.K. Carbon nanotube-polyaniline nanotube core-shell structures for electrochemical applications. *Eur. Polym. J.* **2014**, *57*, 137–142. [CrossRef]
120. Downs, C.; Nugent, J.; Ajayan, P.M.; Duquette, D.J.; Santhanam, K.S.V. Efficient Polymerization of Aniline at Carbon Nanotube Electrodes. *Adv. Mater.* **1999**, *11*, 1028–1031. [CrossRef]
121. Haq, A.U.; Lim, J.; Yun, J.M.; Lee, W.J.; Han, T.H.; Kim, S.O. Direct Growth of polyaniline chains from N-Doped sites of carbon nanotubes. *Small* **2013**, *9*, 3829–3833. [CrossRef]
122. Yu, M.; Ma, Y.; Liu, J.; Li, S. Polyaniline nanocone arrays synthesized on three-dimensional graphene network by electrodeposition for supercapacitor electrodes. *Carbon* **2015**, *87*, 98–105. [CrossRef]
123. Liu, Y.; Ma, Y.; Guang, S.; Ke, F.; Xu, H. Polyaniline-graphene composites with a three-dimensional array-based nanostructure for high-performance super-capacitors. *Carbon N.Y.* **2015**, *83*, 79–89. [CrossRef]
124. Huang, J.; Wang, K.; Wei, Z. Conducting polymer nanowire arrays with enhanced electrochemical performance. *J. Mater. Chem.* **2010**, *20*, 1117–1121. [CrossRef]
125. Li, H.; Lu, X.; Yuan, D.; Sun, J.; Erden, F.; Wang, F.; He, C. Lightweight flexible carbon nanotube/polyaniline films with outstanding EMI shielding properties. *J. Mater. Chem. C* **2017**, *5*, 8694–8698. [CrossRef]
126. Zhou, H.; Zhi, X.; Zhai, H.-J. A facile approach to improve the electrochemical properties of polyaniline-carbon nanotube composite electrodes for highly flexible solid-state supercapacitors. *Int. J. Hydrogen Energy* **2018**, *43*, 18339–18348. [CrossRef]
127. Zhou, W.; Sasaki, S.; Kawasaki, A. Effective control of nanodefects in multiwalled carbon nanotubes by acid treatment. *Carbon* **2014**, *78*, 121–129. [CrossRef]
128. Chang, C.-M.; Weng, C.-J.; Chien, C.M.; Chuang, T.L.; Lee, T.-Y.; Yeh, J.M.; Wei, Y. Polyaniline/carbon nanotube nanocomposite electrodes with biomimetic hierarchical structure for supercapacitors. *J. Mater. Chem. A* **2013**, *1*, 14719–14728. [CrossRef]
129. Zheng, L.; Wang, X.; An, H.; Wang, X.; Yi, L.; Bai, L. The preparation and performance of flocculent polyaniline/carbon nanotubes composite electrode material for supercapacitors. *J. Solid State Electrochem.* **2011**, *15*, 675–681. [CrossRef]
130. Kosynkin, D.V.; Higginbotham, A.L.; Sinitskii, A.; Lomeda, J.R.; Dimiev, A.; Price, B.K.; Tour, J.M. Longitudinal unzipping of carbon nanotubes to form graphene nanoribbons. *Nature* **2009**, *458*, 872–876. [CrossRef] [PubMed]
131. Cano-Marquez, A.G.; Rodriguez-Macias, F.J.; Campos-Delgado, J.; Espinosa-Gonzales, C.G.; Tristan-Lopez, F.; Ramirez-Gonzalez, D.; Cullen, D.A.; Smith, D.J.; Terrones, M.; Vega-Cantu, Y.I. Ex-MWNTs:graphene sheets and ribbons produced by lithium intercalation and exfoliation of carbon nanotubes. *Nano Lett.* **2009**, *9*, 1527–1533. [CrossRef] [PubMed]
132. Elías, A.L.; Botello-Méndez, A.R.; Meneses-Rodríguez, D.; González, V.J.; Ramírez-González, D.; Ci, L.; Muñoz-Sandoval, E.; Ajayan, P.M.; Terrones, H.; Terrones, M. Longitudinal Cutting of Pure and Doped Carbon Nanotubes to Form Graphitic Nanoribbons Using Metal Clusters as Nanoscalpels. *Nano Lett.* **2010**, *10*, 366–372. [CrossRef]
133. Kim, K.; Sussman, A.; Zettl, A. Graphene Nanoribbons Obtained by Electrically Unwrapping Carbon Nanotubes. *ACS Nano* **2010**, *4*, 1362–1366. [CrossRef]
134. Jiao, L.; Zhang, L.; Wang, X.; Diankov, G.; Dai, H. Narrow graphene nanoribbons from carbon nanotubes. *Nature* **2009**, *458*, 877–880. [CrossRef]
135. Sivakkumar, S.R.; Kim, W.J.; Choi, J.-A.; MacFarlane, D.R.; Forsyth, M.; Kim, D.-W. Electrochemical performance of polyaniline nanofibers and polyaniline/multi-walled carbon nanotube composite as an electrode material for aqueous redox supercapacitors. *J. Power Sources* **2007**, *171*, 1062–1068. [CrossRef]
136. Sarker, S.; Lee, K.-S.; Seo, H.W.; Jin, Y.-K.; Kim, D.M. Reduced graphene oxide for Pt-free counter electrodes of dye-sensitized solar cells. *Sol. Energy* **2017**, *158*, 42–48. [CrossRef]
137. Wickramaarachchi, K.; Sundaram, M.M.; Henry, D.J.; Gao, X. Alginate Biopolymer Effect on the electrodeposition of manganese dioxide on electrodes for supercapacitors. *ACS Appl. Energy Mater.* **2021**, *4*, 7040–7051. [CrossRef]
138. Ramkumar, R.; Minakshi, M. Fabrication of ultrathin $CoMoO_4$ nanosheets modified with chitosan and their improved performance in energy storage device. *Dalton Trans.* **2015**, *44*, 6158–6168. [CrossRef] [PubMed]

139. Xia, X.; Lei, W.; Hao, Q.; Wang, W.; Wang, X. One-step synthesis of $CoMoO_4$/graphene composites with enhanced electrochemical properties for supercapacitors. *Electrochim. Acta* **2013**, *99*, 253–261. [CrossRef]
140. Guo, D.; Zhang, H.; Yu, X.; Zhang, M.; Zhang, P.; Li, Q.; Wang, T. Facile synthesis and excellent electrochemical properties of $CoMoO_4$ nanoplate arrays as supercapacitors. *J. Mater. Chem. A* **2013**, *1*, 7247–7254. [CrossRef]
141. Zhang, Q.; Deng, Y.; Hu, Z.; Liu, Y.; Yao, M.; Liu, P. Seaurchin-like hierarchical $NiCo_2O_4$@$NiMoO_4$ core–shell nanomaterials for high performance supercapacitors. *Phys. Chem. Chem. Phys.* **2014**, *16*, 23451–23460. [CrossRef]
142. Ghosh, D.; Giri, S.; Moniruzzaman, M.; Basu, T.; Mandal, M.; Das, C.K. α $MnMoO_4$/graphene hybrid composite: High energy density supercapacitor electrode material. *Dalton Trans.* **2014**, *43*, 11067–11076. [CrossRef]
143. Dambies, L.; Vincent, T.; Domard, A.; Guibal, E. Preparation of Chitosan Gel Beads by Ionotropic Molybdate Gelation. *Biomacromolecules* **2001**, *2*, 1198–1205. [CrossRef]

Disclaimer/Publisher's Note: The statements, opinions and data contained in all publications are solely those of the individual author(s) and contributor(s) and not of MDPI and/or the editor(s). MDPI and/or the editor(s) disclaim responsibility for any injury to people or property resulting from any ideas, methods, instructions or products referred to in the content.

Article

Electrically Conductive Adhesive Based on Thermoplastic Hot Melt Copolyamide and Multi-Walled Carbon Nanotubes

Paulina Latko-Durałek [1,2,*], Michał Misiak [1] and Anna Boczkowska [1]

[1] Faculty of Materials Science and Engineering, Warsaw University of Technology, Wołoska 141 Street, 02-507 Warsaw, Poland
[2] Technology Partners Foundation, Pawińskiego 5A Street, 02-106 Warsaw, Poland
* Correspondence: paulina.latko@technologypartners.pl

Abstract: For the bonding of the lightweight composite parts, it is desired to apply electrically conductive adhesive to maintain the ability to shield electromagnetic interference. Among various solvent-based adhesives, there is a new group of thermoplastic hot melt adhesives that are easy to use, solidify quickly, and are environment-friendly. To make them electrically conductive, a copolyamide-based hot melt adhesive was mixed with 5 and 10 wt% of carbon nanotubes using a melt-blending process. Well-dispersed nanotubes, observed by a high-resolution scanning microscope, led to the formation of a percolated network at both concentrations. It resulted in the electrical conductivity of 3.38 S/m achieved for 10 wt% with a bonding strength of 4.8 MPa examined by a lap shear test. Compared to neat copolyamide, Young's modulus increased up to 0.6 GPa and tensile strength up to 30.4 MPa. The carbon nanotubes improved the thermal stability of 20 °C and shifted the glass transition of 10 °C to a higher value. The very low viscosity of the neat adhesive increased about 5–6 orders of magnitude at both concentrations, even at elevated temperatures. With a simultaneous growth in storage and loss modulus this indicates the strong interactions between polymer and carbon nanotubes.

Keywords: hot melt adhesive; carbon nanotubes; adhesion; rheology; microstructure

Citation: Latko-Durałek, P.; Misiak, M.; Boczkowska, A. Electrically Conductive Adhesive Based on Thermoplastic Hot Melt Copolyamide and Multi-Walled Carbon Nanotubes. *Polymers* **2022**, *14*, 4371. https://doi.org/10.3390/polym14204371

Academic Editor: Xia Dong

Received: 1 September 2022
Accepted: 11 October 2022
Published: 17 October 2022

Publisher's Note: MDPI stays neutral with regard to jurisdictional claims in published maps and institutional affiliations.

Copyright: © 2022 by the authors. Licensee MDPI, Basel, Switzerland. This article is an open access article distributed under the terms and conditions of the Creative Commons Attribution (CC BY) license (https://creativecommons.org/licenses/by/4.0/).

1. Introduction

Electrically conductive adhesives (ECAs) are a group of materials developed to be used as adhesives in electronics or the aviation and automotive industry as a bonding medium of thermoplastic or thermosetting matrix composites. In the first application, the ECAs can replace the traditional Pb-Sn solder used to assemble the components of printed circuit boards. The high interest in ECAs in electronic packaging is due to easier processing and higher resolution printing, lower processing temperature, and environmental friendliness compared to Pb-Sn soldiers [1]. In the second application area, ECAs can improve electromagnetic shielding properties or lightning strike protection by forming the interlayer between the joining lightweight composites used in automotive or aircraft sectors. Here, the ECAs are a promising candidate for eliminating commonly used rivets resulting in lower stress concentration and weight of the final parts [2].

The main components of ECAs are polymer and electrically conductive fillers. The polymer matrix consists of both thermoplastics like polyimide and, mainly, thermosets such as silicone, epoxy, or acrylate resins, and they are responsible for the mechanical properties of the adhesive layer [3]. Because polymers (excluding conductive polymers) are insulators, they need to be modified with electrically conductive fillers, which provide a sufficient level of conductivity and do not decrease the overall mechanical performance. Of the broad range of conductive fillers, micro silver is most commonly used in the form of powder, flakes, spheres, nanowires, or dendrites [4]. However, to achieve a sufficient level of conductivity, it is necessary to add a high amount of silver–between 25 wt% and even

80 wt%. Such an amount causes difficulties during processing, high cost, and impairs the adhesive layer's mechanical properties. Moreover, the electrical conductivity diminishes as an effect of corrosion and oxidation of the metallic filler and localization of the charge carriers [5,6]. To overcome these weaknesses, the new approach focuses on applying carbon-based fillers or nanofillers. The best-known examples include carbon nanotubes (CNTs), graphene, carbon black, and graphite. Due to their high surface area, they form a conductive network in the polymer at lower concentrations than metal fillers [7]. Moreover, they are light and not corroding, and their price is decreasing yearly owing to increased production. However, the critical challenge for the ECAs containing carbonaceous fillers is to achieve the percolated network that allows transferring electrons through the contact points between conductive fillers or by the tunneling effect. Carbon-based nanofillers form agglomerates, which must be destroyed during processing to achieve homogenous dispersion and distribution in the polymer matrix. Ideally, they will form the percolated conductive network at a low concentration [8]. However, it should be noted that the main parameter affecting the overall conductivity of the ECAs is the filler-filler contact resistance, which disturbs the free flow of the electrons. Therefore, agglomerates or microparticles with fewer contact points can positively affect the ECAs' conductivity [1,9]. Hence, researchers analyze the mutual impact of both types of fillers–metallic and carbonaceous because the built network results in more effective conductance. For example, Luo et al. [10] proved that adding 4.5 wt% CNTs together with 50 wt% of silver flakes improved the electrical conductivity by about 255% compared to the sample without CNTs, and the values reached were 2.56×10^5 S/m. The other tested possibility is to modify the surface of CNTs with silver to obtain functionalized CNTs and then add them to the polymer. However, such a solution resulted in a relatively low electrical conductivity of 2.83×10^{-6} S/m [11]. Instead of using conductive fillers, the ECAs can also be produced by modifying the polymer matrix with intrinsically conductive polymers such as polyaniline or polypyrrole [12].

Applying ECAs in electronics requires high electrical conductivity and good mechanical performance, high thermal conductivity, long curing time, resistance to low and high temperatures, and humidity [13]. For the other application of ECAs, such as adhesive bonding of the lightweight composite structures, the essential property is electrical conductivity needed for electromagnetic interference (EMI) shielding (0.1–10 S/m) or lightning strike protection (min. 10 S/m). The other mentioned requirements are high strength, the ability to withstand a wide temperature range, and a low coefficient of thermal expansion [14]. Furthermore, such ECAs should prevent delamination caused by insufficient adhesion between the layers and be easy to apply. Among various polymers used as a matrix in ECAs, the ones that have achieved the highest popularity are solvent-based epoxy, silicones, or acrylates. However, in recent years, solvent-free and thermoplastic hot melt adhesives have attracted great attraction. They are complex materials consisting of: (i) polymer (polyamide, polyurethane, polyolefins, ethylene copolymers) responsible for strength and hot tack; (ii) resin that improves contact with the substrate; (iii) tackifier that adjusts the glass transition and dilutes polymer; (iv) wax that increases setting speed [15]. Hot melts are available in the form of pellets, powder, sticks, and bars, which are solid at room temperature. When the temperature rises, they become liquid and soft, but they solidify during cooling. Depending on the formulation, hot melt adhesives offer a broad range of properties, including adhesion strength, working temperatures, and viscosity level [16]. Hot melts are used primarily where the process speed matters, such as non-wovens in sanitary products, construction, packaging, bottle labeling, bookbinding, or temporary attachments [17]. Hot melt adhesives have also found a place in the composites industry to improve the bonding strength of hybrid parts (different types of glass/carbon composites) [18]; improve the interlaminar toughness in lightweight composites [19]; increase the adhesion between the nanofibrous mat and its supporting woven polyester fabric [20] or as the conductive adhesive to provide an electrically conductive layer between bonding parts. So far, there are only a few publications describing ECAs based on hot melt adhesives, such as polyurethane [21], polyolefines [22], and our previous work on polyamide-based hot melts [23]. Compared to epoxy adhesives, the main advantages of the hot melts in adhesive bonding are lack of solvent, no requirement of special surface treatment; shorter curing time; lack of weight increase compared to rivets, and recyclability [18].

This study aimed to analyze the effect of the high loading of multi-walled carbon nanotubes (MWCNTs) on the properties of commercial non-conductive hot melt polyamide. For this research, a hot melt with very low viscosity was selected to examine how much the electrical conductivity will be improved, but also considering the rheological, thermal, and mechanical properties. Based on the achieved conductivity and the strength of the adhesive bonds, they are a promising candidate to be applied as ECAs for the adhesive bonding that requires EMI shielding.

2. Materials and Methods

2.1. Materials

As the hot melt adhesive, the copolyamide (coPA) hot melt in the form of granules with the trade name Vestamelt®722 was supplied by Evonik (Essen, Germany). According to the datasheet, the melting temperature is 107 °C and melt flow index = 310 g/10 min (2.16 kg/160 °C) classified that polymer as being low viscous. MWCNTs with the trade name NC7000 from Nanocyl (Sambreville, Belgium) synthesized by catalytic carbon deposition from the gas phase were used as the electrically conductive filler. The average diameter of that type of MWCNTs is 9.5 nm, length 1.5 µm, and purity > 95%.

2.2. Composites Fabrication

The selected coPA was mixed with 5% and 10% by weight (wt%) of MWCNTs using an industrial twin-screw extruder by Nanocyl. The pellets of neat coPA and both masterbatches were dried before each process in a vacuum oven at 60 °C for a minimum of 6 h. The specimens for the rheological and mechanical tests were prepared directly from the pellets using the laboratory injection moulding machine HAAKE Mini Jet Pro Piston Injection Molding System (ThermoScientific, Karlsruhe, Germany). The parameters of the injection moulding process are presented in Table 1.

Table 1. Injection moulding parameters for coPA and it's masterbatches containing 5 wt% and 10 wt% MWCNT.

MWCNTs Content	Barrel Temperature (°C)	Mould Temperature (°C)	Injection Pressure (bar)	Injection Time (s)	Post Pressure (bar)	Post Time (s)
0	110	35	650	3	400	6
5 wt%	150	35	800	5	700	10
10 wt%	150	35	800	5	700	10

2.3. Characterization Techniques

Firstly, the dispersion and distribution of MWCNTs in the masterbatches were examined using a Polarized Light Microscope (Bipolar-PL, PZO, Warsaw, Poland). Slides with a thickness of 2–3 µm were cut directly from the masterbatch pellets using an ultramicrotome (EM UC6, Leica, Vienna, Austria). From the obtained images (7 for each material) the number of MWCNT agglomerates was counted using image software (ImageJ). The percentage ratio AA was calculated by dividing the area of all agglomerates by the total area, excluding those agglomerates with a diameter lower than 1 µm. The second method applied to analyze the dispersion of MWCNTs was a Transmission Scanning Electron Microscope (HR STEM S5500, Hitachi, Tokyo, Japan) that allows the observations on a nanometer scale. For the analysis, the slides with a thickness of 80–90 nm were cut with diamond knives at −100 °C by an ultramicrotome (EM UC6, Leica, Vienna, Austria). The observations were performed at 30 kV.

An ARES rheometer (model 4400-0107, TA Instruments, New Castle, DE, USA) was used to analyze the viscoelastic properties of the hot melt adhesives by oscillatory test in the plate-to-plate mode. Firstly, a dynamic strain sweep test as a function of the variable strain γ (0.07–100%) at a constant frequency of 1 Hz was performed to determine the strain within the elastic range. Then a variable frequency test was performed with the determined deformation, in this case, 0.1% and at three temperatures of 180 °C, 200 °C and 220 °C. The specimens for the rheological analysis were produced by injection moulding in the form of rounds 25 mm in diameter and with a thickness of 1 mm.

The thermal stability of the materials was determined by thermogravimetric analysis (TGA). The examination was conducted on a TGA Q500 instrument (TA Instruments, New Castle, DE, USA). For this, samples of 10 ± 0.5 mg were prepared, then transferred to tared platinum pans and heated from 20 °C to 900 °C at a heating rate of 10 °C/min. Two flow rates of nitrogen of 10 mL/min in the chamber and 90 mL/min in the oven were used during the analysis. Thermal stability was determined by degradation temperatures of 5% ($T_{5\%}$) and 10% ($T_{10\%}$) weight loss, as well as by the maximum degradation peak (T_d).

The thermal properties of all materials were examined by Differential Scanning Calorimetry (DSC) using the Q1000 Differential Scanning Calorimeter (TA Instruments, New Castle, DE, USA). The 6.5 ± 0.5 mg samples were placed in an aluminum hermetic crucible and analyzed under the heat-cool-heat program from −80 °C to 220 °C with a heating/cooling rate of 10 °C/min under a nitrogen atmosphere. The curves obtained from the test were analyzed using TA Universal Analysis 2000 software version 4.5A. The glass transition temperature (T_g), melting temperature (T_m), and enthalpy of melting (ΔH_m) were determined from the heating curves, while the crystallization temperature (T_c) and enthalpy of crystallization (ΔH_c) were taken from the cooling curves. Due to the lack of data about the enthalpy of melting of 100% crystalline coPA, the degree of crystallinity was not calculated.

The electrical volume conductivity of unfilled and filled coPA was measured by the Keithley 6221/2182A device equipped with copper electrodes. Test samples with dimensions of 10 cm × 10 cm were prepared by pressing the pellets on a hydraulic press (Hydraulische Werkstattpresse WPP 50 E, Unicraft, Hallstadt, Deutschland) at the temperature of 115 °C and pressure of 30 MPa. Afterwards, small specimens of around 1 cm × 1 cm were cut from different sections of the bigger sample, to determine the homogeneity of the electrical conductivity. To maintain good contact between electrodes and the measured sample silver paste was applied.

Mechanical properties of the studied materials were analyzed by uniaxial tensile tests according to PN EN ISO527 using an MTS QTest 10 (MTS Systems, Eden Prairie, MN, USA) testing machine. Five test specimens represented each material. Tensile experiments were performed at a constant crosshead speed of 10 mm/min using an extensometer with a gauge length of 50 mm for strain measurements. From the stress-strain curves, the tensile modulus of elasticity, ultimate tensile strength (F_{tu}) and elongation at break (ε_b) were determined. Tensile tests were carried out on small dog bone specimens.

The hardness was determined based on the Shore method according to PN-EN ISO 868:2005 standard. The hardness test was carried out using the WHS-180 hardness tester by Wilson-Wolpert (ATM Qness, Mammelzen, Germany). All measurements were made using the Shore D scale for testing hard plastics. For each material, 25 measurements were made. The final result for coPA and its masterbatches containing MWCNT was obtained by calculating the arithmetic mean of the values. A load of 50 N was applied during the test.

The wettability of the surfaces of coPA and its masterbatches was tested by measuring the contact angle using an OCA15 (DataPhysics Instruments, Filderstadt, Germany) goniometer equipped with OCA software. The contact angle hysteresis was determined by calculating the difference between the advancing and receding contact angles. The test was performed on the round specimens from the injection moulding using a 5 µL droplet. The contact angle results are the averaged values of five different measuring points on each surface.

The effect of MWCNTs on adhesive efficiency was studied using a lap shear test. In order to obtain test specimens, two aluminum plates were pressed together at 135 °C, with neat coPA and coPA + MWCNTs between them. The thickness of the adhesive bond was kept at 0.1mm, and the dimensions of the specimens are shown in Figure 1. The lap-shear test was performed using an MTS810 servo-hydraulic testing machine in accordance with ASTM D 1002. The crosshead speed was kept at 1.3 mm/min.

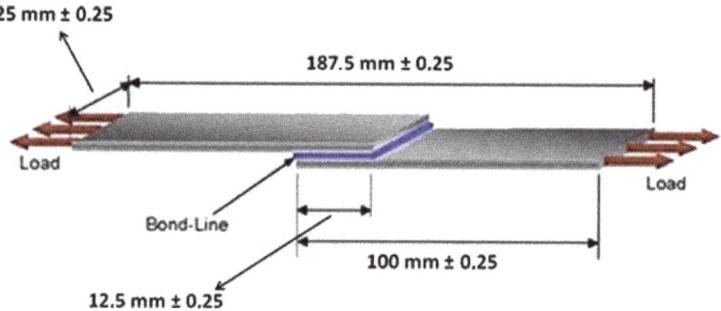

Figure 1. The dimension of the specimen for the lap-shear test [11].

3. Results

3.1. Microstructure

The dispersion and distribution of MWCNTs in the coPA matrix were analyzed in terms of agglomerates (so-called macrodispersion) using a light optical microscope. Sample images are presented in Figure 2a,c. At first glance, it can be seen that the coPA containing 5 wt% of MWCNTs (Figure 2a) has more agglomerates than the masterbatch with 10 wt% of MWCNTs (Figure 2c). The higher MWCNT concentration causes a higher shear force during extrusion and more effective destruction of the primary agglomerates. The quantitative analysis confirms this because the percentage ratio of the MWCNT agglomerates in the nanocomposites equals $A_A = 5.8 \pm 1.1\%$ and $AA = 3.1 \pm 1.0\%$ for 5 wt% and 10 wt% MWCNTs, respectively. The images obtained by TEM are shown in Figure 2b,d present a more detailed analysis of the nanocomposite microstructure. Here, the MWCNTs are visible as single long tubes, loosely connected at some places in the form of bundles. Moreover, they are not arranged in any specific direction. Although MWCNTs are homogeneously dispersed in the copolyamide matrix, there are some empty places where nanotubes do not occur. Such sites are visible as white areas, and they are responsible for decreasing the electrical conductivity of the materials. However, the overall state of MWCNTs dispersion can be classified as good enough.

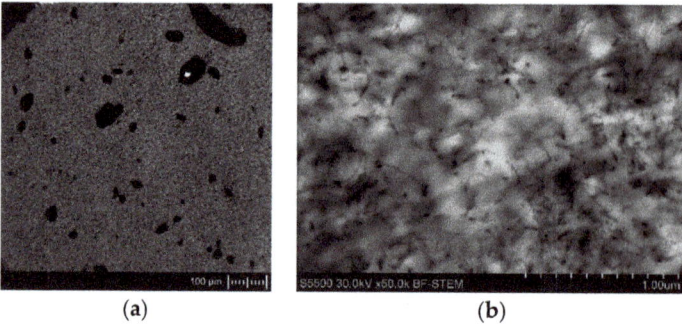

(a)　　　　　　　　　　(b)

Figure 2. Cont.

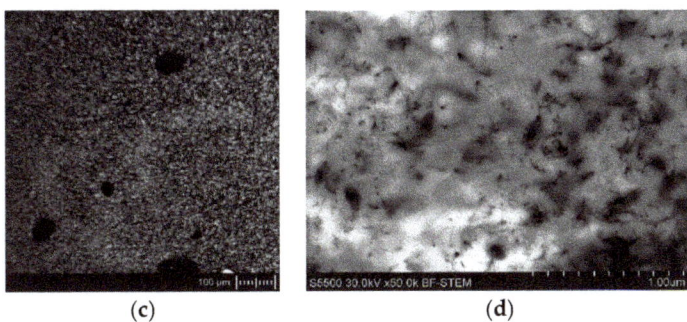

Figure 2. The dispersion of MWCNTs in the coPA: (**a**) macrodispersion in coPA + 5 wt% MWCNTs; (**b**) macrodispersion in coPA + 10 wt% MWCNTs; (**c**) dispersion in coPA + 5 wt% MWCNTs; (**d**) dispersion in coPA + 10 wt% MWCNTs. Scale: 100 µm (**a,c**) and 1 µm (**b,d**).

3.2. Rheological Properties

Oscillatory rheology measurements were performed to examine the effect of the addition of MWCNTs on the viscoelastic properties of coPA. As shown in Figure 3 the complex viscosity of pure coPA increases by about 6 orders of magnitude in the presence of 5 and 10 wt% MWCNTs which is related to the restriction of the polymer chains' movement. It should be noted that there is a negligible difference between 5 and 10 wt% concentration, because a stronger effect is usually visible at low concentrations of CNTs. Here, at 5 wt% the network has already been percolated and further addition of MWCNTs does not change that structure much and does not affect the chains' movement. Such a high increase in viscosity demonstrates the interaction between coPA and nanotubes. Looking into the curves, it can be seen that since neat coPA is a non-Newtonian liquid, its viscosity is not dependent on the frequency. In contrast, nanocomposites behave as Newtonian liquid with the viscosity decreasing together with the frequency [24]. Interestingly, the viscosity of the materials does not decrease at higher temperatures because the curves remain almost unchaged at 180 °C, 200 °C and 220 °C.

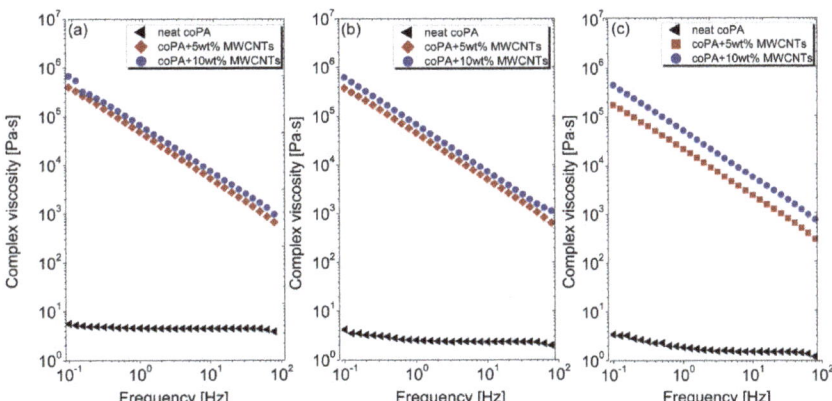

Figure 3. Variation in a complex viscosity as a function of frequency of unfilled and filled coPA measured at: (**a**) 180 °C; (**b**) 200 °C and (**c**) 220 °C.

The formation of a percolated network in the coPA-based nanocomposites is also demonstrated by an increase in storage and loss modulus, which describe the viscous and elastic properties, respectively. As can be seen in Figure 4 both moduli increase after the addition of 5 wt% and 10 wt% MWCNTs in comparison to neat coPA. Similarly to the viscosity, there is a negligible difference between 5 and 10 wt% MWCNTs and, what is more, frequency has no effect on both moduli. It should be also noticed that for neat coPA the loss modulus is higher than the storage modulus within the whole frequency range, however for nanocomposites, this is reversed. This shows that nanocomposites behave more like elastic than viscous material, which has been reported for many thermoplastic nanocomposites modified with CNTs [25].

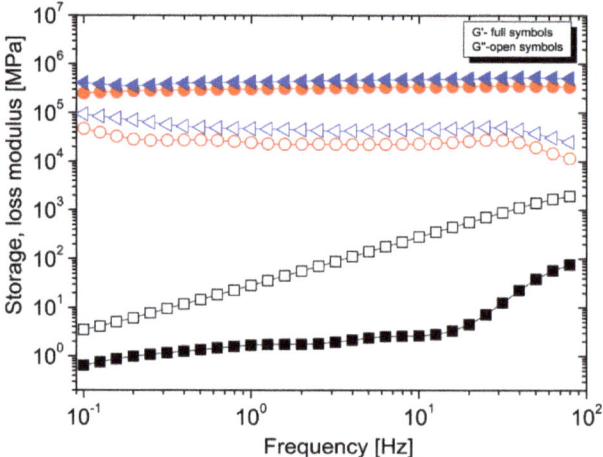

Figure 4. The dependence between storage modulus (full symbols) and loss modulus (open symbols) for neat coPA (black curves), coPA + 5 wt% MWCNTs (red curves) and coPA + 10 wt% MWCNTs (blue curves). Test temperature 180 °C, strain 0.1%.

3.3. Thermal Properties

The effect of the addition of MWCNTs on the thermal stability of coPA was studied by TGA. The obtained results are shown in Table 2, while the curves are shown in Figure 5. It can be seen that for neat coPA the decomposition which corresponds to 5% weight loss ($T_{5\%}$) starts at 310 °C. In the presence of 5 wt% MWCNTs, $T_{5\%}$ increases to 336 °C. This behaviour of the material indicates that the presence of CNTs delays the degradation of the polymer. Interestingly, the presence of 10 wt% MWCNTs causes a slight decrease in $T_{5\%}$ compared to 5 wt% MWCNTs. The same trend can be observed for 10% weight loss of the material. The maximum degradation temperature for pure coPA occurs at about 440 °C. For material with the addition of 5 wt% MWCNTs, this temperature is 465 °C. It can be concluded that a high MWCNTs content in the coPA structure has less effect on the thermal stability of the polymer than a lower concentration. In comparison, Mahmood.N et al. indicated that adding only 0.5 wt% MWCNTs to the PA6 matrix caused a shift in the temperature distribution by 70 °C [26].

The heating and cooling curves of unfilled coPA and filled with 5 wt% and 10 wt% MWCNT are displayed in Figure 6. From the obtained first heating curves (Figure 6a), two characteristic points, the glass transition temperature and the melting temperature was observed. The addition of MWCNTs does not change T_m relative to neat coPA; other researchers have previously observed a similar relationship in polyurethane-based hot melt adhesives [27]. Moreover, the presence of two glass transition peaks in neat coPA indicates that studied adhesive consists of copolymer having segments of PA 6 and PA66 (also confirmed by FTIR analysis). Most likely, the lower T_g value is related to the PA6 segment, while the higher value is related to the presence of the PA66 segment [28]. The addition of 5 wt% MWCNTs shifted the second T_g peak toward higher values of about 15 °C. This confirms a good dispersion of nanotubes in the coPA matrix, which limits the polymer chains' mobility [29]. The second heating curve showed similar behaviour of all tested materials (Figure 6b). However, from cooling curves (Figure 6c), the formation of a more marked crystallization peak is observed, shifted towards higher temperatures after the addition of 5 wt% MWCNTs. Introducing MWCNTs in the polymer matrix induces a nucleation process, which affects the crystalline phase formation. These observations were confirmed by other researchers [30]. In the case of coPA with 10 wt% MWCNTs, no further temperature changes were observed. The formation of a new crystalline phase can also be noticed by the changing character of melting curves as well as decreasing the value of enthalpy of melting.

Table 2. The results of thermal analysis.

Material	TGA			DSC							
				1st Heating			2nd Heating				Cooling
	$T_{5\%}$ (°C)	$T_{10\%}$ (°C)	T_d (°C)	T_g (°C)	T_m (°C)	ΔH_m (J/g)	T_g (°C)	T_m (°C)	T_c (°C)	ΔH_m (J/g)	T_c (°C)
coPA	310	346	437	17 48	101	18	17	103	76	27	-
coPA + 5 wt% MWCNT	336	378	465	18 63	101	13	28	103	-	17	74
coPA + 10 wt% MWCNT	330	373	456	19 66	100	10	30	101	-	15	74

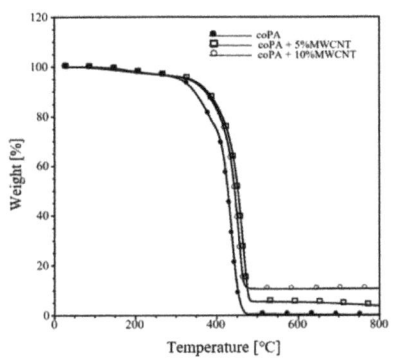

Figure 5. The TGA curves for the neat coPA and its masterbatches containing 5 wt% and 10 wt% MWCNT.

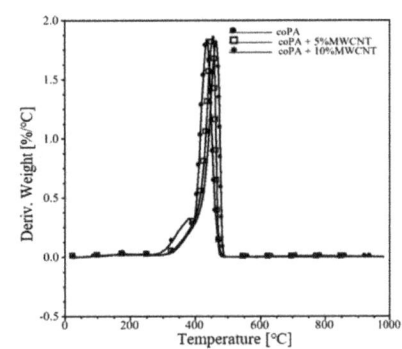

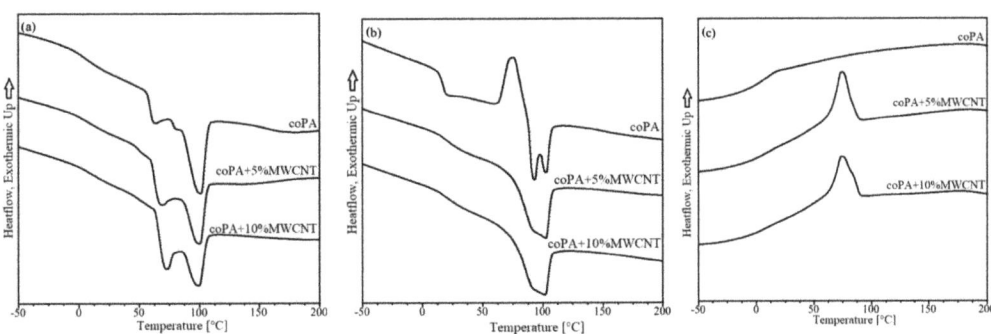

Figure 6. (a) First heating curves; (b) second heating curves; and (c) cooling curves for the unfilled coPA and masterbatches containing 5 wt% and 10 wt% MWCNTs.

3.4. Electrical Properties

Electrical conductivity is the main property since the examined materials based on hot melt coPA and MWCNTs are a candidate for use as ECA's materials. Figure 7 shows the maps of the conductivity of masterbatches measured on hot-pressed square plates divided into small pieces. It can be seen that the conductivity is not the same at every point of the plate, however, the deviation is not significant (except one value for 10 wt% MWCNTs). The average electrical conductivity value for coPA with 5 wt% MWCNTs was 0.57 ± 0.09 S/m, while introducing 10 wt% MWCNTs increased the conductivity to 3.38 ± 1.07 S/m. In comparison to polyolefine hot melt adhesive containing the same type of MWCNTs, the electrical conductivity at 5 wt% was 0.01 S/m, which is much lower than that obtained for coPA nanocomposites [22]. Obviously, the achieved conductivity is much lower than reported for typical ECAs containing silver as a conductive filler, but the manufacturing process is much easier and faster. The electrical conductivity of neat coPA was below the measuring range of the instrument and was <10^{-9}. These results were correlated with the microstructure presented in Section 3.1. The calculated value of A_A affects the value of electrical conductivity. Electrical conductivity increases while the average area of MWCNT agglomerates decreases [4,31]. The higher electrical conductivity with lower agglomerate size was the result of better dispersion of the filler in the polymer matrix, and thus more conductive pathways present in the test material.

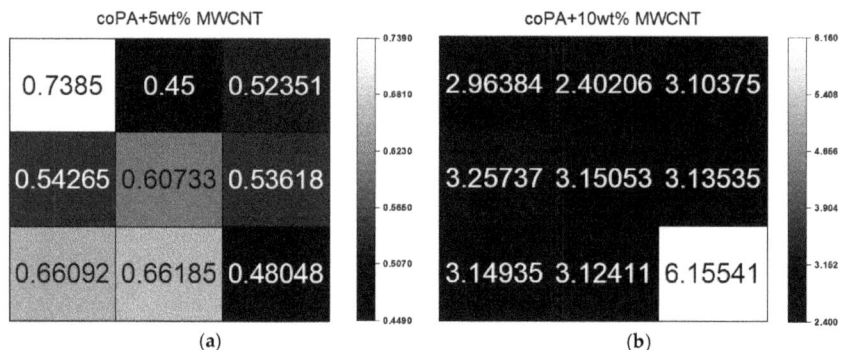

Figure 7. Volume electrical conductivity for masterbatches containing (a) 5 wt% and (b) 10 wt% MWCNTs.

3.5. Mechanical Properties

Figure 8 shows the characteristic stress-strain curves of neat coPA and its composites with MWCNTs. The elastic modulus, tensile strength and elongation at break as a function of the concentration of MWCNTs are presented in Table 3.

Young's modulus increased with the addition of MWCNTs at a concentration of 5 wt% by 205% and at a concentration of 10 wt% by 330%. Similarly, tensile strength improved by 129% at 5 wt% concentration and by 171% with addition of 10 wt% MWCNTs. Only elongation at break decreased, what was expected and reported already for polyolefine hot melt adhesives containing 5 wt% of MWCNTs [22]. The addition of MWCNTs increases the stiffness and strength of thermoplastic adhesive-based materials caused by the uniform dispersion of nanotubes in coPA matrix. There is also a visible effect of MWCNTs on the Shore D hardness, that grown from 54° ShD for neat coPA to 60 and 64° ShD for 5 wt% and 10 wt% MWCNTs, respectively; such a reinforcing effect has also been reported for single-walled carbon nanotubes mixed with rubber [32].

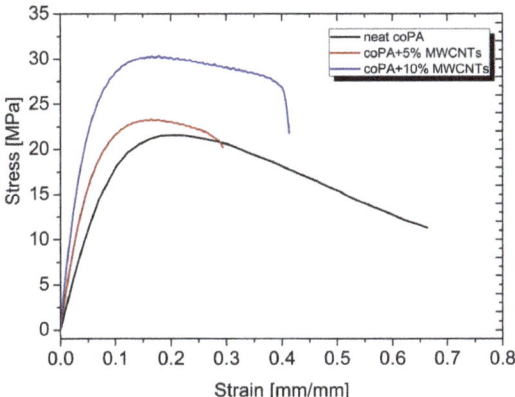

Figure 8. Representative stress-strain curves of coPA and its masterbatches containing 5 wt% and 10 wt% MWCNT.

Table 3. Mechanical properties of neat coPA and coPA/MWCNTs composites.

Material	Tensile Test			Hardness Test
	Elastic Modulus (GPa)	Ultimate Tensile Strength (MPa)	Elongation at Break (%)	Hardness (°ShD)
coPA	0.20 ± 0.02	17.7 ± 1.81	55.9 ± 21.8	54
coPA + 5 wt% MWCNT	0.41 ± 0.02	22.9 ± 0.61	36.7 ± 11.2	60
coPA + 10 wt% MWCNT	0.66 ± 0.03	30.4 ± 0.73	28.6 ± 4.28	65

3.6. Contact Angle Measurements

The main property of the ECAs is high adhesion to different substrates. Therefore, the effect of MWCNTs inclusion into coPA hot melt was analyzed by the measurement of the contact angle. The results (Figure 9) show that the average contact angle of neat coPA was 92°, but in the presence of 5 and 10 wt% MWCNTs (lack of differences between the concentration) it was decreased to 80°. This means that MWCNTs change the hydrophobic character of coPA hot melt to a hydrophilic one. This is related to the modification of the surface of coPA due to the incorporation of MWCNTs. Nevertheless, the contact angle of 80° means that the nanocomposites have hydrophilic properties, so their wettability should be improved. Similar results were reported for the other types of coPA hot melt adhesives [23].

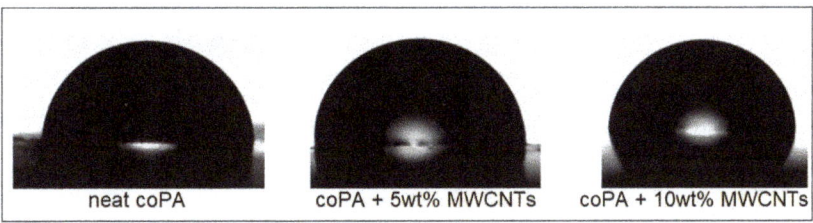

Figure 9. Images taken during contact angle measurement showing the effect of MWCNTs addition on the wettabily of coPA hot melt.

3.7. Adhesive Properties

From the practical point of view, the main challenge is to keep the same or even better adhesive properties of ECAs after the addition of any conductive fillers. To see how MWCNTs affect the adhesive properties of coPA, a lap shear test was performed. The calculated shear strength is presented in Table 4. For the neat coPA hot melt, the shear strength was 8.4 MPa. Adding 5 wt% and 10 wt% MWCNTs decreased shear strength to 5.4 MPa and 4.8 MPa, respectively. In other words, a high concentration of MWCNTs reduces the adhesive properties of the coPA-based ECAs, due to too much carbon. The literature indicates that up to 3 wt% of MWCNTs, the shear strength increases, but further addition of the filler leads to decreasing the adhesive strength [22]. Similar decreasing in the adhesive properties was found for PA6 hot melt adhesive mixed with lithium bromide, which was explained by lack of the bond formation between the adhesive and the aluminum substrate [33]. It is noteworthy, however, that the shear strength values obtained for coPA hot melt modified by even 10 wt% MWCNTs are high and comparable with the literature [34]. The specimens after lap shear test shown in Figure 10 indicate that in all cases, there was adhesive failure between the adherent and the hot melt adhesive. Moreover, the adhesion strength of the tested materials is influenced by many other properties, including surface roughness of the adherent, wettability of the adherent by adhesive, layer thickness, temperature applied during bonding and even concentration of the hot melt in the bonding layers [35]. These factors will be studied in the future.

Table 4. The result of the lap-shear test.

Material	Lap Shear Strength (MPa)
coPA	8.4 ± 0.51
coPA + 5 wt% MWCNT	5.4 ± 0.68
coPA + 10 wt% MWCNT	4.8 ± 0.25

Figure 10. Fracture pattern of joints bonded with MWCNTs modified hot melt at different concentrations.

4. Conclusions

This paper describes a group of electrically conductive adhesives fabricated by melt-blending thermoplastic copolyamide hot melt containing 5 wt% and 10 wt% MWCNTs. The neat coPA was characterized by low viscosity, which, together with the storage modulus and loss modulus, was increased after MWCNTs additions by about 5-6 orders of magnitude due to strong interactions between the coPA chains and nanotubes. At elevated temperatures, the viscosity does not drop because the movement of the polymer chains is hindered due to the high content of MWCNTs. Such behavior is also confirmed by shifting the glass transition temperature of neat coPA by about 15 °C towards higher values at both MWCNT concentrations. However, the nanotubes have a negligible effect of on the melting point, but as the melting curve changes, it was assumed that MWCNTs work as nucleation agents leading to the formation of a new crystal phase. From microscopic images, the area ratio of MWCNT agglomerates calculated by ImageJ was A_A = 5.8% for coPA + 5 wt% MWCNTs and A_A = 3.1% for coPA + 10 wt% MWCNTs. Fewer agglomerates resulted in higher electrical conductivity for 10 wt%–3.38 S/m, while for 5 wt%, it was 0.58 S/m. The addition of MWCNTs improved thermal stability and temperature of decomposition by about 25–30 °C compared to neat coPA indicates the homogenously dispersed nanotubes. Elastic modulus, tensile strength, and hardness were improved in the presence of MWCNTs by a maximum of 330%, 171%, and 20%, respectively. The addition of MWCNTs also changed the hydrophobic character of coPA hot melt to a hydrophilic one, signifying higher wettability. However, it did not affect the adhesion strength, which decreased from 8.5 MPa to 4.8 MPa as the effect of MWCNT addition. Nevertheless, such adhesion can classify coPA filled with MWCNTs as a good adhesive that, together with achieved electrical conductivity, can be applied as ECAs for EMI shielding applications.

Author Contributions: Conceptualization, P.L.-D. and M.M.; methodology, P.L.-D. and M.M.; software, P.L.-D. and M.M.; validation, P.L.-D. and A.B.; formal analysis, P.L.-D.; investigation, P.L.-D. and M.M.; resources, P.L.-D.; data curation, P.L.-D.; writing—original draft preparation, P.L.-D. and M.M.; writing—review and editing, P.L.-D.; visualization, P.L.-D.; supervision, P.L.-D. and A.B.; project administration, P.L.-D.; funding acquisition, P.L.-D. All authors have read and agreed to the published version of the manuscript.

Funding: The research leading to these results has received funding from the Norway Grants 2014–2021 via the National Centre for Research and Development. Grant number: NOR/SGS/3DforCOMP/0171/2020-00.

Institutional Review Board Statement: Not applicable.

Informed Consent Statement: Not applicable.

Data Availability Statement: All data are included in the manuscript.

Acknowledgments: The authors would like to thank Nadir Kchnit from Nanocyl for the fabrication of masterbatches, Rafał Kozera for the microscopic observations and Paulina Kozera for the calculation the agglomerates ratio using ImageJ.

Conflicts of Interest: The authors declare no conflict of interest.

References

1. Aradhana, R.; Mohanty, S.; Nayak, S.K. A review on epoxy-based electrically conductive adhesives. *Int. J. Adhes. Adhes.* **2020**, *99*, 102596. [CrossRef]
2. Nele, L.; Palmieri, B. Electromagnetic heating for adhesive melting in CFRTP joining: Study, analysis, and testing. *Int. J. Adv. Manuf. Technol.* **2020**, *106*, 5317–5331. [CrossRef]
3. Wong, C.P.; Xu, J.; Zhu, L.; Li, Y.; Jiang, H.; Sun, Y.; Lu, J.; Dong, H. Recent advances on polymers and polymer nanocomposites for advanced electronic packaging applications. In Proceedings of the 2005 Conference on High Density Microsystem Design and Packaging and Component Failure Analysis, HDP'05, Shanghai, China, 27–29 June 2005. [CrossRef]
4. Latko-Durałek, P.; Kozera, R.; Macutkevič, J.; Dydek, K.; Boczkowska, A. Relationship between Viscosity, Microstructure and Electrical Conductivity in Copolyamide Hot Melt Adhesives Containing Carbon Nanotubes. *Materials* **2020**, *13*, 4469. [CrossRef] [PubMed]
5. Derakhshankhah, H.; Mohammad-Rezaei, R.; Massoumi, B.; Abbasian, M.; Rezaei, A.; Samadian, H.; Jaymand, M. Conducting polymer-based electrically conductive adhesive materials: Design, fabrication, properties, and applications. *J. Mater. Sci. Mater. Electron.* **2020**, *31*, 10947–10961. [CrossRef]
6. Yim, M.J.; Li, Y.; Moon, K.S.; Paik, K.W.; Wong, C.P. Review of recent advances in electrically conductive adhesive materials and technologies in electronic packaging. *J. Adhes. Sci. Technol.* **2008**, *22*, 1593–1630. [CrossRef]
7. Mantena, K.; Li, J.; Lumpp, J.K. Electrically conductive carbon nanotube adhesives on lead free printed circuit board surface finishes. In Proceedings of the IEEE Aerospace Conference, Big Sky, MT, USA, 1–8 March 2008; pp. 2–6. [CrossRef]

8. Alig, I.; Lellinger, D.; Engel, M.; Skipa, T.; Pötschke, P. Destruction and formation of a conductive carbon nanotube network in polymer melts: In-line experiments. *Polymer* **2008**, *49*, 1902–1909. [CrossRef]
9. Jin, J.; Lin, Y.; Song, M.; Gui, C.; Leesirisan, S. Enhancing the electrical conductivity of polymer composites. *Eur. Polym. J.* **2013**, *49*, 1066–1072. [CrossRef]
10. Luo, J.; Cheng, Z.; Li, C.; Wang, L.; Yu, C.; Zhao, Y.; Chen, M.; Li, Q.; Yao, Y. Electrically conductive adhesives based on thermoplastic polyurethane filled with silver flakes and carbon nanotubes. *Compos. Sci. Technol.* **2016**, *129*, 191–197. [CrossRef]
11. Zhang, Y.; Zhang, F.; Xie, Q.; Wu, G. Research on electrically conductive acrylate resin filled with silver nanoparticles plating multiwalled carbon nanotubes. *J. Reinf. Plast. Compos.* **2015**, *34*, 1193–1201. [CrossRef]
12. Mohammad-Rezaei, R.; Massoumi, B.; Abbasian, M.; Eskandani, M.; Jaymand, M. Electrically conductive adhesive based on novolac-grafted polyaniline: Synthesis and characterization. *J. Mater. Sci. Mater. Electron.* **2019**, *30*, 2821–2828. [CrossRef]
13. Cui, H.W.; Kowalczyk, A.; Li, D.S.; Fan, Q. High performance electrically conductive adhesives from functional epoxy, micron silver flakes, micron silver spheres and acidified single wall carbon nanotube for electronic package. *Int. J. Adhes. Adhes.* **2013**, *44*, 220–225. [CrossRef]
14. Ganesh, M.G.; Lavenya, K.; Kirubashini, K.A.; Ajeesh, G.; Bhowmik, S.; Epaarachchi, J.A.; Yuan, X. Electrically conductive nano adhesive bonding: Futuristic approach for satellites and electromagnetic interference shielding. *Adv. Aircr. Spacecr. Sci.* **2017**, *4*, 729–744. [CrossRef]
15. Li, W.; Bouzidi, L.; Narine, S.S. Current Research and Development Status and Prospect of Hot-Melt Adhesives: A Review. *Ind. Eng. Chem. Res.* **2008**, *47*, 7524–7532. [CrossRef]
16. Petrie, E.M. *Handbook of Adhesives and Sealants*; McGraw Hill Professional: New York, NY, USA, 2007.
17. Paul, C.W. *Hot Melt Adhesives*; Elsevier B.V.: Amsterdam, The Netherlands, 2002.
18. Peng, X.; Liu, S.; Huang, Y.; Sang, L. Investigation of joining of continuous glass fibre reinforced polypropylene laminates via fusion bonding and hotmelt adhesive film. *Int. J. Adhes. Adhes.* **2020**, *100*, 102615. [CrossRef]
19. Latko-Durałek, P.; Dydek, K.; Bolimowski, P.; Golonko, E.; Duralek, P.; Kozera, R.; Boczkowska, A. Nonwoven fabrics with carbon nanotubes used as interleaves in CFRP. 13th International Conference on Textile Composites. *IOP Conf. Ser. Mater. Sci. Eng.* **2018**, *406*, 012033. [CrossRef]
20. Amini, G.; Gharehaghaji, A.A. Improving adhesion of electrospun nanofiber mats to supporting substrate by using adhesive bonding. *Int. J. Adhes. Adhes.* **2018**, *86*, 40–44. [CrossRef]
21. Fernández, M.; Landa, M.; Muñoz, M.E.; Santamaría, A. Tackiness of an electrically conducting polyurethanenanotube nanocomposite. *Int. J. Adhes. Adhes.* **2010**, *30*, 609–614. [CrossRef]
22. Wehnert, F.; Pötschke, P.; Jansen, I. Hotmelts with improved properties by integration of carbon nanotubes. *Int. J. Adhes. Adhes.* **2015**, *62*, 63–68. [CrossRef]
23. Latko-Durałek, P.; Mcnally, T.; Macutkevic, J.; Kay, C.; Boczkowska, A. Hot-melt adhesives based on co-polyamide and multiwalled carbon nanotubes. *J. Appl. Polym. Sci.* **2017**, *1*, 1–15. [CrossRef]
24. Chatterjee, T.; Krishnamoorti, R. Rheology of polymer carbon nanotubes composites. *Soft Matter* **2013**, *9*, 9515. [CrossRef]
25. Nobile, M.R. Rheology of polymer–carbon nanotube composites melts. In *Polymer–Carbon Nanotube Composites Preparation, Properties and Applications*; Woodhead Publishing Limited: Sawston, UK, 2011; pp. 428–481. ISBN 978-1-84569-761-7.
26. Mahmood, N.; Islam, M.; Hameed, A.; Saeed, S. Polyamide 6/multiwalled carbon nanotubes nanocomposites with modified morphology and thermal properties. *Polymers* **2013**, *5*, 1380–1391. [CrossRef]
27. Fernández, M.; Landa, M.; Muñoz, M.E.; Santamaría, A. Thermal and viscoelastic features of new nanocomposites based on a hot-melt adhesive polyurethane and multi-walled carbon nanotubes. *Macromol. Mater. Eng.* **2010**, *295*, 1031–1041. [CrossRef]
28. Greco, R.; Nicolais, L. Glass transition temperature in nylons. *Polymer* **1976**, *17*, 1049–1053. [CrossRef]
29. Xie, X.L.; Mai, Y.W.; Zhou, X.P. Dispersion and alignment of carbon nanotubes in polymer matrix: A review. *Mater. Sci. Eng. R Rep.* **2005**, *49*, 89–112. [CrossRef]
30. Landa, M.; Canales, J.; Fernández, M.; Muñoz, M.E.; Santamaría, A. Effect of MWCNTs and graphene on the crystallization of polyurethane based nanocomposites, analyzed via calorimetry, rheology and AFM microscopy. *Polym. Test.* **2014**, *35*, 101–108. [CrossRef]
31. Socher, R.; Krause, B.; Boldt, R.; Hermasch, S.; Wursche, R.; Pötschke, P. Melt mixed nano composites of PA12 with MWNTs: Influence of MWNT and matrix properties on macrodispersion and electrical properties. *Compos. Sci. Technol.* **2011**, *71*, 306–314. [CrossRef]
32. Bakošov, D.; Bakošov, A. Testing of Rubber Composites Reinforced with Carbon Nanotubes. *Polymers* **2022**, *14*, 3039. [CrossRef] [PubMed]
33. Yamaguchi, M.; Takatani, R.; Sato, Y.; Maeda, S. Moisture-sensitive smart hot-melt adhesive from polyamide 6. *SN Appl. Sci.* **2020**, *2*, 1567. [CrossRef]
34. Tang, X.; Cui, H.W.; Lu, X.Z.; Fan, Q.; Yuan, Z.; Ye, L.; Liu, J. Development and characterisation of nanofiber films with high adhesion. In Proceedings of the Electronic Components and Technology Conference, Florida, FL, USA, 21 May–3 June 2011; pp. 673–677. [CrossRef]
35. Boutar, Y.; Naïmi, S.; Mezlini, S.; Da Silva, L.F.M.; Hamdaoui, M.; Ben Sik Ali, M. Effect of adhesive thickness and surface roughness on the shear strength of aluminium one-component polyurethane adhesive single-lap joints for automotive applications. *J. Adhes. Sci. Technol.* **2016**, *30*, 1913–1929. [CrossRef]

Article

Low Molecular Weight Bio-Polyamide 11 Composites Reinforced with Flax and Intraply Flax/Basalt Hybrid Fabrics for Eco-Friendlier Transportation Components

Claudia Sergi [1,*], Libera Vitiello [2], Patrick Dang [3], Pietro Russo [4], Jacopo Tirillò [1] and Fabrizio Sarasini [1]

1. Department of Chemical Engineering Materials Environment, Sapienza Università di Roma and UdR INSTM, 00184 Rome, Italy
2. Department of Chemical, Materials and Production Engineering, University of Naples Federico II, 80125 Naples, Italy
3. Arkema High Performance Polymers, Research & Development, Cerdato, 27470 Serquigny, France
4. Institute for Polymers, Composites and Biomaterials—National Council of Research, 80078 Pozzuoli, Italy
* Correspondence: claudia.sergi@uniroma1.it

Citation: Sergi, C.; Vitiello, L.; Dang, P.; Russo, P.; Tirillò, J.; Sarasini, F. Low Molecular Weight Bio-Polyamide 11 Composites Reinforced with Flax and Intraply Flax/Basalt Hybrid Fabrics for Eco-Friendlier Transportation Components. *Polymers* **2022**, *14*, 5053. https://doi.org/10.3390/polym14225053

Academic Editor: Kamila Sałasińska

Received: 28 September 2022
Accepted: 18 November 2022
Published: 21 November 2022

Publisher's Note: MDPI stays neutral with regard to jurisdictional claims in published maps and institutional affiliations.

Copyright: © 2022 by the authors. Licensee MDPI, Basel, Switzerland. This article is an open access article distributed under the terms and conditions of the Creative Commons Attribution (CC BY) license (https://creativecommons.org/licenses/by/4.0/).

Abstract: The transportation sector is striving to meet the more severe European legislation which encourages all industrial fields to embrace more eco-friendly policies by exploiting constituents from renewable resources. In this framework, the present work assessed the potential of a bio-based, low molecular weight PA11 matrix reinforced with flax and intraply flax/basalt hybrid fabrics. To this aim, both quasi-static and impact performance were addressed through three-point bending and low-velocity impact tests, respectively. For hybrid composites, the effect of stacking sequence, i.e., [0/0] and [0/90], and fiber orientation were considered, while the effect of temperature, i.e., −40 °C, room temperature and +45 °C, was investigated for laminates' impact response. The mechanical experimental campaign was supported by thermal and morphological analyses. The results disclosed an improved processability of the low molecular weight PA11, which ensured a manufacturing temperature of 200 °C, which is fundamental to minimize flax fibers' thermal degradation. Both quasi-static and impact properties demonstrated that hybridization is a good solution for obtaining good mechanical properties while preserving laminates' lightness and biodegradability. The [0/90] configuration proved to be the best solution, providing satisfying flexural performance, with an increase between 62% and 83% in stiffness and between 19.6% and 37.6% in strength compared to flax-based laminates, and the best impact performance, with a reduction in permanent indentation and back crack extent.

Keywords: polyamide 11; basalt; flax; hybrid composites; low-velocity impact; temperature; polymer matrix composites

1. Introduction

Most industrial sectors are making efforts to design and produce components with a lower carbon footprint and a lower environmental impact to redress the strong environmental pollution and to face the ever-increasing waste disposal dilemma [1]. Greener policies intended to boost the use of biodegradable and bio-based materials, to reduce wastes pressure on landfills and to promote a circular economy perspective focused on materials reuse and recycle [2,3] were forced by the more restrictive European legislations. Among them, the 2000/53/EC regulation was specifically addressed to vehicles' end-of-life and presented a big challenge for the automotive sector, which makes a massive use of polymers and fiber-reinforced polymers (FRP) to ensure high mechanical-property standards while reducing vehicles' weight and the related fuel consumption. In particular, 8.8% of the 55 million tons of plastics destined to Europe in 2020 were meant to cover the automotive industry demand [4].

Considering that the automotive field covers almost 40% of the global polyamide (PA) demand [5], the exploitation of natural resources to synthesize eco-sustainable PA is a strategic way to face the environmental pollution challenge from the polymeric matrix perspective. For example, the bio-PA11 by Arkema is completely synthesized from renewable resources, i.e., castor oil [6], allowing for a reduction in environmental impact while preserving mechanical reliability. Indeed, PA11 displays only a slight decrease in the mechanical performance compared to the fully petroleum-based PA12, while ensuring a reduction of CO_2 emissions from 6.9 $kgCO_{2eq}/kg$ to 4.2 $kgCO_{2eq}/kg$ [7].

If polymers' carbon footprint reduction is a first valid initiative to meet the European legislation requirements, further efforts are necessary to solve the issue even from the fibrous-reinforcement perspective. According to the data reported by AVK for 2021, glass FRPs still account for more than 95% of the overall market [8]. A first attempt to assess the problem was made with vegetable fibers to obtain fully bio-based and partially/totally biodegradable composites [9]. On these heels, many European and North American companies, i.e., BASF, Rieter Automotive, Delphi Interior Systems and Visteon Automotive, adopted this strategy [10,11].

Undoubtedly, vegetable fibers offer many advantages throughout the whole life cycle of the component, i.e., production, in-service and disposal. Their renewable origin grants a significant reduction in the production process emissions. For example, the production of hemp and China reed fibers ensures a reduction of almost 97% and 87% in the CO_2 and SO_x emissions with respect to glass fibers, respectively [12,13]. Their low cost, low density and high specific stiffness allow for the reduction of vehicles' fuel consumption and greenhouse gas emissions, providing a reduction of 30% in components' weight and of 20% in their cost [14]. Finally, their biodegradability ensures an easier disposal at the end of the components' life cycle.

Despite all these benefits, vegetable fibers struggle to become competitive on the market, being unable to ensure the same mechanical performance achievable with glass fibers due to the inherently higher mechanical properties of glass and to vegetable fibers' hydrophilicity, which counteracts polymeric matrixes' hydrophobicity, generating a weak interfacial adhesion [15]. The low thermal stability is another drawback which delays vegetable fibers' spread in the industrial landscape [16]. Indeed, temperature lower than 200 °C must be used to avoid fiber degradation, thus decreasing the number of polymers exploitable and their processability.

All these limitations and the rush for bio-based solutions pushed to find another natural solution which was found in the basalt mineral volcanic rock. It proved to be a perfect solution to obtain mechanical properties' competitiveness with glass ones, a greener manufacturing process where no additional chemicals are required and energy consumption is significantly reduced [17], and an easier recycling and recovery of the fibers at the end of composite life cycle [18].

Notwithstanding all the advantages, basalt fibers are not biodegradable and display a density comparable with glass ones, thus losing all the advantages acquired with vegetable reinforcements. Hybridization is a feasible strategy to produce FRPs characterized by satisfying mechanical properties, a lower weight and cost, an improved eco-friendliness and a partial biodegradability. Hybridization benefits are undeniable, as confirmed by the huge number of works that applied this approach for the glass/vegetable fibers system [19,20], but further environmental advantages may be gained by replacing glass fibers with basalt. This idea was assessed using various hybrid configurations with different vegetable fibers such as hemp [21,22] and jute [23,24], but among them, the flax/basalt one was the most widely investigated [25–27] thanks to flax's superior mechanical properties among vegetable fibers [28].

Considering the strategies available to increase composites' eco-friendliness from both the matrix and fiber perspective, the present study aims to assess the potential of a green composite produced with a bio-PA11 matrix reinforced with an intraply flax/basalt hybrid fabric for automotive components. The effect of fibers' hybridization on a different bio-polyamide was already assessed by Bazan et al. [29], who investigated a bio-polyamide 10.10 reinforced with aramid/basalt fibers, and by Armioun et al. [30], who used a PA11 as a matrix and a wood/carbon combination as a hybrid reinforcement. Some preliminary studies on the hybrid configurations proposed in this work were also performed by Russo et al. [31] and by Sergi et al. [32] from the quasi-static and impact response perspective, respectively, but some important steps forward were taken in this work.

In particular, a low molecular weight bio-PA11 was used as a matrix, allowing for the processing of the composite at lower temperatures, i.e., lower than 200 °C, thus reducing flax fibers' thermal degradation. Russo et al. [31] and Sergi et al. [32] assessed the effect of a butyl-benzene-sulfonamide plasticizer on the quasi-static and impact properties of the PA11 flax/basalt composite, respectively, but none of them evaluated the effect of the stacking sequence or fiber orientation. Moreover, Sergi et al. evaluated the impact response at room temperature and +80 °C, but no information is available about the cryogenic impact response. In light of this, the proposed study investigated the quas-istatic, i.e., three-point bending, and impact properties of the low molecular weight bio-PA11 reinforced with the intraply flax/basalt hybrid woven fabric assessing the effect of stacking sequence and fiber orientation. In particular, the $[0/0]_s$ and the $[0/90]_s$ stacking sequences were considered, and a direct comparison with a flax-reinforced PA11 was provided to disclose the significant improvement offered by hybridization. Considering that impact resistance is a key feature for the transportation field, composite impact response was investigated as a function of both impact energy, i.e., 10 J, 20 J and 30 J, and operating temperature, i.e., −40 °C, room temperature and +45 °C. The mechanical characterization was accompanied by a morphological and structural characterization through scanning electron microscopy and X-ray diffraction; by a thermal characterization through thermogravimetric analysis, differential scanning calorimetry and dynamic mechanical analysis; and by a post-impact analysis through profilometry. Moreover, a rheological study aimed at evaluating matrix processability was also carried out.

2. Materials and Methods

2.1. Materials and Manufacturing Process

The low molecular weight bio-PA11 FMNO Rilsan® by Arkema (Serquigny, France) was selected as a polymer matrix, while a flax and intraply flax/basalt hybrid fabric were selected as reinforcements. In particular, a 2/2 flax twill supplied by Composites Evolution Ltd. (Chesterfield, UK) with an areal weight of 300 g/m² and a 2/2 interwoven flax/basalt hybrid twill provided by Lincore® with an areal density of 360 g/m² and a 50 wt% of basalt and flax fibers were used to produce the laminates.

Composites were manufactured by hot compression molding using a P400E hydraulic press by Collin GmbH (Maitenbeth, Germany) applying the film-stacking technique. In particular, ten plies of reinforcement were alternated with 100 μm-thick PA11 films and were compressed at 200 °C, increasing the applied pressure from 0.5 to 3 MPa with a 0.5 MPa increment every two minutes. Before composites manufacturing, the PA11 matrix was vacuum oven dried at 70 °C overnight. Three different laminate configurations were considered: PA11 reinforced with flax (Flax), PA11 reinforced with the intraply hybrid fabric using a $[0/0]_s$ stacking sequence (Hybrid [0/0]) and PA11 reinforced with the intraply hybrid fabric using a $[0/90]_s$ stacking sequence (Hybrid [0/90]). The main characteristics of the three configurations investigated are reported in Table 1.

Table 1. Main characteristics of the three composite configurations under study.

Composite	Reinforcement Type	Stacking Sequence	Thickness (mm)	Fiber Volume Fraction
PA11_Flax	Flax	$[0/0]_s$	5.30 ± 0.05	0.40
PA11_Hybrid $[0/0]_s$	50% Flax/50% Basalt	$[0/0]_s$	4.40 ± 0.05	0.44
PA11_Hybrid $[0/90]_s$	50% Flax/50% Basalt	$[0/90]_s$	3.90 ± 0.05	0.49

2.2. Rheological Characterization and Molecular Weight Assessment

Rheological analyses were performed on the polyamide 11 films provided by Arkema using a stress-controlled rotational rheometer AR-G2 by TA Instruments (New Castle, DE, USA) equipped with 25 mm diameter parallel plates. Consecutive frequency scan tests were carried out on the same sample over a time interval of approximately 16 min, from the frequency ω = 100 up to 1 rad s^{-1} at three temperatures: 200 °C, 210 °C and 220 °C. Before the analyses, PA11 films were dried at T = 70 °C under vacuum overnight (i.e., 12 h).

To simulate the environment of the compression molding process, the experiments were run in air atmosphere. The complex viscosity (η^*) was recorded as a function of frequency in the linear elastic regime (strain amplitude of 5%), in turn identified by preliminary strain amplitude tests.

Size exclusion chromatography (SEC) was performed to check if there was an evolution of the M_n or M_w during processing of the composites. The samples were dissolved for 24 h at room temperature in hexafluoroisopropanol (HFIP). The molecular weights are given in PMMA equivalent (g/mol).

2.3. Thermal and Structural Characterization

Thermal and microstructural characterizations were performed to analyze and interpret material mechanical response. Thermogravimetric analysis (TGA), Differential Scanning Calorimetry (DSC) and Dynamic Mechanical Analysis (DMA) were used to for thermal characterization, while X-ray diffraction (XRD) was employed for structural characterization. Three samples for each characterization technique and each configuration were tested.

TGA was performed on both PA11 neat matrix and the related composites with a Setsys Evolution by Setaram according to ISO 11358. Tests were performed in an inert nitrogen atmosphere from room temperature to 800 °C, employing a heating ramp of 10 °C/min. DSC was carried out on the PA11 matrix with a DSC 214 Polyma by Netzsch (Selb, Germany), according to ISO 11357. Specimens were tested in a nitrogen atmosphere and in a temperature range from 20 °C to 250 °C using a heating/cooling ramp of 10 °C/min. The specimens were subjected to a first heating to remove the thermal history, and the data resulting from the cooling and second heating were used for the analysis. By applying Equation (1), where ΔH_m is the melting enthalpy evaluated from the DSC curve and ΔH_{m0} is the melting enthalpy of the PA11 matrix fully crystallized (ΔH_{m0} = 226.4 J/g) [33], it was also possible to evaluate matrix crystallinity X_c:

$$X_c = \frac{\Delta H_m}{\Delta H_{m0}} * 100 \qquad (1)$$

DMA was performed on the neat matrix and on all laminates' configurations along both flax and basalt directions with a DMA 242 E Artemis by Netzsch according to ISO 6721. Samples with a length of 60 mm and a width of 10 mm were tested using a 3-point bending configuration, a temperature range from 20 °C to 100 °C, a heating ramp of 2 °C/min, an amplitude of 30 µm and a frequency of 1 Hz. Finally, XRD analysis was performed with a Philips X'Pert PRO diffractometer on flax and flax/basalt hybrid composites collecting the spectra in the range of 2θ = 10°–40° using a step size of 0.02°, a time per step of 3 s and a CuK$_\alpha$ monochromatic radiation (40 kV–40 mA).

2.4. Quasi-static Characterization: Flexural Properties

Flax and flax/basalt PA11 composites were subjected to quasi-static characterization in 3-point bending with a Zwick/Roell Z010 universal testing machine equipped with a 10 kN load cell. Tests were carried out on rectangular specimens with a 16:1 span-to-thickness ratio according to ASTM D790 and a test speed of 2.5 mm/min. For the hybrid composite configurations, tests were performed along both flax and basalt directions to disclose potential variations and anisotropy in composite mechanical behavior.

After mechanical testing, specimens were subjected to fracture surface analysis through an FE-SEM MIRA 3 by Tescan. All specimens were sputter-coated with a thin layer of gold to prevent charging because of their low electrical conductivity. Sputter coating was carried out with an Edwards S150B sputter coater.

2.5. Impact Characterization

The impact properties of the materials under study were evaluated through low-velocity impact tests using an instrumented drop weight tower Instron/Ceast 9340. Tests were performed with a hemispherical impactor with an overall mass of 3.055 kg and a diameter of 12.7 mm, using a sample holder with a circular unsupported area of 40 mm in diameter. A pneumatic system assured specimen clamping to sample holder and an anti-rebound system ensured impactor block after the rebound to prevent a second impact. Operating temperature effect was evaluated performing tests at room temperature (25 °C), −40 °C and +45 °C after conditioning the specimens for two hours to ensure a homogeneous temperature profile throughout the material. The combinations of temperature and impact energies investigated for each composite configuration are reported in Table 2.

Table 2. Impact energy and temperature combinations investigated for each PA11 composite configuration.

Laminate	Flax	Hybrid [0/0]$_s$	Hybrid [0/90]$_s$
−40 °C	10 J, 20 J	10 J, 20 J, 30 J	20 J
Room Temperature	10 J, 20 J, 30 J	10 J, 20 J, 30 J	20 J
+45 °C	10 J, 20 J, 30 J	10 J, 20 J, 30 J	20 J

Impacted specimens were also subjected to a post-impact analysis to evaluate damage extent as a function of impact energy and operating temperature. The residual indentation depth was assessed through profilometry using a laser profilometer Talyscan 150 by Taylor Hobson and a scan speed of 8500 µm/s. The resulting scanned surface was analyzed with the software TalyMap 3D. The back crack extent was also quantified with the image processing software Image J by analyzing the impacted specimen photographs.

3. Results and Discussion

3.1. Rheological Characterization and Molecular Weight Assessment

The evolution over time of the complex viscosity of the PA11 matrix subjected to consecutive frequency scan tests at three different temperatures is shown in Figure 1. As expected, the higher the test temperature, the lower the viscosity, which, among other things, at the lowest temperature examined (200 °C), shows an almost Newtonian trend apparently not influenced by time. On the contrary, at temperatures of 210 and 220 °C, by repeatedly testing the same sample, an increase in the complex viscosity is evident, with an increasingly marked upward trend of the curves as the test temperature increases: the effect is particularly accentuated in the low-frequency region, in correspondence of which, notoriously, the response of large portions of chains is detected.

This behavior, typical of condensation polymers such as polyamides and polyesters, has been the subject of previous research, according to which the increasing evolution of viscosity and of some viscoelastic parameters is due to the concomitant occurrence of various structural changes. These phenomena are related to both the intrinsically hygroscopic character of these polymers, which inevitably influences the rheological behavior and the optimal process modalities of the same, and the environment in which the material is tested (inert or air).

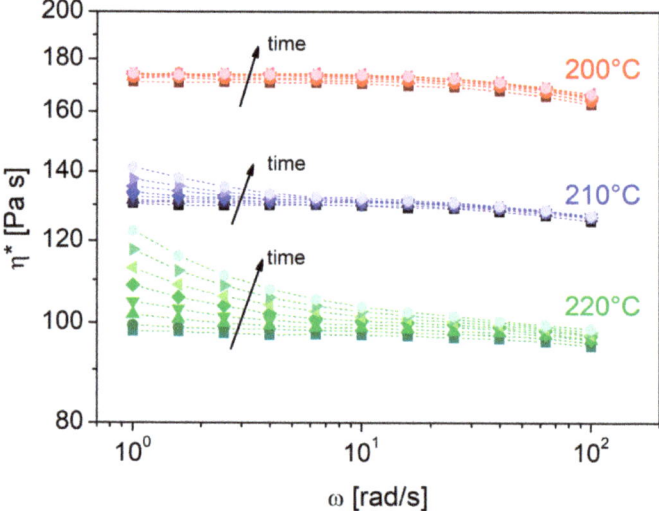

Figure 1. Variation of the complex viscosity (η^*) of polyamide 11 in air at three different temperatures: from bottom to top, 220 °C, 210 °C and 200 °C.

In general, if the moisture content of the material is less than that of thermodynamic equilibrium, it is reasonable to attribute the increase in viscosity to an increase in the molecular weight of the material due to the occurrence of post-condensation phenomena. In practice, this mechanism is prevalent if we consider the behavior of the melt in an inert environment (nitrogen) [34]. Conversely, in the air, it is very likely that simultaneous thermo-degradative phenomena such as chain scissions and cross-linking occur. It is reasonable to believe that these phenomena begin at the edge of the disk-shaped sample and then propagate over time in the bulk of the sample between the plates thanks to the combination of air and humidity of the material [35]. Filippone et al. [35], combining rheological tests and molecular measurements such as size exclusion chromatography and mass-chromatography, have shown that the increase in complex viscosity over time is essentially related to the extension of cross-linked polymer fractions (insoluble gels) and, therefore, the hindering of the macroscopic flows.

In light of the previous considerations, to limit the occurrence of the aforementioned structural changes and avoid an excessive increase in the viscosity of the matrix that could compromise the impregnation of the reinforcing fabrics, the lowest temperature explored (200 °C) was chosen for the laminate production. Results from SEC (Table 3) appear to be in line with the rheological characterization. Compared to the initial PA11 pellet, the composite samples showed higher M_n and M_w and also a higher M_w/M_n. This indicates that in the case of flax-based samples, there is an increase in the molecular weight, especially on the high-M_w end, while hybrid samples displayed an enlargement on both sides of molecular weight. The very high M_w of flax-based composites might come from some branching reactions due to process conditions.

Table 3. Molecular weights obtained by size-exclusion chromatography.

Composite	M_n (g/mol)	M_w (g/mol)	M_z (g/mol)	M_w/M_n	M_z/M_w
PA11_Matrix	25,900	47,200	69,900	1.8	1.5
PA11_Flax	30,600	109,900	497,600	3.6	4.5
PA11_Hybrid	23,400	54,400	309,900	2.3	5.7

Branching can be caused by oxidation; therefore, the higher fiber areal weight of hybrid fabric might reduce the permeability of air/oxygen during processing which, coupled with lower thickness and higher thermal conductivity, resulted in a shorter cooling time, so the residence time at high temperature was reduced. This would explain the higher degree of branching observed in flax-based laminates.

3.2. Thermal and Structural Characterization

3.2.1. Thermogravimetric Analysis (TGA)

The thermal stability of the PA11 matrix and the resulting composites, i.e., flax and hybrid, was studied by TGA. The resulting mass loss and derivative curves are shown in Figure 2, while the mass drops and the related degradation temperatures are summarized in Table 4. The neat PA11 matrix is characterized by a single degradation step, with a maximum degradation rate at around 475 °C, while the flax and hybrid composites are characterized by two degradation steps related to flax and PA11 matrix degradation, respectively. The mass loss connected with flax fibers' degradation is 42.6% for pure flax composite and 22% for hybrid composites and takes place at 368 °C and 354 °C, respectively. These results are in perfect agreement with the ones provided by Lafranche et al. [36], who reported flax degradation at 365 °C, and by Kannan et al. [37], who identified flax degradation in the 333–375 °C temperature range.

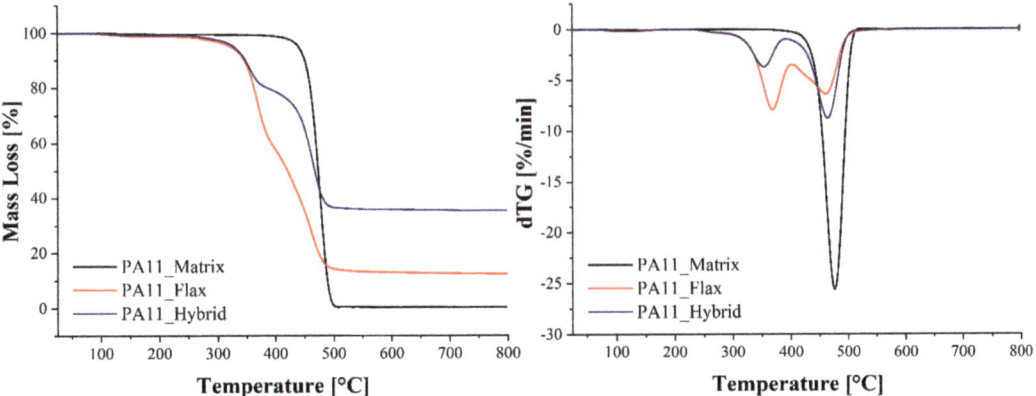

Figure 2. Mass loss and derivative curves of the PA11 matrix and of the flax and flax/basalt hybrid composites.

Table 4. Mass drops and related degradation temperatures for the PA11 matrix and the flax and flax/basalt hybrid composites.

	PA11_Matrix	PA11_Flax	PA11_Hybrid
1st drop temperature (°C)	-	368.5	354.0
1st mass loss (%)	-	42.6	22.0
2nd drop temperature (°C)	475.6	459.6	463.3
2nd mass loss (%)	100	44.5	42.7

3.2.2. Differential Scanning Calorimetry (DSC) and X-ray Diffraction (XRD)

DSC analysis provided further proof of the 200 °C eligibility as a manufacturing temperature. From the DSC curves of the low molecular weight PA11 shown in Figure 3, it was possible to evaluate the crystallization and melting temperatures of the matrix. In particular, a crystallization temperature of 162.6 °C and a double melting peak at 182.3 °C and 191 °C were detected. This means that 200 °C is the best compromise to achieve full polymer melting while preventing polymer structural changes and deterioration over time.

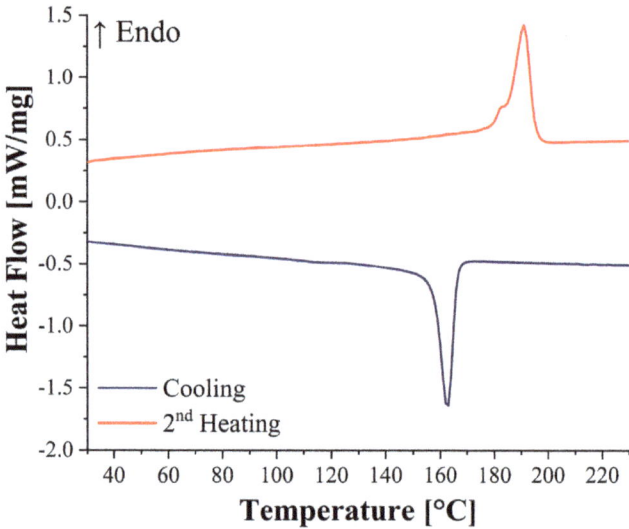

Figure 3. DSC curves of the low molecular weight PA11 matrix.

A crystallinity degree of 30.3% was calculated, and it is slightly higher than the 27% reported by Russo et al. for a regular PA11 matrix by Arkema [31]. This can be ascribed to the lower molecular weight of the matrix, which ensures an easier macromolecular mobility, thus promoting matrix crystallization [38].

As previously mentioned, the PA11 matrix is characterized by a double melting peak ascribable to a two-phase transition resulting from two different crystalline phases [39]. An XRD analysis of the flax and hybrid composites was carried out to confirm this hypothesis, and the resulting XRD spectra are reported in Figure 4. Both flax and hybrid composites display two diffraction peaks at 20.2°–22.7° and 20.8°–22.6°, respectively.

The peaks at around 20° can be ascribed to the (200) plane of the triclinic α structure [40] and to the (100) plane of the α' structure, i.e., a pseudohexagonal δ phase [41]. The peak at around 23° must be ascribed to the (010) and (210) of the triclinic α structure [40]. These results are coherent with the ones reported by Sergi et al. [32] for a regular PA11 matrix. The observation of the two peaks in the XRD spectra resulting from the two different crystalline phases is a valid support to explain the two-stage melting detected in the DSC. Two more peaks at 15.0° and 16.4° for flax composites and at 15.0° and 16.7° for hybrid composites were also identified and must be ascribed to the crystalline component of cellulose I contained in the flax fibers [42]. Indeed, cellulose I also presents a diffraction peak at around 22°–22.8°, which, added to the diffraction resulting from PA11, explains the strong intensity detected at that diffraction angle.

3.2.3. Dynamic Mechanical Analysis (DMA)

DMA allowed us to evaluate the glass transition temperature and the evolution of storage modulus as a function of temperature for the neat PA11 matrix and for both flax and hybrid configurations accounting also for fiber orientation. Figure 5 shows the DMA curves of PA11 matrix, flax composite and hybrid [0/0] composites tested along the flax and basalt direction and hybrid [0/90] composites tested along the flax and basalt direction, while Table 5 summarizes the glass transition temperature values evaluated from tanδ and the values of E' at 25, 45, 60 and 80 °C.

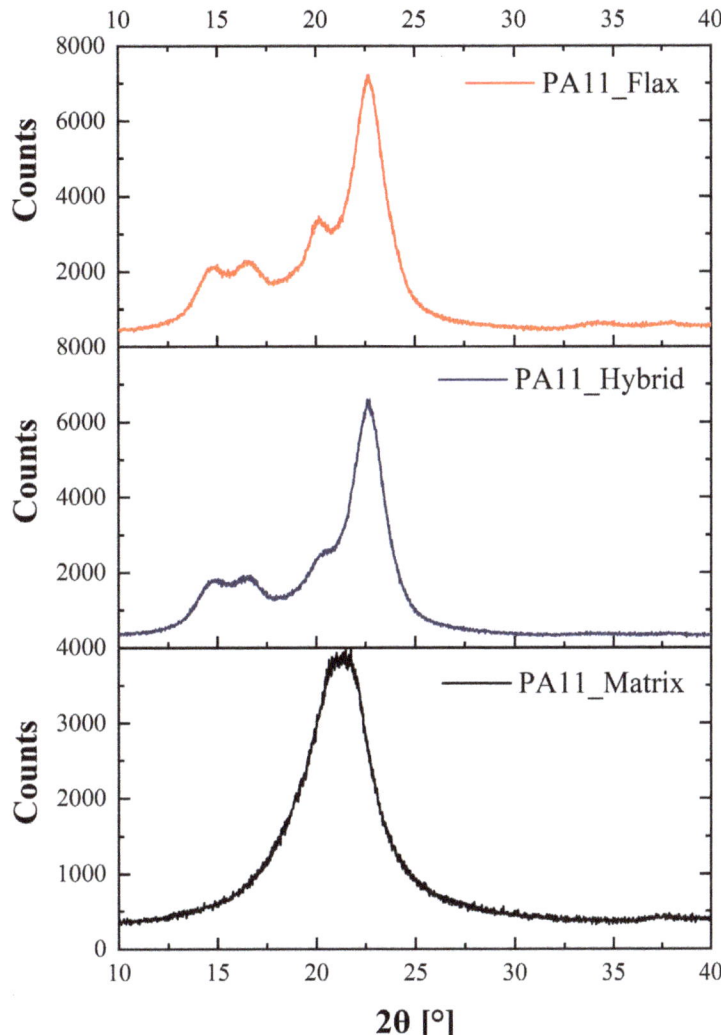

Figure 4. XRD spectra of the as-received PA11_flax, PA11_hybrid composites and PA11_matrix.

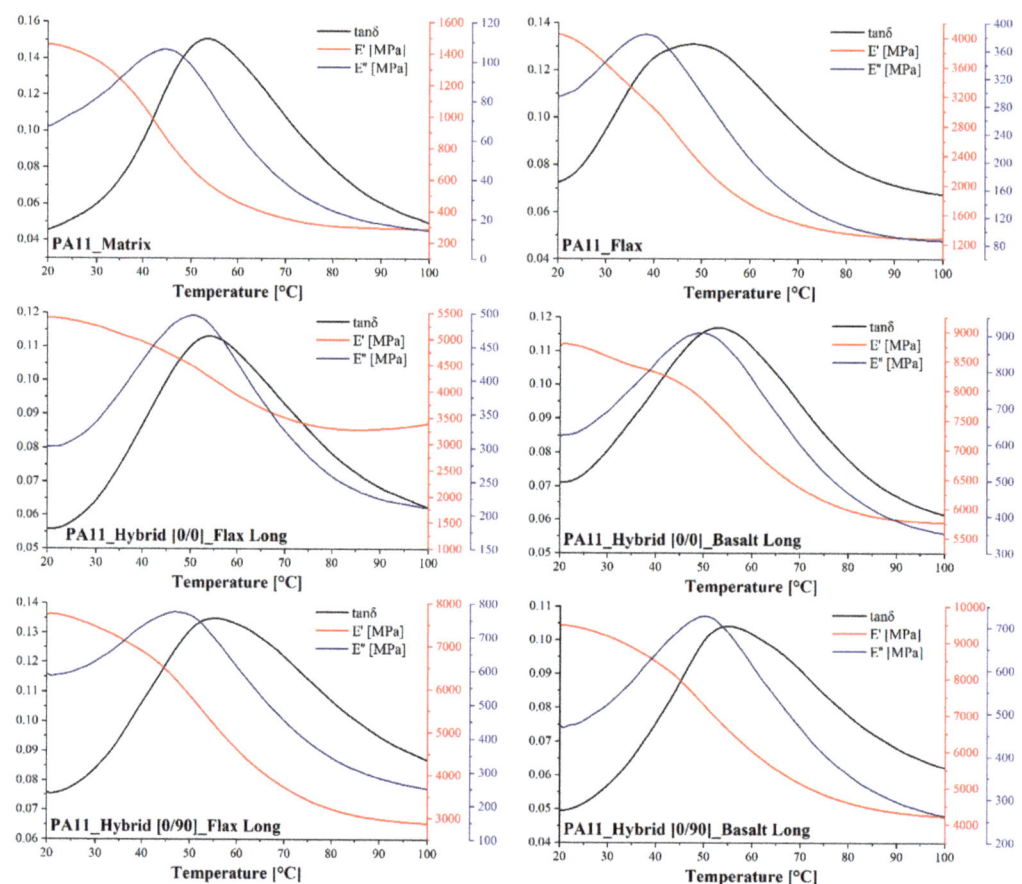

Figure 5. DMA curves of PA11 matrix, flax composite and hybrid [0/0] composites tested along flax and basalt direction and hybrid [0/90] composites tested along flax and basalt direction.

Table 5. Glass transition temperature and E' evolution as a function of temperature of PA11 matrix, flax composite and hybrid [0/0] composites tested along flax and basalt direction and hybrid [0/90] composites tested along flax and basalt direction.

	Glass Transition Temperature [°C]	E'(25 °C) [MPa]	E'(45 °C) [MPa]	E'(60 °C) [MPa]	E'(80 °C) [MPa]
PA11_Matrix	53.4	1420.9	847.0	454.9	303.8
PA11_Flax	48.0	3899.6	2650.5	1748.3	1353.0
PA11_Hybrid [0/0]_Basalt_Long	52.9	8730.4	8125.4	7006.6	5981.5
PA11_Hybrid [0/0]_Flax_Long	53.9	5353.2	4739.3	3924.8	3296.2
PA11_Hybrid [0/90]_Basalt_Long	55.2	9378.0	7951.6	6012.8	4603.3
PA11_Hybrid [0/90]_Flax_Long	55.0	7657.4	6437.0	4590.8	3204.2

All hybrid configurations are characterized by a glass transition temperature comparable with the neat PA11 matrix one at around 53–55 °C. On the contrary, flax laminates exhibit a decrease of 6 °C in this value, probably as a consequence of the plasticizing effect played by moisture, which is absorbed at a larger extent in the composite because of flax fibers' higher content.

3.3. Mechanical Characterization

3.3.1. Flexural Characterization

The quasi-static performance of the bio-based composites under study was evaluated in three-point bending, also assessing the effect of fiber orientation for hybrid configurations. The flexural modulus and strength values are reported in Table 6. The introduction of flax fibers in the PA11 matrix provides an increase of 195% in the bending stiffness and of 78% in the flexural strength, but a further improvement between 380 and 464% in the stiffness and between 112 and 280% in the strength can be achieved by exploiting the hybrid configurations. As expected, flax laminates are characterized by lower flexural properties than hybrids because of the inherent superior mechanical performance of basalt, thus proving that hybridization is a viable solution to produce components with satisfying mechanical requirements while maintaining their partial biodegradability and lightness.

Table 6. Flexural modulus and strength values for flax and hybrid configurations along flax and basalt directions.

	Flexural Modulus [GPa]	Flexural Strength [MPa]
PA11_Matrix	1.74 ± 0.14	72.33 ± 3.35
PA11_Flax	5.13 ± 0.46	128.47 ± 6.34
PA11_Hybrid [0/0]_Basalt_Long	9.82 ± 0.36	276.14 ± 9.38
PA11_Hybrid [0/0]_Flax_Long	8.34 ± 0.29	175.03 ± 5.95
PA11_Hybrid [0/90]_Basalt_Long	9.4 ± 0.31	153.69 ± 7.27
PA11_Hybrid [0/90]_Flax_Long	9.79 ± 0.20	218.55 ± 6.33

Among hybrids, the [0/0] configuration is characterized by the highest and the lowest flexural stiffness when tested along basalt and flax directions, respectively. Again, the intrinsic higher mechanical performance of basalt fibers ensures the best outcomes when the laminate is tested along their directions, determining a decrease of 15% in stiffness and of 36.6% in strength when tested transversally to their orientation. These results are coherent with the E' evolution as a function of temperature reported in Table 5. In fact, the [0/0] configuration tested along the basalt direction shows the highest values along the whole temperature range, while the same configuration tested along the flax direction is characterized by the lowest ones. The [0/90] configuration featured intermediate quasi-static properties with a flexural stiffness closer to the [0/0] upper limit. The obtained results are promising, considering that Vitiello et al. [43] reported a 15.7 GPa flexural stiffness and a 174 MPa flexural strength for a regular PA11 matrix reinforced with a basalt 2 × 2 twill, and Vitiello et al. [44] acknowledged a bending stiffness of around 13 GPa and a strength of around 130 MPa for a PA11 reinforced with a plain-weave basalt fabric. The results compare favorably even with more traditional composite configurations, such as polypropylene (PP) glass-reinforced ones. In particular, Russo et al. [45] and Simeoli et al. [46] report a flexural modulus between 15.1 and 15.7 GPa and a flexural strength between 112 and 183 MPa working with fiber volume fractions around 0.5 and 0.54, which are higher than the ones reported in this work, respectively.

Tested specimens were further analyzed through SEM to investigate composite fracture surface, and Figure 6 shows the fracture surface of both flax (A) and hybrid (B and C) laminates. Figure 6A,B highlight some interfacial adhesion issues arising between the extremely hydrophilic flax and the PA11 matrix. In particular, Figure 6A shows a partial detachment of the matrix from the fibers, proving that further improvements can be achieved to increase matrix/fiber load transfer. Figure 6B, in turn, underlines that PA11/flax compatibility is better compared to a more hydrophobic matrix such as polypropylene and polyethylene, but there is certainly further room for improvement. In particular, the flax fiber displays a surface with some residual traces of matrix proving some degree of compatibility. Better results in terms of compatibility were achieved between PA11 and basalt fibers, as shown in Figure 6C. Despite the presence of fiber pull out, basalt fibers display many residual traces of PA11 on their surface, which prove to successfully transfer the load from the matrix to the fibers. If the improvement of the quasi-static properties of the proposed bio-based composites is the main goal from the design perspective, many physical and chemical treatments can be exploited to improve flax and PA11 adhesion and the resulting load bearing capabilities. Quiles-Carrillo et al. [47] demonstrated that the introduction of glycidyl-silane and amino-silane in bio-PA10-10/slate composites allows for the increase in their tensile strength and stiffness. Moreover, a good interfacial adhesion was achieved by Armioun et al. [30] in PA11/wood fiber composites by adding a maleic anhydride grafted polypropylene as coupling agent.

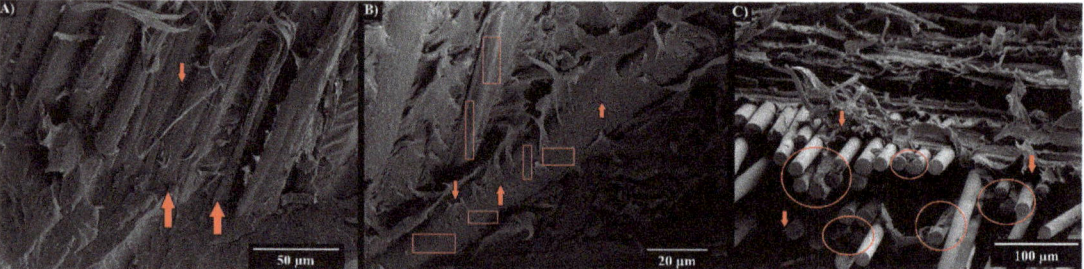

Figure 6. SEM micrographs detailing the fracture surface of flax (**A,B**) and hybrid composites (**C**).

3.3.2. Impact Characterization

Quasi-static mechanical properties are important to ensure composites use, but their impact response is also crucial to ensure their applicability in the transportation sector. The main impact properties of flax, [0/0] and [0/90] composites for different impact energies at room temperature are summarized in Table 7, while Figure 7 shows the force-displacement curves of the aforementioned laminates impacted at 20 J.

Table 7. Main impact properties, i.e., maximum force, maximum displacement and damage degree, of flax, [0/0] and [0/90] hybrid composites at room temperature at different impact energies.

		PA11_Flax	PA11_Hybrid [0/0]	PA11_Hybrid [0/90]
10 J	Max Force [N]	4244.2 ± 89.20	4200.1 ± 0.70	-
	Max Displacement [mm]	3.25 ± 0.09	3.55 ± 0.02	-
	Damage Degree	0.76 ± 0.03	0.74 ± 0.01	-
20 J	Max Force [N]	4879.8 ± 97.70	5342.2 ± 148.60	6793.1 ± 61.00
	Max Displacement [mm]	5.26 ± 0.05	5.38 ± 0.08	5.01 ± 0.01
	Damage Degree	0.84 ± 0.03	0.79 ± 0.01	0.74 ± 0.01
30 J	Max Force [N]	5179.1 ± 70.00	5614.8 ± 18.20	-
	Max Displacement [mm]	12.36 ± 0.04	7.43 ± 0.02	-
	Damage Degree	1.00 ± 0.01	0.84 ± 0.01	-

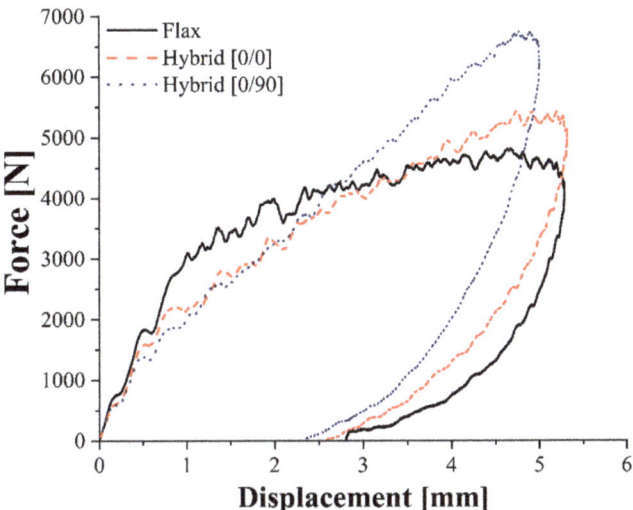

Figure 7. Force-displacement curves of flax and hybrid composites for a 20 J impact at room temperature.

As already acknowledged in flexural tests, both hybrid configurations considerably outperform flax–PA11 composites displaying a higher maximum force, a lower maximum displacement and therefore a higher stiffness. Moreover, flax composites are always characterized by a higher damage degree, defined as the ratio of absorbed to impact energy, which can compromise their integrity and their residual mechanical properties. The higher damage extent is also confirmed by laminates' damage mode shown in Figure 8 and permanent indentation and back crack extent data summarized in Figure 9 for a 20 J impact.

Concerning biocomposite damage mode, hybrid configurations display a single crack on the rear side which runs perpendicularly to flax fibers due to the lower strength of these fibers, which act as a damage preferential path requiring a lower energy to propagate. On the contrary, PA11–flax composites are characterized by a cross-shaped rear crack which derives from their symmetric nature and the circularity of the sample holder where no preferential paths occur. Focusing on the damage extent, flax composites display a much higher back crack extent than hybrids and a higher permanent indentation than hybrid [0/90], thus confirming that the higher amount of energy absorbed, and the higher damage degree, can significantly jeopardize their residual structural performance.

Among the two hybrid configurations, [0/90] laminates respond better to the impact than [0/0] ones exhibiting the highest maximum force and the lowest maximum displacement and damage degree. This must be attributed to the lower anisotropy of the [0/90] configuration resulting from fabric orientation alternation which involves basalt fibers orientation along two different directions. This ensures satisfying load-bearing capabilities along two perpendicular directions rather than one, thus delaying flax fibers' overloading and breakage. This is confirmed by the damage-extent analysis. In fact, [0/90] hybrids are characterized by a decrease of 38.6% in the permanent indentation and of 17.7% in back crack extent.

The effect of low and high temperatures on the impact response of the bio-based composites under study was assessed, and Figure 10 shows the impact response curves, i.e., force against displacement, of flax, [0/0] and [0/90] hybrid laminates impacted at 20 J at −40 °C, room temperature and +45 °C. Regardless of the operating temperature, flax-reinforced PA11 is always characterized by the worst impact response displaying the lowest maximum force and the highest maximum displacement.

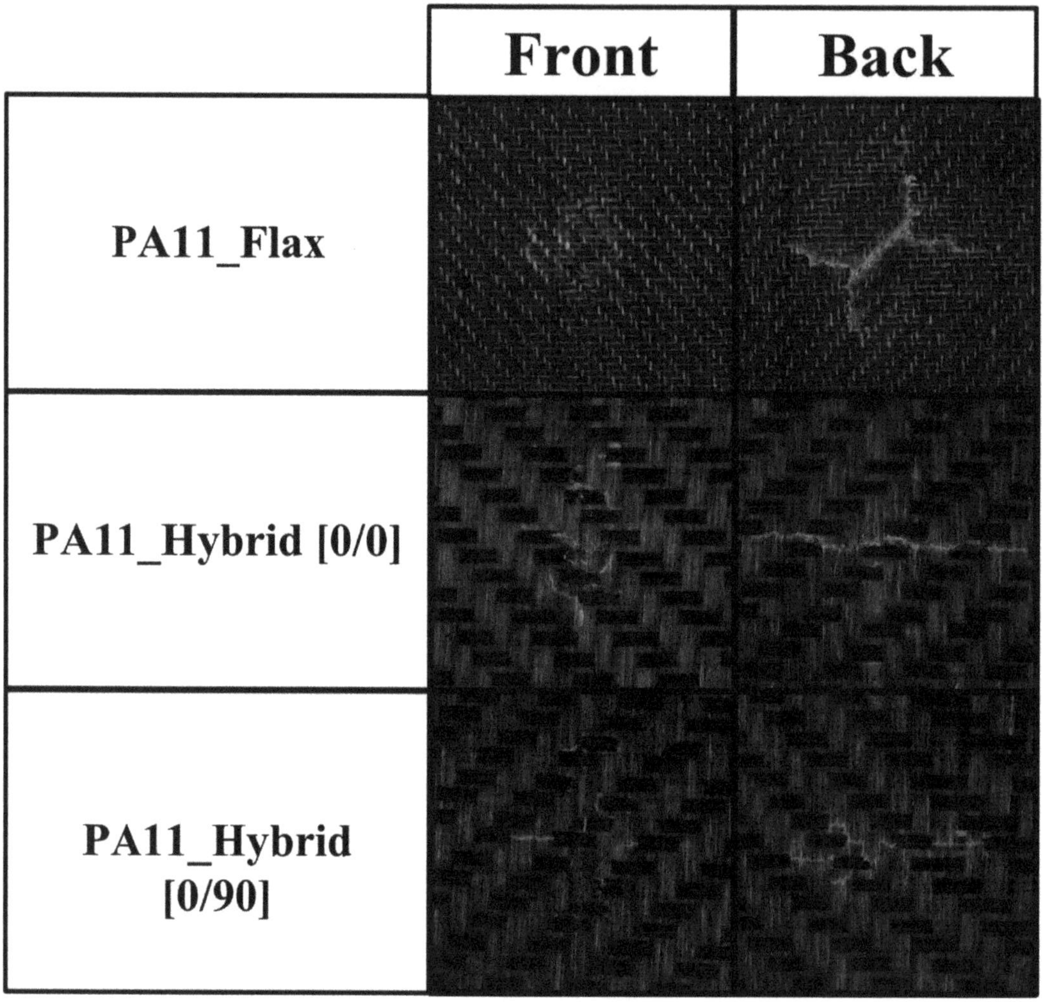

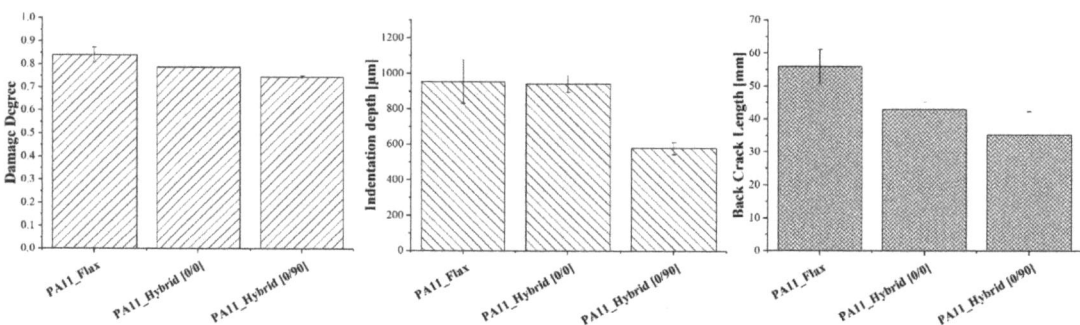

Figure 8. Front and back damage mode of flax, hybrid [0/0] and hybrid [0/90] laminates impacted at room temperature at 20 J.

Figure 9. Damage degree, permanent indentation and back crack extent of flax and hybrid laminates after a 20 J impact at room temperature.

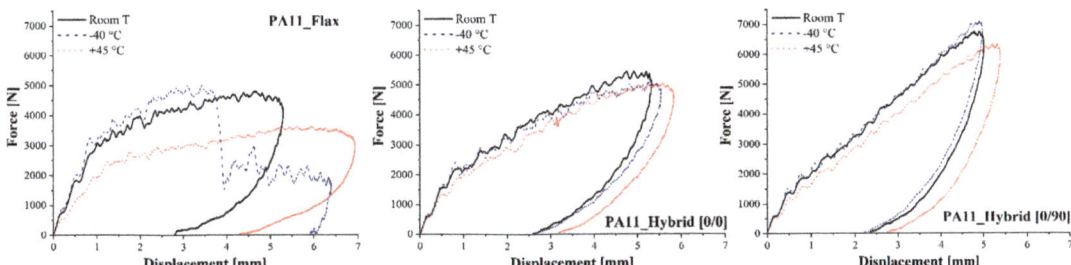

Figure 10. Force-displacement curves of flax and hybrid composites for a 20 J impact at −40 °C, room temperature and +45 °C.

For all composite configurations, the progressive increase in operating temperature induces a decrease in laminates' maximum force and an increase in the maximum displacement, which derives from a progressive decrease in laminates' stiffness. This effect becomes particularly evident at +45 °C, which is extremely close to the PA11 matrix glass transition temperature, and a transition from the glassy to the rubbery state can significantly modify composite response. The stiffer, even if more brittle, response of the laminates for decreasing operating temperature is also confirmed by flax composites, which undergo significant penetration phenomena when impacted at 20 J and −40 °C while remaining in the elastic rebound region when tested at room temperature and +45 °C.

The progressive approach of matrix glass transition temperature determines its progressive softening, and the resulting reduction in stiffness makes the laminate more compliant toward the impact, allowing it to involve a wider area of the specimen, thus preventing damage localization and penetration phenomena. All these considerations are supported by the composite damage mode shown in Figure 11 and the damage extent analysis reported in Figure 12. Considering flax composites, the highest damage degree is achieved at −40 °C due to the brittle response of the laminate, which leads to a strong penetration of the impactor and appears in the form of a pronounced permanent indentation, i.e., six times higher than the one of specimens impacted at room temperature and +45 °C.

Different is the trend for hybrid laminates, where a progressive increase in damage degree can be observed for increasing operating temperatures while keeping the permanent indentation almost constant and decreasing the back crack extent. Furthermore, hybrids impacted at −40 °C display some brittle cracks arising on the front side of the specimen near the contact area with the impactor. These last items become progressively less branched and pronounced when working at higher operating temperatures. This can be explained considering that the laminates are approaching PA11 glass transition temperature, thus making the matrix more prone to plastic deformation and activating further energy dissipation mechanisms. This allows for an increase in the amount of energy that the laminate is able to store, keeping constant the local indentation and delaying the formation of the back crack resulting from the approach of the bending elastic limit of the laminate.

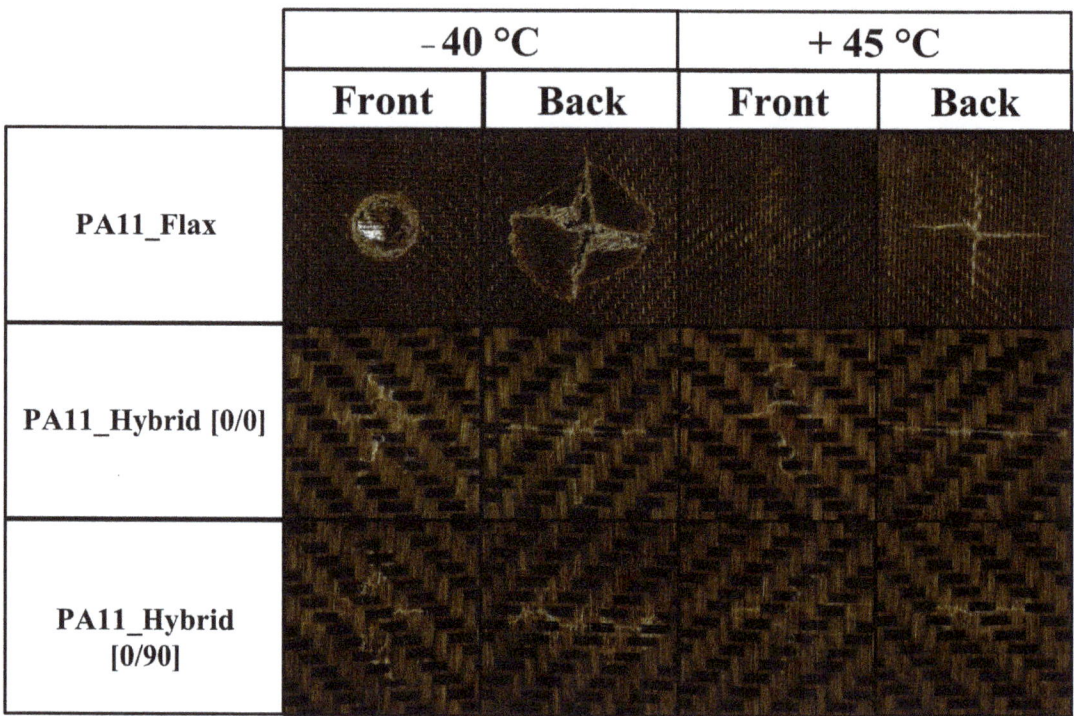

Figure 11. Front and back damage mode of flax, hybrid [0/0] and hybrid [0/90] laminates impacted at 20 J at −40 °C and +45 °C.

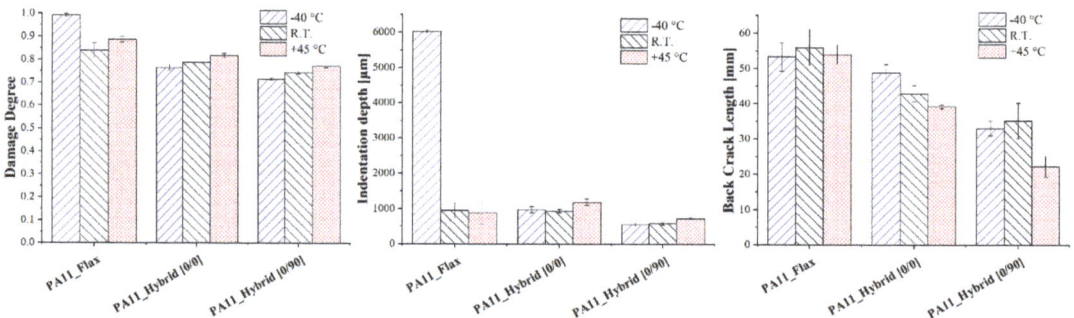

Figure 12. Damage degree, permanent indentation and back crack extent of flax and hybrid laminates after a 20 J impact at −40 °C, room temperature and +45 °C.

4. Conclusions

The present research work investigated the thermal, quas-istatic and impact properties of a low molecular weight bio-based PA11 reinforced with flax and intraply flax/basalt hybrid fabrics to assess the feasibility of these bio-based composites for the automotive sector. For hybrid composites, the effect of stacking sequence, i.e., [0/0] and [0/90], and fiber orientation were considered, while the effect of temperature, i.e., −40 °C, room temperature and +45 °C, was investigated for laminates' impact properties. The main outcomes of the study are as follows:

- The low molecular weight of the PA11 matrix ensures an easier manufacturing of the composites displaying a good processability already at 200 °C. This allows one to obtain components of good quality while preventing matrix and, above all, flax fibers' thermal degradation.
- Both quasi static and impact tests proved that hybrid laminates significantly outperform flax ones, thus proving that hybridization is an effective way to improve the mechanical performance of bio-based composites while preserving their lightness.
- Among the hybrid configurations under study, the [0/90] is the best at ensuring a satisfying bending stiffness and an improved impact resistance deriving from the higher isotropy in basalt fibers' orientation, which allows for a better distribution of the impact load and delays flax fibers' overloading and breakage.
- Operating temperature strongly influences composite impact properties, determining a progressive decrease in matrix, and therefore in laminate, stiffness, which makes the component more compliant toward the impact, thus involving a wider area of the specimen and preventing damage localization.

As already discussed, the results obtained for the [0/90] configuration compare favorably with PA11 fully basalt-reinforced composites, but also with more traditional composite configurations such as PP/glass-reinforced ones, thus validating the viability of this bio-based solution for the manufacturing of more eco-friendly and sustainable automotive components. In addition, this material combination might provide a better recyclability compared to thermoset-based composites, as PA11/flax composites have already proven to be very efficient after recycling [48], a behavior that can be even improved by the presence of basalt fibers.

Author Contributions: Conceptualization, F.S., P.R. and P.D.; methodology, F.S., P.R. and P.D.; validation, F.S., C.S. and P.D.; formal analysis, F.S., P.R., C.S. and P.D.; investigation, J.T., P.R., P.D., L.V. and C.S.; resources, F.S., J.T., P.R. and P.D.; data curation, F.S., C.S. and P.R.; writing—original draft preparation, C.S. and F.S.; writing—review and editing, P.R., P.D. and J.T.; visualization, C.S.; supervision, F.S., J.T. and P.R.; funding acquisition, F.S., P.R. and P.D. All authors have read and agreed to the published version of the manuscript.

Funding: The work was developed in the framework of the project Thalassa (PON "R&I" 2014–2020, grant ARS01_00293, Distretto Navtec) funded by MUR (Italian Ministry for University and Research).

Institutional Review Board Statement: Not applicable.

Informed Consent Statement: Not applicable.

Data Availability Statement: All data are contained within the article.

Conflicts of Interest: The authors declare no conflict of interest.

References

1. Oladele, I.O.; Adelani, S.O.; Agbabiaka, O.G.; Adegun, M.H. Applications and Disposal of Polymers and Polymer Composites: A Review. *Eur. J. Adv. Eng. Technol.* **2022**, *9*, 65–89.
2. Shamsuyeva, M.; Endres, H.-J. Plastics in the context of the circular economy and sustainable plastics recycling: Comprehensive review on research development, standardization and market. *Compos. Part C Open Access* **2021**, *6*, 100168. [CrossRef]
3. Shogren, R.; Wood, D.; Orts, W.; Glenn, G. Plant-based materials and transitioning to a circular economy. *Sustain. Prod. Consum.* **2019**, *19*, 194–215. [CrossRef]
4. PlasticEurope. Association of Plastics Manufacturers. In *Plastics-the Facts 2021 An Analysis of European Plastics Production, Demand and Waste Data*; PlasticEurope: London, UK, 2021.
5. Fortune Business Insights. Nylon market size by type, by application and regional forecast, 2020-2027. In *Market Research Report*; Fortune Business Insights: Pune, India, 2020.
6. Winnacker, M.; Rieger, B. Biobased Polyamides: Recent Advances in Basic and Applied Research. *Macromol. Rapid Commun.* **2016**, *37*, 1391–1413. [CrossRef] [PubMed]
7. Armioun, S.; Pervaiz, M.; Sain, M. Biopolyamides and High-Performance Natural Fiber-Reinforced Biocomposites. In *Handbook of Composite from Renewable Materials*; Wiley: New York, NY, USA, 2017; Volume 3, pp. 253–270.
8. Witten, E.; Mathes, V. *The European Market for Fibre Reinforced Plastics/Composites in 2021*; AVK, (Industrievereinigung Verstärkte Kunststoffe): Galten, Denmark, 2022.

9. Mohanty, A.K.; Misra, M.; Drzal, L.T. Sustainable Bio-Composites from Renewable Resources: Opportunities and Challenges in the Green Materials World. *J. Polym. Environ.* **2002**, *10*, 19–26. [CrossRef]
10. Boland, C.S.; De Kleine, R.; Keoleian, G.A.; Lee, E.C.; Kim, H.C.; Wallington, T.J. Life Cycle Impacts of Natural Fiber Composites for Automotive Applications: Effects of Renewable Energy Content and Lightweighting. *J. Ind. Ecol.* **2015**, *20*, 179–189. [CrossRef]
11. Li, M.; Pu, Y.; Thomas, V.M.; Yoo, C.G.; Ozcan, S.; Deng, Y.; Nelson, K.; Ragauskas, A.J. Recent advancements of plant-based natural fiber–reinforced composites and their applications. *Compos. Part B Eng.* **2020**, *200*, 108254. [CrossRef]
12. Shahzad, A. Hemp fiber and its composites—A review. *J. Compos. Mater.* **2011**, *46*, 973–986. [CrossRef]
13. Joshi, S.V.; Drzal, L.T.; Mohanty, A.K.; Arora, S. Are natural fiber composites environmentally superior to glass fiber reinforced composites? *Compos. Part A Appl. Sci. Manuf.* **2004**, *35*, 371–376. [CrossRef]
14. Huda, M.S.; Drzal, L.T.; Ray, D.; Mohanty, A.K.; Mishra, M. Natural-fiber composites in the automotive sector. In *Properties and Performance of Natural-Fibre Composites*; Pickering, K.L., Ed.; Woodhead Publishing: Cambridge, UK, 2008; pp. 221–268, ISBN 9781845692674.
15. Pickering, K.L.; Aruan Efendy, M.G.; Le, T.M. A review of recent developments in natural fibre composites and their mechanical performance. *Compos. Part A Appl. Sci. Manuf.* **2016**, *83*, 98–112. [CrossRef]
16. Monteiro, S.N.; Calado, V.; Rodriguez, R.J.S.; Margem, F.M. Thermogravimetric behavior of natural fibers reinforced polymer composites—An overview. *Mater. Sci. Eng. A* **2012**, *557*, 17–28. [CrossRef]
17. Militký, J.; Mishra, R.; Jamshaid, H. Basalt Fibers. In *Handbook of Properties of Textile and Technical Fibres*; Elsevier: Amsterdam, The Netherlands, 2018; pp. 805–840, ISBN 9780081012727.
18. Balaji, K.V.; Shirvanimoghaddam, K.; Rajan, G.S.; Ellis, A.V.; Naebe, M. Surface treatment of Basalt fiber for use in automotive composites. *Mater. Today Chem.* **2020**, *17*, 100334. [CrossRef]
19. Amico, S.C.; D'Almeida, J.R.M.; De Carvalho, L.H.; Cioffi, M.O.H. Hybrid Vegetable/Glass fiber composites. In *Lignocellulosic Polymer Composites: Processing, Characterization, and Properties*; Wiley: New York, NY, USA, 2014; pp. 63–81.
20. Nurazzi, N.M.; Asyraf, M.R.M.; Fatimah Athiyah, S.; Shazleen, S.S.; Rafiqah, S.A.; Harussani, M.M.; Kamarudin, S.H.; Razman, M.R.; Rahmah, M.; Zainudin, E.S.; et al. A Review on Mechanical Performance of Hybrid Natural Fiber Polymer Composites for Structural Applications. *Polymers* **2021**, *13*, 2170. [CrossRef]
21. Kumar, C.S.; Arumugam, V.; Dhakal, H.; John, R. Effect of temperature and hybridisation on the low velocity impact behavior of hemp-basalt/epoxy composites. *Compos. Struct.* **2015**, *125*, 407–416. [CrossRef]
22. Dhakal, H.N.; Sarasini, F.; Santulli, C.; Tirillò, J.; Zhang, Z.; Arumugam, V. Effect of basalt fibre hybridisation on post-impact mechanical behaviour of hemp fibre reinforced composites. *Compos. Part A Appl. Sci. Manuf.* **2015**, *75*, 54–67. [CrossRef]
23. Prasath, K.A.; Krishnan, B.R. Mechanical Properties of Woven Fabric Basalt/Jute Fibre Reinforced Polymer Hybrid Composites. *Int. J. Mech. Eng. Robot. Res.* **2013**, *2*, 279–290.
24. Amuthakkannan, P.; Manikandan, V.; Jappes, J.T.W.; Uthayakumar, M. Influence of stacking sequence on mechanical properties of basalt-jute fiber-reinforced polymer hybrid composites. *J. Polym. Eng.* **2012**, *32*, 547–554. [CrossRef]
25. Almansour, F.; Dhakal, H.; Zhang, Z. Investigation into Mode II interlaminar fracture toughness characteristics of flax/basalt reinforced vinyl ester hybrid composites. *Compos. Sci. Technol.* **2018**, *154*, 117–127. [CrossRef]
26. Sarasini, F.; Tirillò, J.; Ferrante, L.; Sergi, C.; Russo, P.; Simeoli, G.; Cimino, F.; Ricciardi, M.R.; Antonucci, V. Quasi-Static and Low-Velocity Impact Behavior of Intraply Hybrid Flax/Basalt Composites. *Fibers* **2019**, *7*, 26. [CrossRef]
27. Ferrante, L.; Sergi, C.; Tirillò, J.; Russo, P.; Calzolari, A.; Sarasini, F. Temperature effect on the single and repeated impact responses of intraply flax/basalt hybrid polypropylene composites. *Polym. Compos.* **2021**, *42*, 4397–4411. [CrossRef]
28. Peças, P.; Carvalho, H.; Salman, H.; Leite, M. Natural Fibre Composites and Their Applications: A Review. *J. Compos. Sci.* **2018**, *2*, 66. [CrossRef]
29. Bazan, P.; Nosal, P.; Wierzbicka-Miernik, A.; Kuciel, S. A novel hybrid composites based on biopolyamide 10.10 with basalt/aramid fibers: Mechanical and thermal investigation. *Compos. Part B Eng.* **2021**, *223*, 109125. [CrossRef]
30. Armioun, S.; Panthapulakkal, S.; Scheel, J.; Tjong, J.; Sain, M. Biopolyamide hybrid composites for high performance applications. *J. Appl. Polym. Sci.* **2016**, *43595*, 1–9. [CrossRef]
31. Russo, P.; Simeoli, G.; Vitiello, L.; Filippone, G. Bio-Polyamide 11 Hybrid Composites Reinforced with Basalt/Flax Interwoven Fibers: A Tough Green Composite for Semi-Structural Applications. *Fibers* **2019**, *7*, 41. [CrossRef]
32. Sergi, C.; Vitiello, L.; Russo, P.; Tirillò, J.; Sarasini, F. Toughened Bio-Polyamide 11 for Impact-Resistant Intraply Basalt/Flax Hybrid Composites. *Macromol* **2022**, *2*, 154–167. [CrossRef]
33. Tencé-Girault, S.; Lebreton, S.; Bunau, O.; Dang, P.; Bargain, F. Simultaneous SAXS-WAXS Experiments on Semi-Crystalline Polymers: Example of PA11 and Its Brill Transition. *Crystals* **2019**, *9*, 271. [CrossRef]
34. Acierno, S.; Van Puyvelde, P. Rheological behavior of polyamide 11 with varying initial moisture content. *J. Appl. Polym. Sci.* **2005**, *97*, 666–670. [CrossRef]
35. Filippone, G.; Carroccio, S.; Mendichi, R.; Gioiella, L.; Dintcheva, N.; Gambarotti, C. Time-resolved rheology as a tool to monitor the progress of polymer degradation in the melt state—Part I: Thermal and thermo-oxidative degradation of polyamide 11. *Polymer* **2015**, *72*, 134–141. [CrossRef]
36. Lafranche, E.; Oliveira, V.M.; I Martins, C.; Krawczak, P. Prediction of injection-moulded flax fibre reinforced polypropylene tensile properties through a micro-morphology analysis. *J. Compos. Mater.* **2013**, *49*, 113–128. [CrossRef]

37. Kannan, T.G.; Wu, C.M.; Cheng, K.B.; Wang, C.Y. Effect of reinforcement on the mechanical and thermal properties of flax/polypropylene interwoven fabric composites. *J. Ind. Text.* **2012**, *42*, 417–433. [CrossRef]
38. Chivers, R. The effect of molecular weight and crystallinity on the mechanical properties of injection moulded poly(aryl-ether-ether-ketone) resin. *Polymer* **1994**, *35*, 110–116. [CrossRef]
39. Xiao, X.; Cai, Z.; Qian, K. Structure evolution of polyamide (11)'s crystalline phase under uniaxial stretching and increasing temperature. *J. Polym. Res.* **2017**, *24*, 1–8. [CrossRef]
40. Ricou, P.; Pinel, E.; Juhasz, N. Temperature Experiments for Improved Accuracy in the Calculation of Polyamide-11 Crystallinity by X-Ray Diffraction. *Denver X-ray Conf. Appl. X-ray Anal.* **2005**, *48*, 170–175.
41. Latko, P.; Kolbuk, D.; Kozera, R.; Boczkowska, A. Microstructural Characterization and Mechanical Properties of PA11 Nanocomposite Fibers. *J. Mater. Eng. Perform.* **2015**, *25*, 68–75. [CrossRef]
42. Terinte, N.; Ibbett, R.; Schuster, K.C. Overview on native cellulose and microcrystalline cellulose I structure studied by X-ray diffraction (WAXD): Comparison between measurement techniques. *Lenzinger Berichte* **2011**, *89*, 118–131.
43. Vitiello, L.; Russo, P.; Papa, I.; Lopresto, V.; Mocerino, D.; Filippone, G. Flexural Properties and Low-Velocity Impact Behavior of Polyamide 11/Basalt Fiber Fabric Laminates. *Polymers* **2021**, *13*, 1055. [CrossRef]
44. Vitiello, L.; Papa, I.; Lopresto, V.; Mocerino, D.; Filippone, G.; Russo, P. Manufacturing of bio-polyamide 11/basalt thermoplastic laminates by hot compaction: The key-role of matrix rheology. *J. Thermoplast. Compos. Mater.* **2022**, 1–16. [CrossRef]
45. Russo, P.; Acierno, D.; Simeoli, G.; Iannace, S.; Sorrentino, L. Flexural and impact response of woven glass fiber fabric/polypropylene composites. *Compos. Part B Eng.* **2013**, *54*, 415–421. [CrossRef]
46. Simeoli, G.; Acierno, D.; Meola, C.; Sorrentino, L.; Iannace, S.; Russo, P. The role of interface strength on the low velocity impact behaviour of PP/glass fibre laminates. *Compos. Part B Eng.* **2014**, *62*, 88–96. [CrossRef]
47. Quiles-Carrillo, L.; Boronat, T.; Montanes, N.; Balart, R.; Torres-Giner, S. Injection-molded parts of fully bio-based polyamide 1010 strengthened with waste derived slate fibers pretreated with glycidyl- and amino-silane coupling agents. *Polym. Test.* **2019**, *77*, 105875. [CrossRef]
48. Gourier, C.; Bourmaud, A.; Le Duigou, A.; Baley, C. Influence of PA11 and PP thermoplastic polymers on recycling stability of unidirectional flax fibre reinforced biocomposites. *Polym. Degrad. Stab.* **2017**, *136*, 1–9. [CrossRef]

Article

Modification of Polyvinyl Chloride Composites for Radiographic Detection of Polyvinyl Chloride Retained Surgical Items

Martina Polaskova [1], Tomas Sedlacek [1,*], Zdenek Polasek [2] and Petr Filip [3,*]

[1] Centre of Polymer Systems, Tomas Bata University in Zlín, Trida Tomase Bati 5678, 760 01 Zlín, Czech Republic
[2] Department of Food Technology, Faculty of Technology, Tomas Bata University in Zlín, Vavreckova 275, 760 01 Zlín, Czech Republic
[3] Institute of Hydrodynamics, Czech Academy of Sciences, Pod Patankou 5, 166 12 Prague, Czech Republic
* Correspondence: sedlacek@utb.cz (T.S.); filip@ih.cas.cz (P.F.)

Abstract: The ever-present risk of surgical items being retained represents a real medical peril for the patient and potential liability issues for medical staff. Radiofrequency scanning technology is a very good means to substantially reduce such accidents. Radiolucent medical-grade polyvinyl chloride (PVC) used for the production of medical items is filled with radiopaque agents to enable X-ray visibility. The present study proves the suitability of bismuth oxychloride (BiOCl) and documents its advantages over the classical radiopaque agent barium sulfate ($BaSO_4$). An addition of BiOCl exhibits excellent chemical and physical stability (no leaching, thermo-mechanical properties) and good dispersibility within the PVC matrix. As documented, using half the quantity of BiOCl compared to $BaSO_4$ will provide a very good result. The conclusions are based on the methods of rotational rheometry, scanning electron microscopy, dynamic mechanical analysis, atomic absorption spectroscopy, and the verification of zero leaching of BiOCl out of a PVC matrix. X-ray images of the studied materials are presented, and an optimal concentration of BiOCl is evaluated.

Keywords: retained surgical items; polyvinyl chloride; bismuth oxychloride; barium sulfate; radiopacity

Citation: Polaskova, M.; Sedlacek, T.; Polasek, Z.; Filip, P. Modification of Polyvinyl Chloride Composites for Radiographic Detection of Polyvinyl Chloride Retained Surgical Items. *Polymers* **2023**, *15*, 587. https://doi.org/10.3390/polym15030587

Academic Editors: Kamila Sałasińska and Cornelia Vasile

Received: 15 December 2022
Revised: 20 January 2023
Accepted: 20 January 2023
Published: 23 January 2023

Copyright: © 2023 by the authors. Licensee MDPI, Basel, Switzerland. This article is an open access article distributed under the terms and conditions of the Creative Commons Attribution (CC BY) license (https://creativecommons.org/licenses/by/4.0/).

1. Introduction

Unintentionally retained surgical items (RSI) or retained foreign objects (RFO) represent an everlasting problem where the impact—apart from non-negligible financial costs—is significant for both medical staffs and is a danger to patients' lives. The true RSI incidence is unknown as some patients will be initially unaware of hosting such items. The data introduced in the literature do not differ significantly: from 1 case in 1000–1500 abdominal cavity operations [1,2] to 1 case in 2000 operations under general anesthesia [3]. It should be pointed out that these data are relatively new.

The thorough analysis presented in the literature [3–7] documents that the human factor that participates in a standard counting protocol increasing vigilance and communication among staff [3,8] cannot fully eliminate hard RSI events. It implies that the classical methods (standard counting, presence of a trainee—however, reducing an occurrence of RSI by two-thirds [6]) cannot prevent these adverse patient outcomes [9]. It substantiates that a parallel implementation of more sophisticated methods would be required.

Radiofrequency scanning technology represents a very efficient method for the non-invasive identification of radiopaque items [5,7–10]. The problem is that not all medical items are able to absorb the radiant energy of an electromagnetic field due to their radiolucency (electromagnetic radiation passes through such particular materials) caused by their low electron density. Such items include sponges, swabs, textile materials, and plastic items [1,4,8]. In principle, there are two possibilities of how to adapt radiolucent materials for their detection by radio-frequency identification. The first is to use tags attached to the materials (e.g., textile materials), which ensures their tracking by changes in an electromagnetic field. The alternative to this external method is the application of so-called radiopaque agents within medical plastic items. The composition of these agents is frequently based on iodine, barium, bismuth, gold, and tungsten [11–14]. The agents -quite often in the form of nanoparticles- can be either mixed with the polymer during the manufacturing process at a specific ratio or infused into the manufactured polymer by organic solvent treatment [13,15].

The radiopacity of such composite materials is then subject to the thickness of the final product and the concentration and nature of radiopaque materials used. Radiopaque materials must be chemically and physically stable during processing and under physiological conditions within the human body. In addition, factors such as particle morphology (shape and size), concentration, degree of dispersion, and interface effects between the particles and surrounding polymeric matrix can significantly influence the processability of a compound and the final properties of the composite [10,16–20]. High-loading filler fractions up to 60 wt.% are often required to achieve good X-ray visibility, although such amounts of radiopaque filler can strongly influence mechanical properties [21]. Therefore, the type and quantity of radiopaque material are subject to the specific properties of the individual medical devices.

Barium sulfate ($BaSO_4$) represents the most widely used radiopaque agent [13,22–26]. Nevertheless, parallel to radiopaque efficiency, it is also necessary to analyze other aspects, such as detrimental effects on the mechanical properties of the device or biological reactions [27,28]. As a consequence, other radiopaque agents are intensively studied. Based on atomic number and handling risk, nine potential filler compounds have recently been evaluated [29]. With the development of the radiographical techniques [30], the choice of potential radiopaque candidates is gradually enlarging. In the case of polyurethanes as a basic material, radiopacity was introduced by incorporating an iodinated chain extender resulting in a favorable replacement of barium sulfate [31]. The agents based on bismuth [24,32] also proved to be very competitive. Specifically, greater attention is paid to bismuth oxychloride (BiOCl) [20,29]. Bismuth oxychloride exhibits twice the density resulting in a doubled level of radiopacity using the same filler loading as barium sulfate [33–35]. This significantly contributes to the maintenance of sufficient mechanical properties of a carrier polymer, ensuring no risk of overloading (causing mechanical properties deterioration) and surface smoothness. This reflects in the easier insertion of medical aids and a reduction in the likelihood of thrombus formation [16].

From the above, it is obvious that the overall characteristics of radiopaque materials (the degree of radiodensity/radiopacity, the mechanical properties, and the patient´s comfort) are subject to the specific combination of such radiolucent materials and radiopaque agents, and as such cannot be generalized. The goal of this contribution is to demonstrate the advantage of bismuth oxychloride as a radiopaque agent over classical barium sulfate for frequently used medical items made of medical-grade polyvinyl chloride (PVC). As far as the authors are aware, this compounded material is yet to be studied. Among other things, this substitution substantially helps to improve the thermo-mechanical properties of the resulting material, and better adhesion between bismuth oxychloride and the PVC matrix is exhibited when compared to barium sulfate. To this aim, radiographic, rheological, thermo-mechanical, and morphological characteristics will be determined, including verification of zero leaching of BiOCl out of a PVC matrix.

2. Materials and Methods

2.1. Materials

In the experiments, the medical grade polyvinyl chloride RB 3 (PVC) produced by Modenplast Medical S.R.L. (Fiorano Modenese, Italy) with a density of 1.23 g/cm^3 (test method ISO 1183), hardness Shore A 75 (test method ISO 868), was used. Bismuth oxychloride was supplied by Sigma-Aldrich (Prague, Czech Republic) with a density of 7.70 g/cm^3. It exhibits approximately platelet structure (see Figure 1), ensuring a smooth and shiny surface of composites. Barium sulfate powder provided by Sigma-Aldrich (Prague, Czech Republic) with a density of 4.50 g/cm^3 exhibits–in comparison with bismuth oxychloride- a rather grainy structure (see Figure 1). Prior to mixing with the polymer matrix, it was necessary to dry BaSO$_4$ powder for 15 h at a temperature of 80 °C.

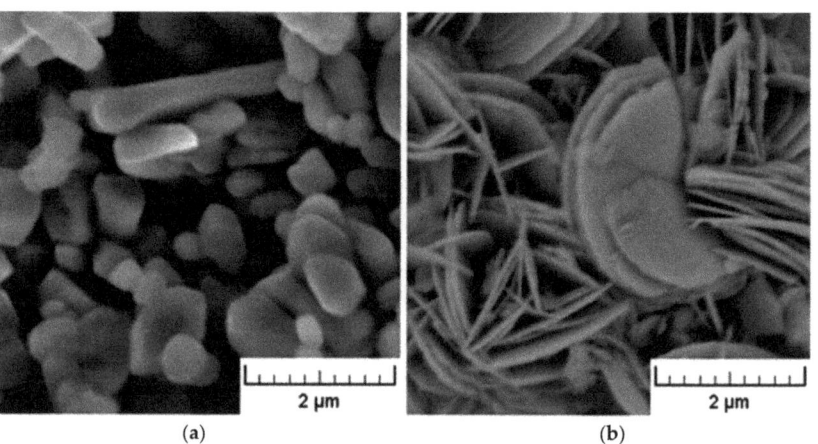

Figure 1. SEM pictures of BaSO$_4$ powder (**a**) and BiOCl powder (**b**).

2.2. Preparation of Compounded Samples

PVC pellets were melt-mixed with various amounts of additives, as summarised in Table 1. Since the density of the material is directly linked to its absorption ability of X-ray photons, expressed through an attenuation coefficient, the comparative concentration ratio of the chosen fillers of 1:1.7 was derived from the different densities of BiOCl and BaSO$_4$ powders. Melt-mixing process was performed via a MiniLab II Micro Compounder (ThermoHaake, Karlsruhe, Germany) at a temperature of 160 °C and screw speed rotation of 35 rpm. The compounding time of 8 min was sufficient for all compounded samples to achieve constant torque of the drive motor. Pure PVC resin was treated in the same way to ensure the identical temperature and shear history for all samples. Finally, the prepared polymer compounds and processed PVC were consequently compressions molded at a temperature of 160 °C into 0.1 and 1.0 mm thick sheets that were further employed for the preparation of testing samples. The area of these sheets was 125 mm × 125 mm. The prepared compounded material is shown in Figure 2.

Figure 2. SEM pictures of the compounded material (from left: 34Ba, 20Bi, pure PVC).

Table 1. Composition of prepared samples.

Sample Label	BaSO$_4$ (ρ = 4.5 g/cm^3)		BiOCl (ρ = 7.7 g/cm^3)	
	wt.%	vol.%	wt.%	vol.%
Pure PVC	-	-	-	-
17Ba	17	5.3	-	-
34Ba	34	12.3	-	-
10Bi	-	-	10	1.7
20Bi	-	-	20	3.8

2.3. Rheometrical Measurements

The temperature-dependent linear viscoelastic properties of the prepared compounds were determined using a rotational rheometer Gemini II (Malvern Panalytical, Malvern, U.K.) with 25 mm parallel-plate geometry and oscillatory mode in the frequency range from 0.015 to 15 s^{-1} and at a temperature of 160 °C. The flow behaviour of the tested samples of 1 mm thickness was then evaluated from the values of complex viscosity as a function of angular frequency. Each measurement (also for the other subsections) was carried out in triplicate and exhibited negligible deviations.

2.4. Scanning Electron Microscopy

A scanning electron microscope (SEM) Vega II LMU (Tescan, Brno, Czech Republic) was utilized to provide images of both the cross-section morphology of the prepared composites and the pristine powders. Composite samples were broken in liquid nitrogen and sputtered coated with a palladium gold alloy prior to the SEM experiments. Imaging was carried out at an accelerating voltage of 10.0 kV.

2.5. Dynamic Mechanical Analysis

Dynamic mechanical analysis (DMA) as a versatile thermal method was performed to obtain both the storage modulus and glass transition temperature of the prepared samples. While the storage modulus as a function of temperature provides information about the material's thermal stability, the glass transition temperature represents the boundary between polymers' elastic and viscoelastic behavior. Specimens with an overall diameter of 2.5 mm and thickness of 1 mm were placed in a DMA instrument—DMA/SDTA 861e (Mettler Toledo GmbH, Greifensee, Switzerland)—and oscillated at a frequency of 10 Hz in shear mode. The specimens were heated at a rate of 3 °C/min in the temperature region from −65 to 80 °C. The linear viscoelastic region of radiopaque composites was determined, prior to the testing regime, by means of a dynamic deformation sweep test at a fixed frequency of 10 Hz and temperature of −65 °C. At this temperature, the linear region of mechanical response was determined for the displacement amplitude up to 0.5 µm.

2.6. Atomic Absorption Spectroscopy

The stability of BiOCl filler within the PVC matrix under a physiological condition of gastric environment was evaluated with the help of atomic absorption spectroscopy. For these purposes, hydrochloric acid (HCl) of pH 2, simulating the gastric juice, was used as a leaching media. Samples of 20 mm × 20 mm × 0.1 mm were immersed into 20 mL of HCl and kept in sealed glass bottles for 7 and 30 days in a shaking incubator Cole-Parmer Stuart Orbital SI500 (Cole-Parmer Instrument Co., Ltd., St. Neots, UK) at a temperature of 37 °C and circular motion speed of 80 rpm. After chosen time periods (7 and 30 days), the samples were removed, and the presence of bismuth within the leachates was examined with the help of an atomic absorption spectrometer GBC 933 AA (GBC Scientific Equipment Pty Ltd., Braeside, Australia).

2.7. X-ray Measurements

Radiopaque properties of PVC composites filled with BiOCl and BaSO$_4$ were recorded using a high-frequency X-ray device Gierth HF 100 Plus (Gierth X-ray International GmbH, Riesa, Germany). X-ray imaging was performed at conditions equal to a tube voltage of 70 kV, an electric charge of 1.2 mAs, and an exposure time of 0.05 s, which are within the X-ray energy range of 20–100 keV commonly used for diagnostic radiology.

3. Results and Discussion

3.1. Rheological Characterization

The complex viscosity data of the studied polymeric systems as a function of angular frequency at a temperature of 160 °C are presented in Figure 3.

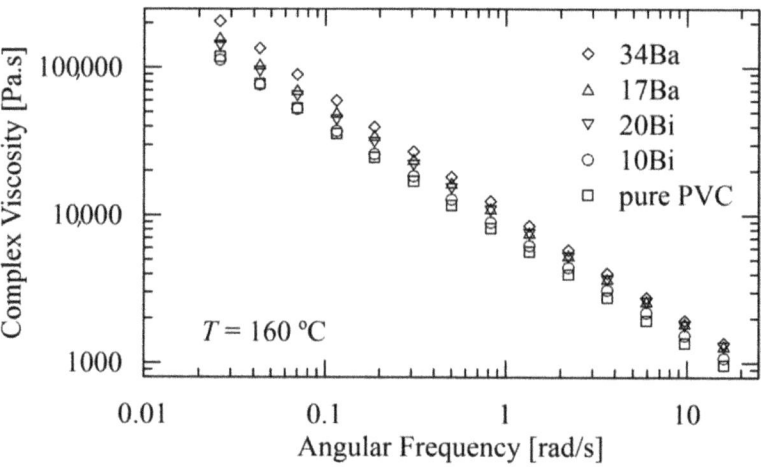

Figure 3. Complex viscosities of the composites and pure PVC matrix as a function of angular frequency at a temperature of 160 °C.

The complex viscosity for the individual composites and pure PVC matrix can be approximated by the relation:

$$\log(\eta) = a - 0.74 \times \log(\omega) \tag{1}$$

where η [Pa.s] is the complex viscosity, and ω [rad/s] represents the angular frequency. The slope of all straight lines is identical (−0.74), and only the additive term a [10^a has units Pa.s$^{1.74}$.rad$^{-0.74}$] (increasing with increasing contents of particular filler) diversifies between the individual composites (see Table 2). The mean deviation of the approximated values $\log(\eta)$ using Equation (1) attains 0.4%, which—for the real non-transformed values of the complex viscosity η—corresponds to 5% for ω = 0.026 rad/s and gradually decreases to 3% for ω = 16 rad/s. These deviations are within the experimental errors. Equation (1), in combination with Table 2, can be used for an estimate of the behavior of the composites with different percentage participation of BaSO$_4$ or BiOCl.

Generally, addition of inorganic particulate fillers into the polymer matrix results in an increase in viscosity due to the two factors. First, the mobility of macromolecules is more restricted, and second, with an increase in filler concentration, the polymer-particle interaction becomes stronger, resulting in a further increase in viscosity [18]. While the flow, permanent competition between the creation and destruction of polymer-particle interaction takes place. While at low shear rates, interactions could be easily restored, leading to higher viscosity values, at higher shear rates, destruction of polymer-particle interaction predominates, which results in minor viscosity variation [19], as documented in Figure 2.

Note that in accordance with the expectation [18], the higher radiopaque filler loadings of $BaSO_4$ (5.3 vol.% for sample 17Ba and 12.3 vol.% for sample 34Ba) were bounded to more significant viscosity increase compared to the flow behavior of BiOCl/PVC compounds with a filler content of 1.7 and 3.8 vol.% for samples 10Bi and 20Bi, respectively. However, these findings are not in compliance with the results of McNally and Rudy [16] and Godinho et al. [22], concluding that the incorporation of $BaSO_4$ particles into the PVC matrix leads to a compound's viscosity decrease. Since the pure PVC resin was not pre-processed in the mentioned studies, the possible explanation of revealed inconsonance can be caused by variation in the temperature history of pure PVC resin and $BaSO_4$/PVC samples due to divergence in their preparation.

Table 2. The values of an additive constant as for the individual composites and pure PVC matrix.

Material	Notation	Parameter a
Pure PVC matrix	PVC	3.86
PVC matrix + 17 wt.% $BaSO_4$	17Ba	3.99
PVC matrix + 34 wt.% $BaSO_4$	34Ba	4.03
PVC matrix + 10 wt.% BiOCl	10Bi	3.88
PVC matrix + 20 wt.% BiOCl	20Bi	3.96

The graphs depicting load vs. time for PVC with the highest filler loading 34Ba and for pure PVC are introduced in the Supplement (Figure S1). As apparent, after an initial increase, the torque values are stabilized after approximately 7 min. The value corresponding to pure PVC is higher by approximately one-eighth of that for 34Ba.

3.2. SEM Analysis

Since a sufficient degree of dispersion of filler particles within the polymeric matrix is one of the factors significantly influencing the extent of polymer-particle interactions and, as such, closely related to compounding process efficiency, the resulting structures of the prepared compounds were evaluated using Scanning Electron Microscopy (SEM) technique. As obvious from Figures 4 and 5, relatively uniform dispersion of the fillers within the PVC matrix was achieved for all tested filler loadings, in spite of a partial agglomeration tendency of both filler types. However, the average size of agglomerates does not exceed 5 µm.

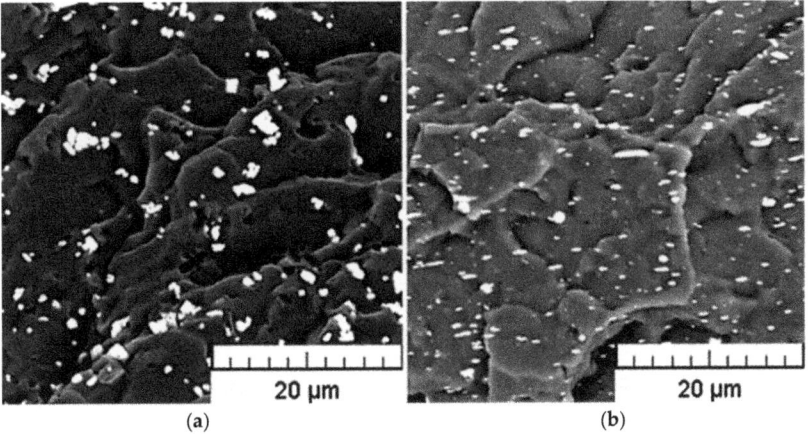

Figure 4. SEM pictures of the cross-sections of 17Ba (**a**) and 10Bi (**b**) samples.

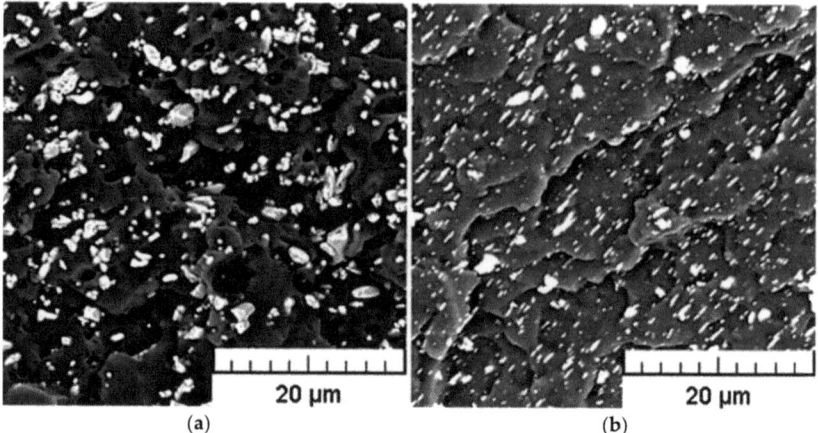

Figure 5. SEM pictures of the cross-sections of 34Ba (**a**) and 20Bi (**b**) samples.

The SEM details of the cross-section areas of BiOCl/PVC and BaSO$_4$/PVC composites are compared in Figure 6. While bulky types of agglomerated particles are seldom found in the BaSO4/PVC samples, an apparent tendency for delamination of originally layered BiOCl particles into nanosheets (see Figure 1) can be observed in BiOCl/PVC materials. Moreover, due to its hydrophobic behavior [12], BiOCl powder is readily dispersed in the polymeric matrix, and BiOCl particles (nanosheets) are completely embedded within the PVC matrix (Figure 6 right). On the other hand, the SEM details of BaSO$_4$/PVC composites revealed rather poor adhesion of BaSO$_4$ filler to the PVC matrix (Figure 6 left). It is apparent that the filler particles are not wetted by the polymeric matrix. Moreover, in the monitored cross-sectional area there can be observed pores or holes probably formed during the breaking of the samples in liquid nitrogen as some particles were pulled out of the polymeric matrix.

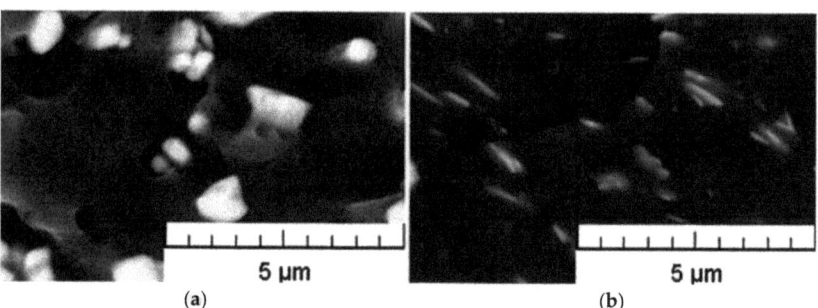

Figure 6. SEM details of the cross-sections of BaSO$_4$/PVC (**a**) and BiOCl/PVC (**b**) samples.

3.3. Impact of Radiopaque Fillers on Thermo-Mechanical Properties

Addition of the radiopaque filler into a polymeric matrix also contributes to the resulting thermo-mechanical properties of the compounds. Since mineral filler has a substantially higher modulus than polymers, its addition usually reflects in a higher modulus of polymer composite [18]. The effect of various amounts of both types of fillers on storage modulus, indicating a change of material elasticity, is presented in Figure 7 in more detail, using the normal coordinates in Figure 8. A comparison of viscoelastic behavior for pure PVC and all four compounds are presented in the Supplement (elastic and storage moduli in Figure S2, elastic and storage moduli in more detail in Figures S3 and S4, and courses of phase angles in Figure S5).

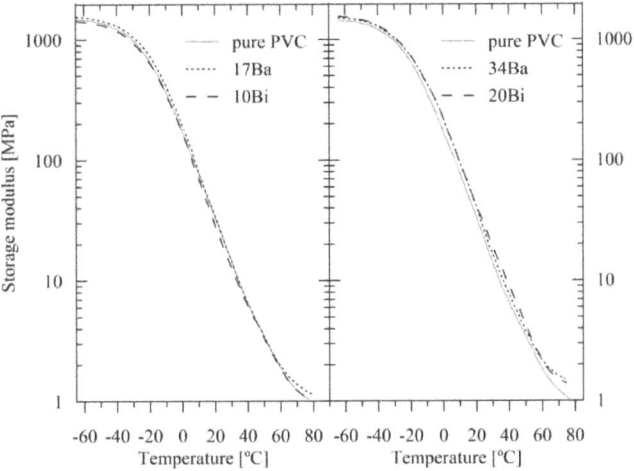

Figure 7. The storage modulus of the samples as a function of temperature (the semi-log coordinates).

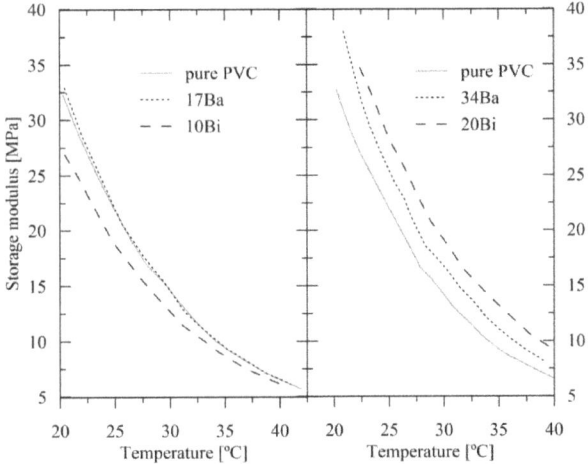

Figure 8. The storage modulus of the samples as a function of temperature—a detail in the normal coordinates.

As apparent for lower temperatures, storage moduli for the individual composites are nearly identical to that of pure PVC. However, for temperatures exceeding zero, there is an increase in storage moduli for the samples 17Ba, 34Ba, and 20Bi in comparison to pure PVC (see Figures 7 and 8). On the other hand, the addition of 10 wt. % of BiOCl (sample 10Bi) leads rather to a storage modulus decrease compared to unfilled PVC (see Figure 7). These results are very favorable as they prove no significant reinforcing of the polymer matrix with the addition of radiopaque fillers.

Glass transition temperature (T_g) is commonly defined as the maximum damping ratio (tan δ) or the maximum loss modulus. Its value can also be derived from the onset of the change in the slope of the storage modulus curve. In this work, T_g was evaluated from a maximum of damping ratios of the prepared composites, as summarised in Table 3. The results document a slight T_g decrease in the samples 17Ba, 34Ba, and 10Bi compared to pure PVC. This course indicates that the radiopaque fillers incorporated into the PVC matrix -at least up to a certain concentration- can work as plasticizers by embedding themselves between the chains of polymers and spacing them apart, allowing thus a higher degree of molecular chain mobility due to reduced chain cooperativity [20]. Nevertheless, it should be noted that the addition of 20 wt. % of BiOCl (sample 20Bi) is assigned with the recovery of pristine T_g and tan δ values (that of PVC), indicating an improvement of adhesion at the filler-polymer interface [36].

Table 3. Effect of radiopaque fillers on the phase transition behavior.

Material	T_g [°C]	tan δ
Pure PVC	24	0.42
17Ba	22	0.46
34Ba	23	0.46
10Bi	22	0.46
20Bi	24	0.42

3.4. Non-Toxicity of BiOCl Compounds

As BiOCl dissolves in concentrated hydrochloric acid, the stability of BiOCl/PVC systems under physiological conditions of a gastric environment [37,38], pretended by HCl of pH 2, was tested. Generally, the solid BiOCl readily dissolves in concentrated hydrochloric acid to give a clear solution of bismuth trichloride. This reaction is reversible, meaning that solid BiOCl and bismuth trichloride in the solution can exist in equilibrium.

Atomic absorption spectrometry, as an adequate method for the determination of bismuth in trace amounts [39,40], was employed. For this purpose, the standard bismuth solution (1 mg/mL) was diluted with nitric acid (65 vol.%) to prepare the calibration curves for 20, 40, and 60 ng/mL. As no peak response was registered during the measurement, it can be concluded that the potential amount of Bi leached out from BiOCl/PVC composites into HCl leaching media is below the lowest limit of the calibration curves, hence lower than 20 ng/mL.

It can be explained either by the proper embedding of BiOCl particles within the PVC matrix, to an extent preventing BiOCl particles from direct contact with leaching media, or that the pH level in the gastric environment is not low enough to dissolve BiOCl particles. By virtue of this analysis, the stability of the BiOCl/PVC system in HCl leaching media for the duration of 30 days was proven, and from this point of view, the utilization of this material, for instance, in the form of catheters can be considered as safe. This is an analogy to the compound poly(glycerol sebacate) acrylate/ bismuth oxychloride, as studied in [24].

In the case of the composites containing $BaSO_4$, there is no leaching out of the polymeric matrix, reasoning the long tradition of $BaSO_4$ utilization in medical applications. Moreover, $BaSO_4$ is usually used as so-called "barium meal" as a means of imaging the digestive tract because, in this case, no toxic side products originate inside the body environment [23].

3.5. Radiopaque Detection

When radiopaque material is inserted into a human body, a light-dark contrast appears on X-ray images, similar to in Figure 9. A lighter area in this picture corresponds to a smaller number of transmitted photons, while a darker area represents the transmission of a larger number of photons. This contrast is essential for an accurate location of medical aids during critical procedures inside the human body. Hence, during the final inspection and after the surgical procedure, erroneously retained aids made of radiopaque material are detectable with X-ray images. Their location is indicated by the lighter shades. Due to the different shades and the radiologist's experience, they cannot be confused with individual bones. For comparison, an X-ray image of a bone is presented in the top part of Figure 9.

Figure 9 shows the prepared samples Ba and Bi with thicknesses of 1 mm and approximate dimensions of 10 mm × 60 mm (manually cut out from pressed sheets, the size was chosen in accordance with the size of a bone used) can be seen on the X-ray images and can be compared. It should be emphasized that BaSO$_4$/PVC samples 17Ba and 34Ba (labeled as 2 and 4, respectively) were filled with almost two times higher concentrations of the filler compared to that of BiOCl/PVC samples 10Bi and 20Bi (1 and 3, respectively), rending thus comparable ability to block the passage of X-rays. Indeed, the samples with higher filler loading, 34Ba, and 20Bi, exhibit a mutually similar contrast, which is -compared to the samples 17Ba and 10Bi—also much brighter and sharper. In addition, the radiopaque properties of prepared samples can be confronted with the contrast of a bone, which is captured in the top part. The question is, which is the optimal concentration for a radiopaque agent. The dominant criterion should be good visibility of erroneously retained aids and the minimum failure in their recognition, as it is still necessary to keep in mind that the evaluation of X-ray images is based on the human factor. In this regard, the dimensions of the aids, and especially the thickness of the employed radiopaque layer, can be even lower than 1 mm, which is the thickness of the samples presented in Figure 9.

Figure 9. X-ray image of the radiopaque samples. The numbers 1, 2, 3, and 4 stand for the samples 10Bi, 17Ba, 20Bi, and 34Ba, respectively. In contrast, a bone is presented in the top part (above sample 10Bi).

On the other hand, there is financial availability of radiopaque materials. With regards to acquisition costs of individual radiopaque agents, according to a pricelist of Sigma Aldrich (Prague, Czech Rep.), the price of BiOCl has more than doubled compared with BaSO$_4$. However, as shown, only half the quantity of BiOCl is sufficient. The remaining difference is more than adequately balanced by the better properties of the BiOCl-PVC compound: thermo-mechanical parameters, adhesion, and stability. Based on the necessary contents of BiOCl and the overall superior quality of the proposed compound, the price of this radiopaque material is viable.

4. Conclusions

At present, the radiopacity of medical items used within the human body should be an inevitable attribute serving to identify their location and to substantially reduce these unintentionally retained surgical items. Radiopaque agents should not degrade the properties of the basic materials used, such as their thermo-mechanical characteristics and, crucially, their biocompatibility and non-toxic properties. The emphasis should be paid to a limited content of radiopaque agents, however, ensuring their errorless detection.

Bismuth oxychloride proved to be a very good candidate for fulfilling all these demands. Very good biocompatibility and negligible in vivo toxicity are discussed in [41,42].

Further, BiOCl is a water-insensitive material that exhibits extremely low toxicity as a result of its almost complete insolubility in aqueous solutions such as biological fluids [43]. Favorable results concerning thermo-mechanical characteristics are presented in Section 3.3. It was shown that BiOCl exhibits comparable properties to the frequently used barium sulfate, but for the case of the PVC matrix, BiOCl weight can be reduced by nearly one-half in comparison with barium sulfate. It has a favorable impact, for instance, on the thermo-mechanical properties of the composite. Moreover, a better adhesion between bismuth oxychloride and polymeric matrix was observed compared to barium sulfate.

Supplementary Materials: The following supporting information can be downloaded at: https://www.mdpi.com/article/10.3390/polym15030587/s1, Figure S1: Load in dependence on time for pure PVC and the highest filler loading 34Ba; Figure S2: A comparison of viscoelastic behaviour of pure PVC and all four compounds; Figure S3: A comparison of elastic moduli of pure PVC and all four compounds; Figure S4: A comparison of storage moduli of pure PVC and all four compounds; Figure S5: A comparison of phase angles tan δ of pure PVC and all four compounds.

Author Contributions: M.P.: writing—original draft and investigation; T.S.: writing—review and editing; conceptualization, investigation; Z.P.: methodology and investigation; P.F.: writing—editing, conceptualization, formal analysis, and data curation. All authors have read and agreed to the published version of the manuscript.

Funding: The authors (M.P., T.S.) thank the Ministry of Education, Youth, and Sports of the Czech Republic—DKRVO (RP/CPS/2022/003). The author (P.F.) thanks the institutional support of the Czech Academy of Sciences, Czech Republic, RVO 67985874.

Acknowledgments: The authors (M.P., T.S.) thank the Ministry of Education, Youth, and Sports of the Czech Republic—DKRVO (RP/CPS/2022/003). The author (P.F.) thanks the institutional support of the Czech Academy of Sciences, Czech Republic, RVO 67985874.

Conflicts of Interest: The authors declare no conflict of interest.

References

1. Sonarkar, R.; Wilkinson, R.; Nazar, Z.; Gajendra, G.; Sonawane, S. Textiloma presenting as a lump in abdomen: A case report. *Int. J. Surg. Case Rep.* **2020**, *77*, 206–209. [CrossRef]
2. Siebert, H. Unbeabsichtigt vergessene Fremdkorper im Operationsgebiet. Medizinische, organisatorische, prophylaktische und rechtliche Aspekte. *Rechtsmedizin* **2015**, *25*, 194–200. [CrossRef]
3. Takahashi, K.; Fukatsu, T.; Oki, S.; Iizuka, Y.; Otsuka, Y.; Sanui, M.; Lefor, A.K. Characteristics of retained foreign bodies and near-miss events in the operating room: A ten-year experience at one institution. *J. Anesth.* **2022**; *online ahead of print*. [CrossRef]
4. Cima, R.R.; Kollengode, A.; Garnatz, J.; Storsveen, A.; Weisbrod, C.; Deschamps, C. Incidence and characteristics of potential and actual retained foreign object events in surgical patients. *J. Am. Coll. Surg.* **2008**, *207*, 80–87. [CrossRef]
5. Asiyanbola, B.; Etienne-Cummings, R.; Lewi, J.S. Prevention and diagnosis of retained foreign bodies through the years: Past, present, and future technologies. *Technol. Health Care* **2012**, *20*, 379–386. [CrossRef]
6. Stawicki, S.P.A.; Moffatt-Bruce, S.D.; Ahmed, H.M.; Anderson, H.L.; Balija, T.M.; Bernescu, I.; Chan, L.; Chowayou, L.; Cipolla, J.; Coyle, S.M.; et al. Retained surgical items: A problem yet to be solved. *J. Am. Coll. Surg.* **2013**, *216*, 15–22. [CrossRef]
7. Kaplan, H.J.; Spiera, Z.C.; Feldman, D.L.; Shamamian, P.; Portnoy, B.M.J.; Ioannides, P.; Leitman, I.M. Risk reduction strategy to decrease incidence of retained surgical items. *J. Am. Coll. Surg.* **2022**, *235*, 494–499. [CrossRef]
8. Weprin, S.; Crocerossa, F.; Meyer, D.; Maddra, K.; Valancy, D.; Osardu, R.; Kang, H.S.; Moore, R.H.; Carbonara, U.; Kim, F.J.; et al. Risk factors and preventive strategies for unintentionally retained surgical sharps: A systematic review. *Patient Saf. Surg.* **2021**, *15*, 24. [CrossRef]
9. Peng, J.P.; Ang, S.Y.; Zhou, H.; Nair, A. The effectiveness of radiofrequency scanning technology in preventing retained surgical items: An integrative review. *J. Clin. Nurs.* **2022**; *online ahead of print*. [CrossRef]
10. Cochran, K. Guidelines in practice: Prevention of unintentionally retained surgical items. *AORN J.* **2022**, *116*, 427–440. [CrossRef]
11. Houston, K.R.; Brosnan, S.M.; Burk, L.M.; Lee, Y.Z.; Luft, J.C.; Ashby, V.S. Iodinated polyesters as a versatile platform for radiopaque biomaterials. *J. Polym. Sci. Part A Polym. Chem.* **2017**, *55*, 2171–2177. [CrossRef]
12. Rivera, E.J.; Tran, L.A.; Hernandez-Rivera, M.; Yoon, D.; Mikos, A.G.; Rusakova, I.A.; Cheong, B.Y.; Cabreira-Hansen, M.D.; Willerson, J.T.; Perin, E.C.; et al. Bismuth@US-tubes as a potential contrast agent for X-ray imaging applications. *J. Mater. Chem. B* **2013**, *1*, 4792–4800. [CrossRef]
13. Tian, L.; Lee, P.; Singhana, B.; Chen, A.; Qiao, Y.; Lu, L.F.; Martinez, J.O.; Tasciotti, E.; Melancon, A.; Huang, S.; et al. Radiopaque resorbable inferior vena cava filter infused with gold nanoparticles. *Sci. Rep.* **2017**, *7*, 2147. [CrossRef]

14. Kim, S.C. Process technology for development and performance improvement of medical radiation shield made of eco-friendly oyster shell powder. *Appl. Sci.* **2022**, *12*, 968. [CrossRef]
15. Tian, L.; Lu, L.; Feng, J.; Melancon, M.P. Radiopaque nano and polymeric materials for atherosclerosis imaging, embolization and other catheterization procedures. *Acta Pharm. Sin. B* **2018**, *8*, 360–370. [CrossRef]
16. McNally, G.; Ruddy, A.C. Rheological, mechanical and thermal behaviour of radiopaque filled polymers. In Proceedings of the Annual Conference of the Society of Plastics Engineers, ANTEC 2005, Boston, MA, USA, 1–5 May 2005; pp. 2924–2928.
17. Lu, G.; Kalyon, D.; Yilgor, I.; Yilgor, E. Rheology and processing of baso4-filled medical-grade thermoplastic polyurethane. *Polym. Eng. Sci.* **2004**, *44*, 1941–1948. [CrossRef]
18. Shenoy, A.V. *Rheology of Filled Polymer Systems*; Springer Science & Business Media: Berlin/Heidelberg, Germany, 2013.
19. Han, C.D. *Rheology and Processing of Polymeric Materials: Volume 2: Polymer Processing*; Oxford University Press: New York, NY, USA, 2006. [CrossRef]
20. Wang, Y.; Huang, S.-W.; Guo, J.-Y. Dynamic mechanical study of clay dispersion in maleated polypropylene/organoclay nanocomposites. *Polym. Compos.* **2009**, *30*, 1218–1225. [CrossRef]
21. Hampikian, J.M.; Heaton, B.C.; Tong, F.C.; Zhang, Z.Q.; Wong, C.P. Mechanical and radiographic properties of a shape memory polymer composite for intracranial aneurysm coils. *Mater. Sci. Eng. C-Biomim. Supramol. Syst.* **2006**, *26*, 1373–1379. [CrossRef]
22. Godinho, J.S.; Moore, I.R.; Mc Nally, G.M.; Murphy, W.R.; Ruddy, A. Influence of barium sulfate on rheological behaviour and mechanical properties of medical-grade PVCs. In Proceedings of the Annual Conference of the Society of Plastics Engineers, ANTEC 2004, Chicago, IL, USA, 16–20 May 2004; pp. 3395–3399.
23. Kalyon, D.M.; Lu, G.; Yilgor, E.; Yilgor, I. Extrusion of $BaSO_4$ filled medical-grade thermoplastic polyurethane. In Proceedings of the Annual Conference of the Society of Plastics Engineers, ANTEC 2003, Nashville, TN, USA, 4–8 May 2003; pp. 261–265.
24. Chang, C.-T.; Chen, H.-T.; Girsang, S.P.; Chen, Y.-M.; Wan, D.; Shen, S.-H.; Wang, J. 3D-printed radiopaque polymer composites for the in situ monitoring of biodegradable medical implants. *Appl. Mater. Today* **2020**, *20*, 100771. [CrossRef]
25. Liu, H.; Zhang, Z.; Gao, C.; Bai, Y.; Liu, B.; Wang, W.; Ma, Y.; Saijilafu; Yang, H.; Li, Y.; et al. Enhancing effects of radiopaque agent $BaSO_4$ on mechanical and biocompatibility properties of injectable calcium phosphate composite cement. *Mater. Sci. Eng. C* **2020**, *116*, 110904. [CrossRef]
26. Li, J.Y.; Wang, H.; Guo, Q.P.; Zhu, C.H.; Zhu, X.S.; Han, F.X.; Yang, H.L.; Li, B. Multifunctional Coating to Simultaneously Encapsulate Drug and Prevent Infection of Radiopaque Agent. *Int. J. Mol. Sci.* **2019**, *20*, 2055. [CrossRef]
27. van Hooy-Corstjens, C.S.J.; Bulstra, S.K.; Knetsch, M.L.W.; Geusens, P.; Kuijer, R.; Koole, L.H. Biocompatibility of a new radiopaque iodine-containing acrylic bone cement. *J. Biomed. Mater. Res. Part B Appl. Biomater.* **2007**, *80B*, 339–344. [CrossRef]
28. Deb, S.; Abdulghani, S.; Behiri, J.C. Radiopacity in bone cements using an organo-bismuth compound. *Biomaterials* **2002**, *23*, 3387–3393. [CrossRef]
29. Shannon, A.; JO'Sullivan, K.; Clifford, S.; O'Sullivan, L. Assessment and selection of filler compounds for radiopaque PolyJet multi-material 3D printing for use in clinical settings. *Proc. Inst. Mech. Eng. Part H J. Eng. Med.* **2022**, *236*, 740–747. [CrossRef]
30. Sneha, K.R.; Sailaja, G.S. Intrinsically radiopaque biomaterial assortments: A short review on the physical principles, X-ray imageability, and state-of-the-art developments. *J. Mater. Chem. B* **2021**, *9*, 8569–8593. [CrossRef]
31. Kiran, S.; James, N.R.; Jayakrishnan, A.; Joseph, R. Polyurethane thermoplastic elastomers with inherent radiopacity for biomedical applications. *J. Biomed. Mater. Res. Part A* **2012**, *100A*, 3472–3479. [CrossRef]
32. Aviv, H.; Bartling, S.; Grinberg, I.; Margel, S. Synthesis and characterization of Bi_2O_3/HSA core-shell nanoparticles for X-ray imaging applications. *J. Biomed. Mater. Res. Part B Appl. Biomater.* **2013**, *101B*, 131–138. [CrossRef]
33. Jagdale, P.; Rovere, M.; Ronca, R.; Vigneri, C.; Bernardini, F.; Calzetta, G.; Tagliaferro, A. Determination of the X-ray attenuation coefficient of bismuth oxychloride nanoplates in polydimethylsiloxane. *J. Mater. Sci.* **2020**, *55*, 7095–7105. [CrossRef]
34. Huang, S.D. Bismuth-based nanoparticles for CT imaging. In *Design and Applications of Nanoparticles in Biomedical Imaging*; Bulte, J.W.M., Modo, M.M.J., Eds.; Springer: Cham, Switzerland, 2017; pp. 429–444. [CrossRef]
35. Tian, L.; Lu, L.; Singhana, B.; Jacobsen, M.; Melancon, A.; Melancon, M. Novel radiopaque bismuth nanoparticle coated polydioxanone and comparison of attenuation in pre-clinical and clinical CTs. *Med. Phys.* **2017**, *44*, 2820.
36. Vrsaljko, D.; Leskovac, M.; Blagojevic, S.L.; Kovacevic, V. Interphase phenomena in nanoparticulate filled polyurethane/poly(vinyl acetate) polymer systems. *Polym. Eng. Sci.* **2008**, *48*, 1931–1938. [CrossRef]
37. Slikkerveer, A.; Helmich, R.B.; de Wolff, F.A. Analysis for bismuth in tissue by electrothermal atomic absorption spectrometry. *Clin. Chem.* **1993**, *39*, 800–803. [CrossRef]
38. Nordberg, G.F.; Fowler, B.A.; Nordberg, M.; Friberg, L.T. (Eds.) *Handbook on the Toxicology of Metals*, 3rd ed.; Elsevier: Burlington, MA, USA, 2007; p. 433.
39. Ohyama, J.-I.; Maruyama, F.; Dokiya, Y. Determination of bismuth in environmental samples by hydride generation-atomic absorption spectrometry. *Anal. Sci.* **1987**, *3*, 413–416. [CrossRef]
40. Narukawa, T.; Uzawa, A.; Yoshimura, W.; Okutani, T. Effect of cobalt as a chemical modifier for determination of bismuth by electrothermal atomic absorption spectrometry using a tungsten furnace. *Anal. Sci.* **1998**, *14*, 779–784. [CrossRef]
41. Ijaz, H.; Zia, R.; Taj, A.; Jameel, F.; Butt, F.K.; Asim, T.; Jameel, N.; Abbas, W.; Iqbal, M.; Bajwa, S.Z.; et al. Synthesis of BiOCl nanoplatelets as the dual interfaces for the detection of glutathione linked disease biomarkers and biocompatibility assessment in vitro against HCT cell lines model. *Appl. Nanosci.* **2020**, *10*, 3569–3576. [CrossRef]

42. Zelepukin, I.V.; Ivanov, I.N.; Mirkasymov, A.B.; Shevchenko, K.G.; Popov, A.A.; Prasad, P.N.; Kabashin, A.V.; Deyev, S.M. Polymer-coated BiOCl nanosheets for safe and regioselective gastrointestinal X-ray imaging. *J. Control. Release* **2022**, *349*, 475–485. [CrossRef]
43. Repichet, S.; Le Roux, C.; Roques, N.; Dubac, J. BiCl3-catalyzed Friedel–Crafts acylation reactions: Bismuth(III) oxychloride as a water insensitive and recyclable procatalyst. *Tetrahedron Lett.* **2003**, *44*, 2037–2040. [CrossRef]

Disclaimer/Publisher's Note: The statements, opinions and data contained in all publications are solely those of the individual author(s) and contributor(s) and not of MDPI and/or the editor(s). MDPI and/or the editor(s) disclaim responsibility for any injury to people or property resulting from any ideas, methods, instructions or products referred to in the content.

Article

Engineering Polypropylene–Calcium Sulfate (Anhydrite II) Composites: The Key Role of Zinc Ionomers via Reactive Extrusion

Marius Murariu [1,*], Yoann Paint [1], Oltea Murariu [1], Fouad Laoutid [1] and Philippe Dubois [1,2,*]

[1] Laboratory of Polymeric and Composite Materials, Materia Nova Materials R&D Center & UMONS Innovation Center, 3 Avenue Copernic, 7000 Mons, Belgium

[2] Laboratory of Polymeric and Composite Materials, Center of Innovation and Research in Materials and Polymers (CIRMAP), University of Mons (UMONS), Place du Parc 20, 7000 Mons, Belgium

* Correspondence: marius.murariu@materianova.be (M.M.); philippe.dubois@umons.ac.be (P.D.);
Tel.: +32-65-554976 (M.M.); +32-65-373000 (P.D.); Fax: +32-65-373054 (P.D.)

Abstract: Polypropylene (PP) is one of the most versatile polymers widely used in packaging, textiles, automotive, and electrical applications. Melt blending of PP with micro- and/or nano-fillers is a common approach for obtaining specific end-use characteristics and major enhancements of properties. The study aims to develop high-performance composites by filling PP with $CaSO_4$ β-anhydrite II (AII) issued from natural gypsum. The effects of the addition of up to 40 wt.% AII into PP matrix have been deeply evaluated in terms of morphology, mechanical and thermal properties. The PP–AII composites (without any modifier) as produced with internal mixers showed enhanced thermal stability and stiffness. At high filler loadings (40% AII), there was a significant decrease in tensile strength and impact resistance; therefore, custom formulations with special reactive modifiers/compatibilizers (PP functionalized/grafted with maleic anhydride (PP-g-MA) and zinc diacrylate (ZnDA)) were developed. The study revealed that the addition of only 2% ZnDA (able to induce ionomeric character) leads to PP–AII composites characterized by improved kinetics of crystallization, remarkable thermal stability, and enhanced mechanical properties, i.e., high tensile strength, rigidity, and even rise in impact resistance. The formation of Zn ionomers and dynamic ionic crosslinks, finer dispersion of AII microparticles, and better compatibility within the polyolefinic matrix allow us to explain the recorded increase in properties. Interestingly, the PP–AII composites also exhibited significant improvements in the elastic behavior under dynamic mechanical stress and of the heat deflection temperature (HDT), thus paving the way for engineering applications. Larger experimental trials have been conducted to produce the most promising composite materials by reactive extrusion (REx) on twin-screw extruders, while evaluating their performances through various methods of analysis and processing.

Keywords: polypropylene; PP; mineral-filled composites; calcium sulfate; gypsum; anhydrite II; melt mixing; thermal and mechanical properties; reactive extrusion; REx; heat deflection temperature (HDT); injection molding; automotive; engineering applications

Citation: Murariu, M.; Paint, Y.; Murariu, O.; Laoutid, F.; Dubois, P. Engineering Polypropylene–Calcium Sulfate (Anhydrite II) Composites: The Key Role of Zinc Ionomers via Reactive Extrusion. *Polymers* **2023**, *15*, 799. https://doi.org/10.3390/polym15040799

Academic Editor: Kamila Sałasińska

Received: 13 January 2023
Revised: 30 January 2023
Accepted: 31 January 2023
Published: 5 February 2023

Copyright: © 2023 by the authors. Licensee MDPI, Basel, Switzerland. This article is an open access article distributed under the terms and conditions of the Creative Commons Attribution (CC BY) license (https://creativecommons.org/licenses/by/4.0/).

1. Introduction

Polypropylene (PP) is one of the most used thermoplastics available for almost all end-use markets, ranging from packaging and textiles to automotive and engineering components [1–4]. Nowadays, the global markets of PP are showing outstanding growth, driven by technological progress and an increase in the number of applications. This is due to the low density of PP, its good chemical resistance, and the related thermal, mechanical, and electrical properties of the PP-based products, as well as their excellent processability and recyclability [5,6]. Moreover, in many engineering applications, metals have been extensively replaced by PP-based materials to achieve weight reductions, low energy/fuel consumption, and cost savings [7,8]. Still, regarding the current development status of PP, it is important to mention the announcement of the production of bio-based PP made from eco-friendly building blocks derived from natural resources [6]. Taking as one key example the use of PP in the automotive sector, a non-exhaustive list of applications includes various interior and exterior components: pillars, consoles, armrests, air cleaner bodies, carpet fibers, dashboard components, flexible bumpers with high impact resistance, engine fans, wheel covers, instrumental panels and door trims, radiator headers, heater baffles and housings, tanks, electrical parts, battery boxes, etc. [2,9,10].

The addition to the PP matrix (homopolymer or PP-based copolymers) of reinforcing fibers (e.g., short or long glass fibers, various natural fibers) [4,11–13], nano- and microfillers [14], impact modifiers [13,15], other dispersed phases and special additives, such as flame retardants [16,17], is known as an important route to achieving the enhancement of PP properties required for specific end-use applications, and in some cases, improved processing and cost savings. The mineral-filled (MF) PP compounds can offer enhancements over the unfilled PP, in properties such as stiffness, thermal and dimensional stability, heat deflection at higher temperatures, better processing, and cost-effectiveness [18]. Moreover, they can be designed for a wide range of applications: engineering parts, automotive components, packaging, production of cups and containers, transport pallets, and others.

PP has been melt blended with clays [19–21], $CaCO_3$ [8,22–25], talc [26–29], mica [30,31], kaolin [28,32–34], and other mineral fillers [35,36] typically used in the polymer composites industry. Talc and $CaCO_3$ are among the most preferred fillers used to produce PP compounds. At low loadings, talc can act as a nucleating agent, reducing PP spherulite size and shortening the processing time, while at higher loadings (up to 40 wt.%) it behaves as a reinforcing filler, significantly increasing the tensile modulus and stiffness of PP [37–39]. On the other hand, there is a dramatic decrease in the strain at break and impact resistance [39], especially at high talc percentage, with the latter parameter being of great importance for many PP applications. To improve the proprieties of MF PP composites, different experimental strategies have been applied: using fillers with various surface treatments [40–42]; compatibilizing agents [43–45] such as PP functionalized/grafted with maleic anhydride (PP-g-MA) [46,47]; using elastomers [26]; special modifiers/additives [48]; following the addition and synergies between micro- and/or nano-fillers [49–52], and so on.

Nonetheless, few laboratory-scale studies have considered the use of synthetic $CaSO_4$ (CS) particles to reinforce PP [53–56]. Unfortunately, the information on the use of stable CS for the reinforcement/filling of PP is scant by comparison to other fillers (talc, $CaCO_3$, kaolin, etc.), while the nature of CS (available in different forms), or the necessity of specific thermal treatments, has been much less reported on. In addition, we believe there is a misunderstanding related to the rapid water absorption or high moisture sensitivity that are specific to CS hemihydrate and to "soluble" β-anhydrite III (AIII), both derived from $CaSO_4 \cdot 2H_2O$ (gypsum), as obtained by thermal treatments above 100 °C and at ca. 200 °C, respectively [57]. On the contrary, compared to these not stable forms, calcination of gypsum at higher temperatures (e.g., at 500–700 °C in an industrial process), allows us to obtain highly stable β-anhydrite II (AII), also known as "insoluble" anhydrite [57–59].

Related to this study, it is worth mentioning that nowadays the producers of natural gypsum look for new markets by offering CS derivatives, such as stable AII, for new applications with higher added value, e.g., for the industry of polymer composites, paints, and coatings. In this respect, CS anhydrite AII, which is characterized by high thermal stability and whiteness, is unfortunately not well known by the polymer composite industry. Consequently, further prospects are required to demonstrate its beneficial value with different polymer matrices and by considering the needs of a large range of applications. It is also important to note that AII has been used with promising results in the manufacturing of polylactic acid (PLA) composites even though this aliphatic (bio)polyester is known for its high thermal and hydrolytic sensitivity [57,60]. Accordingly, the use of natural gypsum, and particularly of its derivatives (i.e., AII), which are available in massive quantities, could be an interesting option to produce new PP-based composites and an answer to the current industrial requests.

Starting from the state of the art, the main goal of this paper is to present the experimental pathways followed to produce, characterize, and design the properties of new MF PP–AII composites containing up to 40 wt.% AII (thus issued from natural gypsum). Regarding the choice of PP as the matrix, it is well known that PP homopolymers are considered for engineering applications because they are characterized by higher crystallinity and improved chemical resistance compared to PP-based copolymers. Nevertheless, they are stiffer and have better strength at high temperatures, but unfortunately, their impact resistance is somewhat limited [61], so this parameter must be carefully considered. Interestingly, it has been reported elsewhere that the filling of PP with rigid fillers ($CaCO_3$, talc, others), results in increased rigidity/stiffness, but in some cases, better impact properties can also be achieved [62,63]. Therefore, it is of great interest to have information about the properties of PP–AII composites and to identify their most relevant features, linked to the requests of end-use applications (i.e., from packaging to automotive/engineering sectors).

In this study, first, to obtain small quantities of materials, PP was filled by melt compounding with 20–40 wt.% AII (β-$CaSO_4$) using internal mixers. Then, the PP–AII composites were characterized to obtain primary information on their properties. These composites (without any modifier) have shown quite good filler distribution within the PP matrix, enhanced thermal stability, and stiffness. To compensate for the decrease in tensile and impact strength, after the high filling of PP with AII (e.g., at 40 wt.% filler), it was necessary to propose custom formulations. For compatibilization and to obtain enhancements of proprieties, the PP/AII compositions have been modified with selected reactive modifiers, such as PP functionalized/grafted with maleic anhydride (PP-g-MA), and an ionomeric monomeric additive, zinc diacrylate. In fact, fillers such as talc, calcium carbonate, and AII are polar in nature; therefore, for better interfacial adhesion (PP–filler), the modification of PP can be accomplished by attaching polar groups onto the PP backbone, e.g., via the addition of PP-g-MA [64,65], or using metallic ionomers (i.e., zinc diacrylate to lead to Zn ionomer) [66,67], experimental methods often considered in prior studies.

The effects of filling PP with AII and reactive modification were deeply evaluated in terms of morphology, mechanical and thermal properties, to allow the selection of key compositions. The formation of clusters/ionic crosslinks of Zn ionomer and the favorable interactions with the filler allowed an impressive increase in the properties of PP–AII composites. Moreover, in the frame of current prospects, the most interesting composites in terms of properties were extrapolated on twin-screw extruders (TSE) to produce larger quantities of materials by reactive extrusion (REx) and to make them available for processing by injection molding (IM). The PP–AII composites produced by REx have been characterized using several methods of analysis to highlight their performances: high tensile and flexural strength, stiffness, enhanced storage modulus under dynamic mechanical solicitations, and superior HDT. By considering the overall features of these new PP–AII composites, it is expected that they will show great promise as potential novel products designed especially for technical/engineering applications requiring rigidity, heat resistance, dimensional stability, and improved processing.

2. Materials and Methods

2.1. Materials

PP homopolymer (Moplen HP400R, produced by LyondellBasell), is a PP grade for IM, suitable for food contact applications, characterized by high fluidity and good stiffness. The characteristics of interest according to the technical sheet of the product are: density: 0.9 g/cm^3, melt flow rate (MFR): 25 g/10 min (230 °C, 2.16 kg), tensile strength (yield): 32 MPa, tensile strain at break: >50%, and Charpy impact strength (notched): 2 kJ/m^2.

$CaSO_4$ β-anhydrite II (AII) mineral filler, delivered as "TOROWHITE Ti-ExR04", was kindly supplied by Toro Gips S.L. (Spain). According to the information provided by the supplier, AII is obtained from high-purity natural gypsum. It is characterized by high whiteness/lightness (L *), AII being an alternative of choice as a white pigment (TiO$_2$) extender. Color measurements performed in the CIELab mode (illuminate D65, 10°) with a SpectroDens Premium (TECHKON GmbH, Königstein, Germany) proved the high lightness of the AII sample, i.e., L * of about 98.0.

PP functionalized/grafted with maleic anhydride (PP-g-MA), modified with 8–10% maleic anhydride (MA), was supplied by Sigma-Aldrich. The molecular weights (M_w and M_n) of PP are, respectively, 9100 and 3900 Da. This product, evaluated only in the primary stage of the study, is abbreviated hereinafter as PPMA1.

Polybond 3200 (supplier Chemtura, abbreviated as PPMA2) is a PP homopolymer modified with maleic anhydride (MA) with the following characteristic features: MFR (190 °C, 2.16 Kg): 115 g/10 min, melting point: 157 °C, and MA content: 0.8–1.2%. PPMA2 was used only as an alternative for producing reference samples by REx.

Zinc diacrylate (produced as Dymalink® 9200 by Total Cray Valley) is an ionomeric monomeric additive recommended for the modification of polyolefins that allows resin producers and compounders to impart ionomeric character to their materials [68,69]. In unreinforced, filled, and thermoplastic polyolefin (TPO) systems, this reactive modifier can lead to an increase in melt strength, enhanced mechanical properties, and high-temperature performances. Dymalink® 9200 was made available as an off-white powder (100% active) with the following characteristic features: molecular weight: 207, specific gravity: 1.68, and functionality: 2. The product will be abbreviated as ZnDA. For more clarity, the chemical structures of the two modifiers tested in this study are shown in Scheme 1.

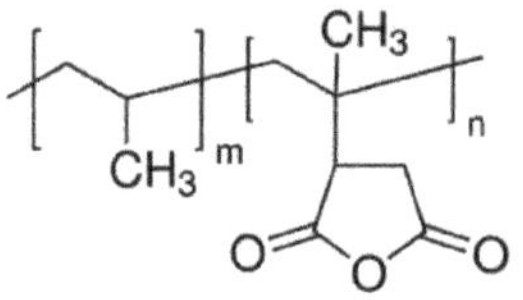

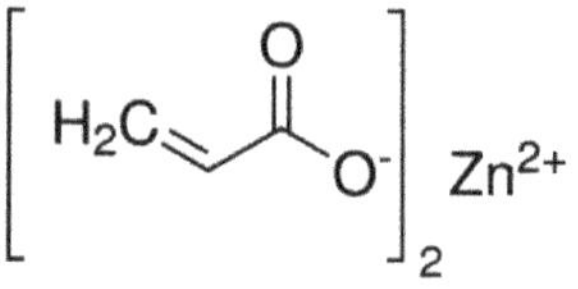

Scheme 1. Chemical structure of reactive modifiers evaluated in the frame of this study: (**a**) PP-g-MA (here a simplified chemical structure is presented; MA end-functionalized PP is likely present, as well as oligo-MA grafted on PP chains) and (**b**) zinc diacrylate.

2.2. Production of PP–AII Composites

2.2.1. Melt Compounding with Internal Mixers

Before melt compounding, all materials were dried overnight (PP and all modifiers at 70 °C, and the AII filler was dried at 100 °C). PP–AII composites were obtained by melt compounding at 220 °C, using a Brabender bench scale internal mixer (W50EHT, Plastograph EC, Brabender GmbH &. Co. KG, Duisburg, Germany) equipped with "came" blades. Conditions of processing: feeding at 30 rpm for 4 min, followed by melt mixing at 100 rpm for 10 min. For the sake of clarity, the codes and compositions of filled PP composites produced with internal mixers are shown in Table 1. Neat PP processed under similar conditions has been used as one key reference. The so-obtained PP composites were processed by compression molding (CM) at 210 °C, using an Agila PE20 hydraulic press to obtain plates (100 mm × 100 mm~3 mm thickness). More specifically, the material was first maintained at low pressure for 180 sec (3 degassing cycles), followed by a high-pressure cycle at 150 bars for 120 sec. Then, the cooling was realized under pressure (50 bars) for 300 sec using tap water (temperature slightly > 10 °C). The plates produced by CM were used to obtain specimens for mechanical characterizations. Throughout this contribution, all percentages are given as weight percent (wt.%).

Table 1. Codes and the compositions of PP samples produced with internal mixers.

Sample Code	PP, wt.%	AII, wt.%	Modifier, wt.%
PP	100	-	-
PP–20AII	80	20	-
PP–20AII–PPMA1	75	20	5
PP–20AII–ZnDA	78	20	2
PP–40AII	60	40	-
PP–40AII–PPMA1	55	40	5
PP–40AII–ZnDA	58	40	2

2.2.2. Reactive Extrusion (REx) Using Twin-Screw Extruders (TSE)

In the subsequent experimental step, larger quantities of the most promising formulations/composites were produced by REx, using a Leistritz twin-screw extruder (TSE) as the equipment (ZSE 18 HP-40, L (length)/D = 40, diameter (D) of screws = 18 mm). Before melt compounding, PP and the AII filler were dried overnight at 70 °C and 100 °C, respectively, using drying ovens with recirculating hot air. The experimental setup used to produce PP–AII–ZnDA composites in larger quantities is shown in Figure 1.

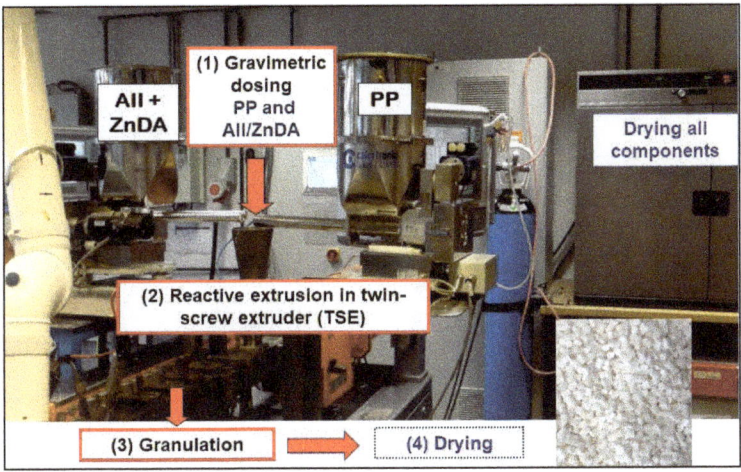

Figure 1. Experimental setup for the production by REx of modified PP–AII composites.

The filler (AII) was previously mixed with ZnDA into a Zeppelin Reimelt Henschel FML4 mixer (Zeppelin Systems GmbH—Henschel Mixing Technology, Kassel, Germany). Basically, to produce PP filled with 20% and 40% AII, the filler was premixed with ZnDA powder (AII/ZnDA weight ratio of 20/2 or 40/2, respectively) for 15 min at 1500 rpm. Two separate gravimetric feeders were used for the dosing of PP (granules) and AII/ZnDA powders. It is noteworthy that to evaluate PPMA2, two controlled feeders were used, a first one to feed the blends of PP and PPMA2, after a previous premixing, and a second one for AII. The parameters of REx/compounding were as follows: (a) temperatures on the heating zones of TSE: Z1 = 180 °C, Z2 = 190 °C, Z3-Z5 = 215 °C, Z6 = 200 °C, and Z7 = 195 °C; (b) die of extrusion = 180 °C; (c) speed of the screws = 170 rpm; (d) throughput = 3 kg/h.

The samples produced by REx as granules have been dried overnight at the temperature of 70 °C. Specimens for tensile, flexural, impact, and heat deflection temperature (HDT) characterizations were produced with a Babyplast 6/10 P injection molding (IM) machine, using adapted processing temperatures (e.g., Z1 = 200 °C; Z2 = 205 °C; and Z3 (die) = 200 °C; temperature of the mold = 30–40 °C).

2.3. Methods of Characterization

(a) Thermogravimetric analyses (TGA) were performed using a TGA Q50 (TA Instruments, New Castle, DE, USA) by heating the samples under air or nitrogen from room temperature (RT) up to maximum of 800 °C (platinum pans, heating ramp of 20 °C/min, 60 cm^3/min gas flow). In TGA, the initial degradation temperature (i.e., $T_{5\%}$, °C), was considered as the temperature at which the weight loss is 5 wt.%, whereas the temperature of the maximum degradation (T_d) was defined as the temperature at which the samples present the maximal mass loss rate (data obtained from D-TG curves).

(b) Differential scanning calorimetry (DSC) measurements were conducted using a DSC Q200 from TA Instruments (New Castle, DE, USA) under nitrogen flow. The traditional DSC procedure was as follows: first, heating scan (10 °C/min) from 0 °C to 220 °C, isotherm at this temperature for 2 min, and then cooling by 10 °C/min to −50 °C, and, finally, a second heating scan from −50 to 220 °C at 10 °C/min. The first scan was used to erase the prior thermal history of the polymer samples. The events of interest linked to the crystallization of PP during the DSC cooling scan, i.e., the peaks of crystallization temperature (T_c) and the enthalpies of crystallization (ΔH_c), were quantified using TA Instruments Universal Analysis 2000 software—Version 3.9A (TA Instruments—Waters LLC, New Castle, DE, USA). It is noteworthy that all data are normalized to the amounts of PP from the samples. The thermal parameters were also evaluated in the second DSC heating scan and abbreviated as follows: melting peak temperature (T_m), melting enthalpy (ΔH_m), and final DC (χ). The DC (degree of crystallinity) was determined using the following general equation:

$$\chi = \frac{\Delta H_m}{\Delta H_m^0 \times W_{PP}} \times 100 \ (\%)$$

where ΔH_m is the enthalpy of melting, W is the weight fraction of PP in composites, and ΔH_m^0 is the melting enthalpy of 100% crystalline PP considered 207 J/g [70].

(c) Mechanical testing: Tensile tests were performed with a Lloyd LR 10K bench machine (Lloyd Instruments Ltd., Bognor Regis, West Sussex, UK), according to ASTM D638-02a norm on specimens-type V, at a typical speed of 10 mm/min (specimens of 3.0–3.2 mm thickness). The flexural properties were determined using a three-point bending test and the NEXYGEN program (Lloyd Instruments Ltd.). The measurements were performed on rectangular specimens (80 mm × 10 mm × 4 mm) at a testing speed of 10 mm/min, by adapting the Lloyd LR 10K tensile bench with bending grips (span = 60 mm), in accordance with ISO 178. For the characterization of Izod impact resistance, rectangular specimens (63 mm × 12 mm × 3.2 mm), a Ray-Ran 2500 pendulum impact tester, and a Ray-Ran 1900 notching apparatus (Ray-Ran Test Equipment Ltd., Warwickshire, UK) were used, according to ASTM D256 norm (method A, 3.46 m/s impact speed, 0.668 kg hammer). All mechanical tests were performed on specimens previously conditioned for at least 48 h at 20 ± 2 °C, under a relative humidity of 50 ± 3%, and the values were averaged over at least five measurements.

(d) Scanning electron microscopy (SEM) analyses on previously cryo-fractured PP samples at a liquid nitrogen temperature were performed using an FE-SEM SU-8020 Hitachi instrument equipped with triple detectors (Hitachi, Tokyo, Japan), at various accelerated voltages and magnitudes. For better insight and easy interpretation, the SEM analyses were performed with detectors for both secondary and backscattered electron imaging (SE and BSE, respectively). Reported microphotographs represented typical morphologies as observed at, at least, three distinct locations. SEM analyses (SE mode) were also performed on the surfaces of selected specimens fractured by tensile or impact testing to have more information about their behavior under mechanical solicitations. SEM combined with energy-dispersive X-ray spectroscopy (EDX) was also used to obtain information about the elemental composition of selected samples and/or to reveal the elemental atomic distribution within the PP matrix.

(e) Rheological measurements: The melt volume/flow rate (MVR/MFR) was determined only on samples produced by REX (as granules) following the procedure described in ASTM D1238, using a Davenport 10 Melt Flow Indexer (AMETEK Lloyd Instruments Ltd., West Sussex, UK) at a temperature of 190 °C, with a 2.16 kg load.

(f) Heat deflection temperature (HDT) measurements were performed according to ISO 75 norm, using an HDT/Vicat 3–300 Allround A1 (ZwickRoell GmbH & Co, Ulm, Germany) equipment. The specimens for testing (80 mm × 10 mm × 4 mm), were produced by IM. All measurements were evaluated under a load of 0.45 MPa and at a heating rate of 120 °C/h using at least three specimens.

(g) DMA (dynamic mechanical analysis) was performed on rectangular specimens obtained by IM (63 mm × 12 mm × 3 mm) using a DMA 2980 apparatus (TA Instruments, New Castle, DE, USA), in dual cantilever bending mode. The dynamic storage and loss moduli (E' and E'', respectively) were determined at a frequency of 1 Hz and amplitude of 20 µm, in the range of temperature from −75 °C to −150 °C at a heating rate of 3 °C/min.

(h) Color measurements (L *, a *, b *) were performed in the CIELab mode (illuminate D65, 10°) with a SpectroDens Premium (TECHKON GmbH, Königstein, Germany).

3. Results and Discussion

3.1. Preliminary Considerations

It is important to note that CS is available in several forms: dihydrate—$CaSO_4 \cdot 2H_2O$ (commonly known as gypsum), hemihydrate—$CaSO_4$ $0.5H_2O$ (Plaster of Paris, stucco, or bassanite), and different types of anhydrites [57,59,71]. The CS phases obtained by progressive dehydration and calcination of gypsum are in the following order [59]: dihydrate → hemihydrate → anhydrite III → anhydrite II → anhydrite I (the last one obtained at temperatures higher than 1180 °C). Dehydration of gypsum above 100 °C at low pressure (vacuum) or under air (at atmospheric pressure) favors the formation of β-CS hemihydrate. Increasing the temperature to around 200 °C enables the production of β-anhydrite III—which is not stable, whereas as mentioned in the introductory part, the calcination at higher temperature (e.g., at 500–700 °C) allows obtaining "insoluble" β-anhydrite II (AII), characterized by extremely slow rates of rehydration [57,59].

Figure 2a,b shows selected SEM pictures to illustrate the morphology of AII microparticles used in this study. The granulometry of AII was characterized by dynamic light scattering, using a Mastersizer 3000 laser particle size analyzer (Malvern Panalytical Ltd., Malvern, UK), the microparticles having a Dv50 of 3.6 µm and a Dv90 of 12.9 µm.

(a)

(b)

Figure 2. (**a,b**) SEM images (SE mode) at (**a**) low and (**b**) high magnification of AII microparticles.

The particulate filler is characterized by a low aspect ratio, whereas a shared morphology, i.e., particles with irregular shape and a fibrillar/flaky aspect, due to the cleavage of CS layers [59] during the production/grinding process, are typically evidenced by SEM analysis. It is worth mentioning that the rough surfaces can increase the number of anchorage points with the polymer matrix, thus offering good filler–resin mechanical interlocking, which can influence the interfacial adhesion and mechanical properties of composites. To allow the use of CS in the production of polymer composites, we will retell that it is of prime importance to dry (dehydrate) the CS dihydrate or hemihydrate prior melt compounding, or it is required to use stable anhydrite forms, such as AII. Indeed, it was reported elsewhere that β-AII (made from natural or synthetic gypsum), is much better suited for melt blending with a polymer sensitive to the degradation by hydrolysis, e.g., the case of polyamides or polyesters, such as polylactic acid (PLA) [57,58,72].

The high purity of the AII sample was proven by SEM-EDX analysis, which confirms the presence of Ca, S, and O as the main atomic elements, and only of few traces of carbon (EDX spectra shown in Figure 3a). Furthermore, the excellent thermal stability of the filler up to 700 °C is confirmed by TGA (Figure 3b). Hence, by its addition to the PP matrix, it is expected to have beneficial effects on the thermal properties of PP composites.

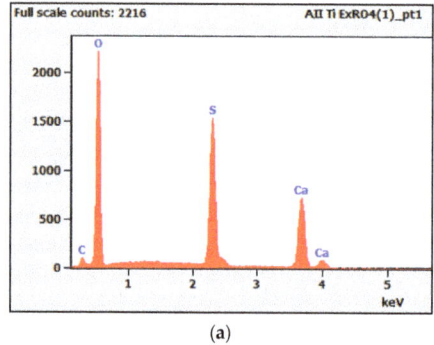
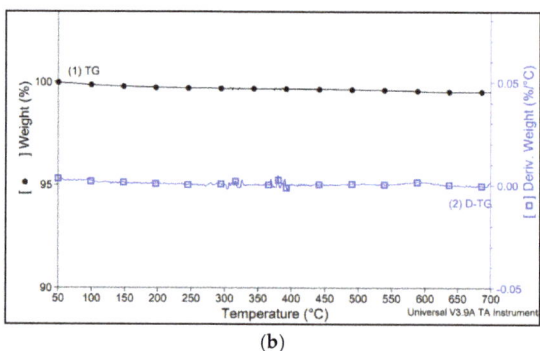

(a) (b)

Figure 3. (a,b) Characterization of AII filler: (a) EDX spectra and (b) TGA (under N_2, 20 °C/min).

For a first evaluation, the PP–AII compositions from Table 1 (see the experimental part), have been produced in small quantities using internal mixers, whereas, in the next experimental step, the most representative composites have been extrapolated on twin-screw extruders (TSE) and deeply characterized for the evidence of key properties.

3.2. PP–AII Composites Produced within Internal Mixers

After the addition of mineral fillers to PP, it is expected that a great part of properties will be improved (e.g., stiffness, thermal stability, aesthetic), while, on the other hand, other properties might decrease to some extent (tensile strength, impact resistance). To maximize the benefits of MF PP composites, it is necessary to understand and combine the relationship between the properties of the matrix and characteristics of dispersed phases (fillers, additives, etc.), to improve their compatibility and interactions, to control the stabilizing or degradation effects, and the influence of the manufacturing process on the final product characteristic features, and so forth.

3.2.1. Morphology of Composites

For better evidence of the filler distribution state through the PP matrix, SEM imaging was performed on all composites (20% and 40% filler) using backscattered electrons (BSE) to obtain a higher phase contrast. Traditionally, it is expected to obtain better individual particles dispersion at lower filler loading than at high filling [58,72], and this assumption is confirmed by the SEM analyses of PP composites filled with 20% AII (SEM micrographs shown in the Supplementary Material, Figure S1). Figure 4a–d show representative SEM-BSE images at different magnifications of cryo-fractured surfaces of PP and of highly filled PP–AII composites containing 40% filler, a loading at which the filler could show some poorer distribution/dispersion.

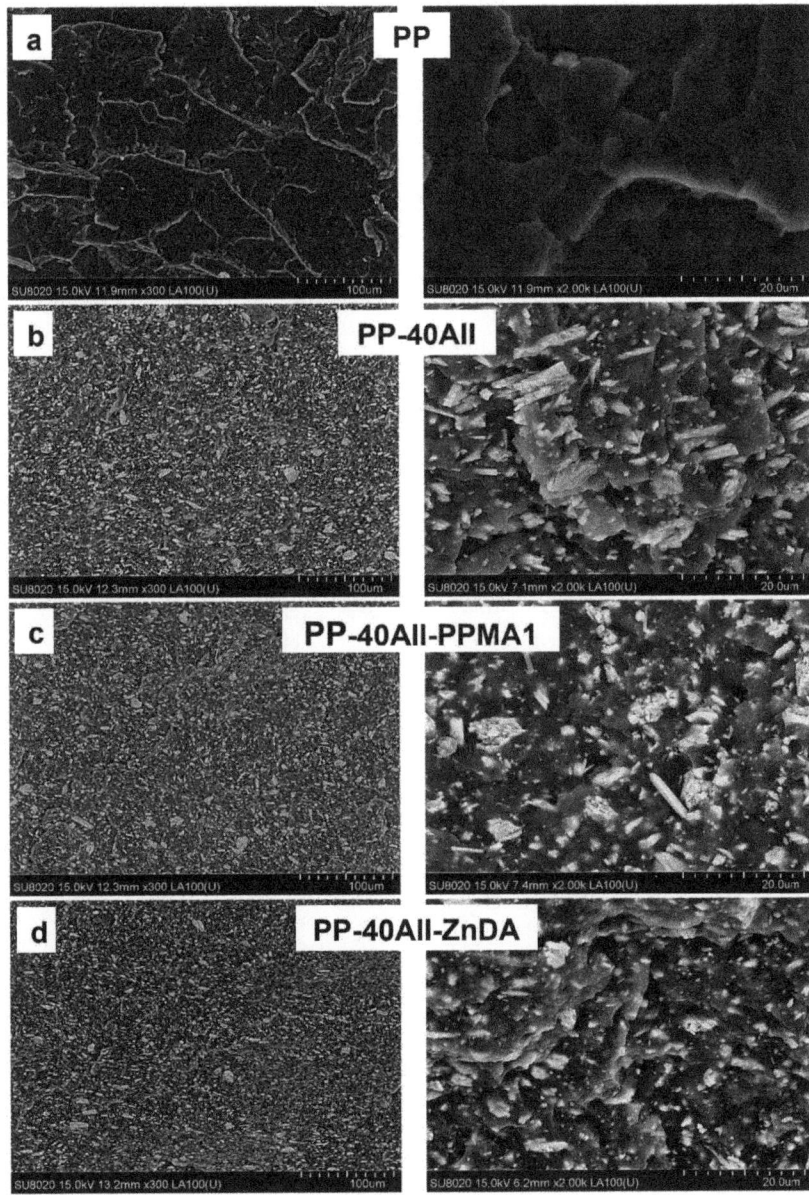

Figure 4. (**a**–**d**). Representative SEM micrographs (BSE) at different magnifications on the cryo-fractured surfaces of PP and highly filled PP composites: (**a**) unfilled PP, (**b**) PP–40AII, (**c**) PP–40AII–PPMA1, and (**d**) PP–40AII–ZnDA composites.

The BSE technique has a high sensitivity to the differences in atomic number, giving information about the distribution of CS (i.e., the presence of Ca atoms is evidenced by brighter zones). Well-distributed particles, with various geometries and quite broad size distribution, are evidenced at the surface of cryo-fractured PP composites, even though the amount of filler is high (i.e., 40%). It is worth mentioning that large aggregates are not observed, whereas such a quality of distributive mixing within the hydrophobic PP matrix is obtained without any previous surface treatment of the hydrophilic filler. Still, from the SEM pictures at higher magnification, it is evident that the AII particles are characterized by a low aspect ratio and irregular shape of micrometric size. On the other hand, the morphology of PP–40AII–ZnDA appears slightly different compared to that of other composites, even though all blends are performed using internal mixers. In fact, the SEM images of this sample suggest better wetting/surrounding by the PP matrix, of well-distributed and even well-dispersed AII microparticles. The improved morphology can be ascribed to a more distinct compatibilizing effect (PP–AII), following the reactive modification.

3.2.2. Thermal Properties

Obviously, AII is a mineral filler characterized by high thermal stability. In the absence of undesirable catalytic degradation effects, its incorporation into different polymers is expected to result in composites characterized by similar, or even better thermal stability, with respect to the neat polymer matrix. Indeed, as shown from the comparative TGA traces shown in Figure 5a,b, the filling with, 20% and 40% AII, respectively, clearly result in a delay in the thermal degradation of PP, well determined by the amounts of filler. Furthermore, a more pronounced stabilization effect is achieved for PP–AII–ZnDA composites, i.e., after the co-addition of AII and ZnDA (chemically grafted as Zn ionomer). In fact, the TGA data (Table 2) confirm the boosting of both thermal parameters, i.e., $T_{5\%}$ and T_d.

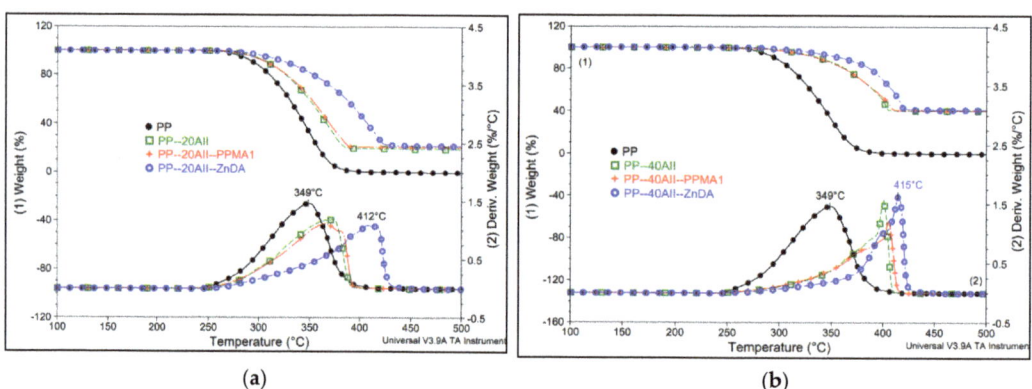

Figure 5. (a,b) Comparative TG and D-TG curves of PP and PP–AII composites filled with different amounts of filler: (a) 20% AII and (b) 40% AII (TGA under air, 20 °C/min).

The significant rise in the onset of thermal degradation ($T_{5\%}$) and of maximum decomposition temperature (T_d, from max. D-TG), is assigned to the presence of filler, with a mention for the PP–AII composites modified with ZnDA, which exhibits the best thermal properties. Indeed, the PP–AII–ZnDA samples show the shift of T_d above 60 °C compared to the neat PP processed under similar conditions. On the other hand, there is no evidence of an additional effect linked to the use of PP-g-MA (PPMA1), because the composites PP–AII and PP–AII–PPMA1 display similar thermal behavior. Still, the increase in thermal stability of PP following the co-addition of AII and ZnDA ($T_{5\%}$ of PP–20AII–ZnDA and PP–40AII–ZnDA is increased by 25 °C and 49 °C, respectively, compared to PP), is considered as a key property in the perspective of the application of such materials. These enhancements can be ascribed at least in part to a better filler dispersion (as was revealed by SEM images), but the influence of other factors is not excluded, e.g., a stabilizing effect linked to the formation of Zn ionomer.

Table 2. Thermal parameters of PP and PP–AII composites as determined by TGA (under air).

Sample Code	$T_{5\%}$, °C	T_d, °C	Residual Product at 600 °C, wt.%
PP	283	349	0.1
PP–20AII	295	374	19.4
PP–20AII–PPMA1	293	365	20.6
PP–20AII–ZnDA	308	412	21.8
PP–40AII	316	400	40.3
PP–40AII–PPMA1	314	404	40.5
PP–40AII–ZnDA	332	415	41.2

Figure 6a,b show the comparative DSC traces of the neat PP and those of PP–AII composites obtained during DSC cooling from the molten state (a), and following the second DSC heating by 10 °C/min (b). From the analysis of DSC results obtained during non-isothermal crystallization (Figure 6a, DSC data shown in Table 3), it emerges that the nucleation/kinetics of crystallization of PP is improved in presence of AII (T_c of PP–AII composites is at about 120 °C, higher than that of neat PP, i.e., 116 °C). On the other hand, when using PPMA1 as a modifier, the crystallization of PP is slightly delayed (e.g., PP–20AII–PPMA1 has a T_c of 115 °C). In contrast, the composites reactively modified with 2% ZnDA (i.e., PP–AII–ZnDA) exhibit a T_c at 126 °C that is significantly higher compared to that of neat PP. A similar behavior was reported elsewhere for PP modified with Surlyn ionomers, by Ma et al. [73]. The crystallization rate was accelerated by the ionic aggregates/clusters of ionomers which were reported to initiate the heterogeneous nucleation of PP. As a result, T_c increased, and the crystallization process was faster for the PP/ionomer systems with respect to the neat PP.

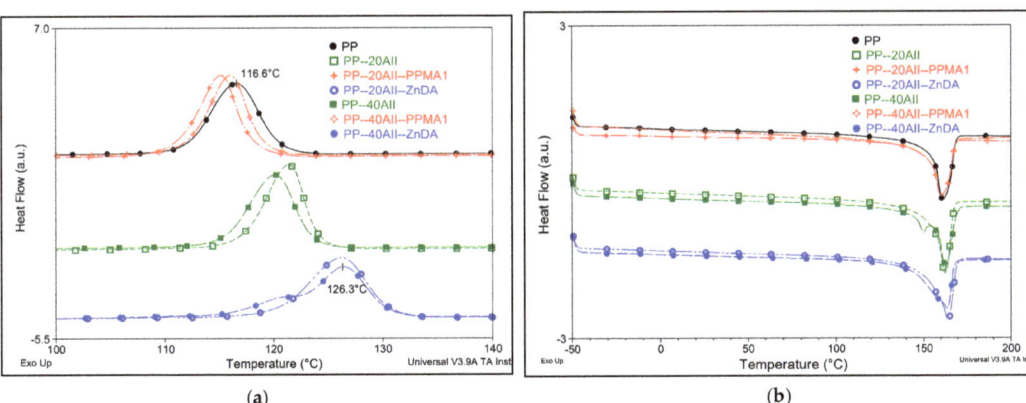

Figure 6. (a,b) Comparative DSC curves of neat PP and PP–(20–40%)AII composites (with/without modifiers) obtained during (a) DSC cooling and (b) second DSC heating (rate of 10 °C/min).

Still, regarding the second DSC heating (Figure 6b, data gathered in Table 3), a slight increase in T_m is noted for the composites that have shown higher T_c (i.e., PP–AII and PP–AII–ZnDA), and that are finally characterized by somewhat higher DC. Moreover, because PP homopolymer inherently has high rates of crystallization, the values of DC for all samples are high and remarkably close to each other, i.e., in the range of 45–48%, which makes it difficult to identify the glass transition temperatures (T_g) under the specific conditions of DSC analysis. However, the crystallization mechanisms of these composites require more understanding (e.g., insight into the crystal nucleation/structure and its growth, α-nucleation, and occurrence with β-nucleation), via different techniques of investigation (DSC, polarized light microscopy, X-ray diffraction, etc.); therefore, we will refrain here from more comments. Further studies are under consideration.

Table 3. Comparative DSC data of PP and PP–AII composites obtained during non-isothermal crystallization from the molten state and following the succeeding heating scan (DSC cooling and second heating by 10 °C/min).

Sample	DSC Cooling		Second DSC Heating		
	T_c, °C	ΔH_c, J/g	T_m, °C	ΔH_m, J/g	χ, %
PP	116	93.2	161	93.6	45.2
PP–20AII	121	97.2	163	98.2	47.4
PP–20AII–PPMA1	115	94.6	160	94.8	45.8
PP–20AII–ZnDA	126	96.4	164	99.5	48.1
PP–40AII	120	94.5	163	96.6	46.7
PP–40AII–PPMA1	119	91.2	161	92.5	44.7
PP–40AII–ZnDA	126	93.2	164	99.9	48.3

Abbreviations: DSC cooling scan: crystallization temperature (T_c) and enthalpy of crystallization (ΔH_c); second DSC heating: peak of melting temperature (T_m); enthalpy of fusion (ΔH_m), and final degree of crystallinity (χ).

3.2.3. Mechanical Properties

In general, it is expected that the reinforcement of PP with mineral fillers will lead to improved mechanical properties, such as increased stiffness (i.e., higher tensile and flexural modulus, enhanced flexural strength), whereas, on the other hand, the reduction of tensile strength, of strain at break and impact resistance, is often reported [39].

The evolution of the tensile and impact properties of PP and PP–AII composites (reactively modified or not) is shown in Figure 7a–d. By analyzing the effects of AII addition into PP (without any modifier), is noted that the ultimate (maximum) tensile strength (σ_t) of the PP (37 MPa) is gradually diminished to 29 MPa and 25 MPa, adding 20% and 40% filler, respectively (Figure 7a). On the other hand, when using PP-g-MA as a compatibilizer, only limited improvements are seen, especially for the highly filled composites (i.e., PP–40AII–PPMA1), characterized by a σ_t of about 29 MPa. On the contrary, the most interesting properties are revealed by PP–AII–ZnDA composites (σ_t of 32–34 MPa). However, the stress–strain curves (Figure 7b) give more insight into the mechanical behavior during tensile testing of PP composites. Accordingly, a stronger reinforcing effect by filling is obtained for PP–AII composites using ZnDA than PPMA1, which was found to be less effective as a compatibilizer, under the specific experimental conditions used in this study. However, it is not excluded that the low molecular weight of PPMA1 can explain the decrease in mechanical parameters; therefore, the testing of other PP-g-MA products was also considered (more information shown elsewhere). Although the elongation at the break (ε_b) of specimens obtained by CM decreased drastically, especially at high filling, from 11% (neat PP) to 3–6% for composites containing 40% AII. Nevertheless, currently there is still a need for a better understanding of the mechanisms that can explain the enhancements of properties of PP–AII composites after the reactive modification with ZnDA (i.e., interactions between Zn ionomer and filler, formation of clusters or ionic crosslinking networks, liable for better thermo-mechanical properties). Moreover, the interfacial adhesion (polymer filler), is considered to be one of the main factors affecting the strength of the composites [62]. In connection with the interfacial properties (PP–AII) in the different composites, it is worth mentioning that for better insight, supplementary SEM analyses have been performed on the surfaces of fractured specimens by mechanical testing. The SEM micrographs suggest a stronger interfacial adhesion between PP and filler in the case of PP–AII–ZnDA composites (numerous regions of contact and better wetting of AII by the PP matrix, the particles are more deeply lodged within the polymeric matrix, etc.), rather than the debonding at the interface, which is more specifically seen for the other composites (SEM images shown as Supplementary Material—Figure S2, and representative examples in Section 3.3.1.).

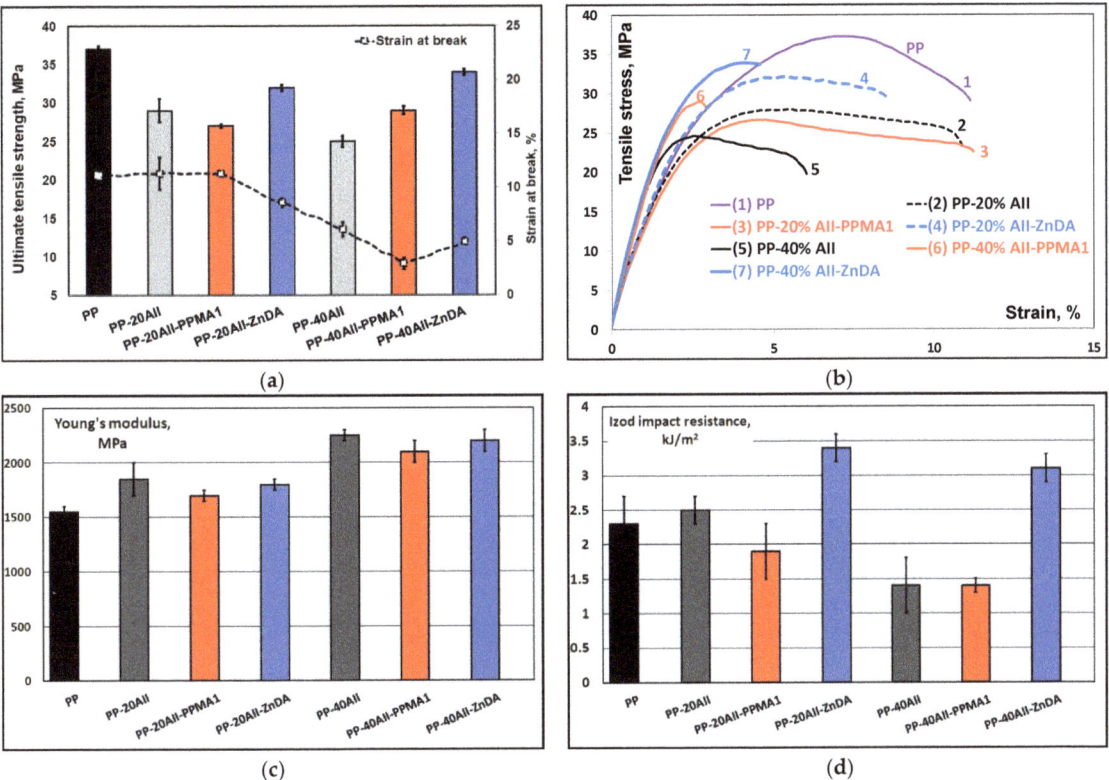

Figure 7. (**a**–**d**) Comparative mechanical properties of PP and PP–AII composites with/without modifiers: (**a**) tensile strength and strain at break; (**b**) tensile stress–strain curves; (**c**) Young's modulus; (**d**) impact resistance (Izod).

Regarding the evolution of rigidity/Young's modulus (Figure 7c), it is reasonable to consider that this parameter is determined by the amounts of filler, e.g., a progressive increase of 19% and 42% compared to the neat PP is obtained for PP–AII composites containing, respectively, 20% and 40% AII. Here, the effect of ZnDA is less evident, maybe because this parameter (Young's modulus) is determined in the limit of elasticity at low strain/deformation.

In terms of impact properties (Figure 7d), unfortunately, PP homopolymers are relatively "brittle", requiring low crack propagation energy for breaking (impact resistance of 2.3 kJ/m^2, on specimens performed by CM). In several cases, it has been reported that the addition of rigid fillers can have positive effects on the impact resistance of polymers, and PP is included [62]. This improvement was also stated for PLA composites filled with 20% AII, which are characterized by higher impact resistance than the neat PLAs [57]. Interestingly, the PP–20AII sample shows a little bit higher impact resistance than the neat PP matrix. At higher filler amounts (40% AII), a dramatic decrease in this parameter is observed for the unmodified composite (PP–40AII). On the other hand, notable enhancements are seen after the reactive modification with ZnDA (impact resistance in the range of 3–3.5 kJ/m^2, much higher than that of neat PP). Accordingly, ZnDA leads to the most noteworthy enhancement of impact/toughness, whereas the modification with PPMA1 was found, again, to be less effective. The increase in impact resistance of PP modified with ionomers is elsewhere ascribed to the decrease in the spherulite size of PP, to a specific/unique structure and morphology [73]. The ionomers allow the formation of crosslink points and networks of molecular chains, which could absorb the impact energy and prevent crack initiation.

3.2.4. Key Considerations and Findings

The overall experimental results reveal the key role of the reactive modification with metallic acrylate monomer (ZnDA) for the enhancement of thermal and mechanical properties of PP–AII composites. On the contrary, PP-g-MA (i.e., PPMA1) was found to be less effective in improving the compatibility/interfacial properties between the PP matrix and filler, and the final characteristics of the composites. To the best of our knowledge, the PP–AII–ZnDA composites/formulations are entirely new; therefore some more comprehension will be required to explain their performances using additional methods and techniques of characterization, not concerned by this paper (i.e., Fourier transform infrared (FTIR) and X-ray photoelectron spectroscopy (XPS), rheometric analyses, and the use of predictive mathematical models for composite mechanical properties). It is also noteworthy to mention that the SEM-EDX analyses performed on PP–AII–ZnDA composites give evidence for the fine distribution/dispersion of elemental Zn within the PP matrix (an illustrative example is shown in Figure 8).

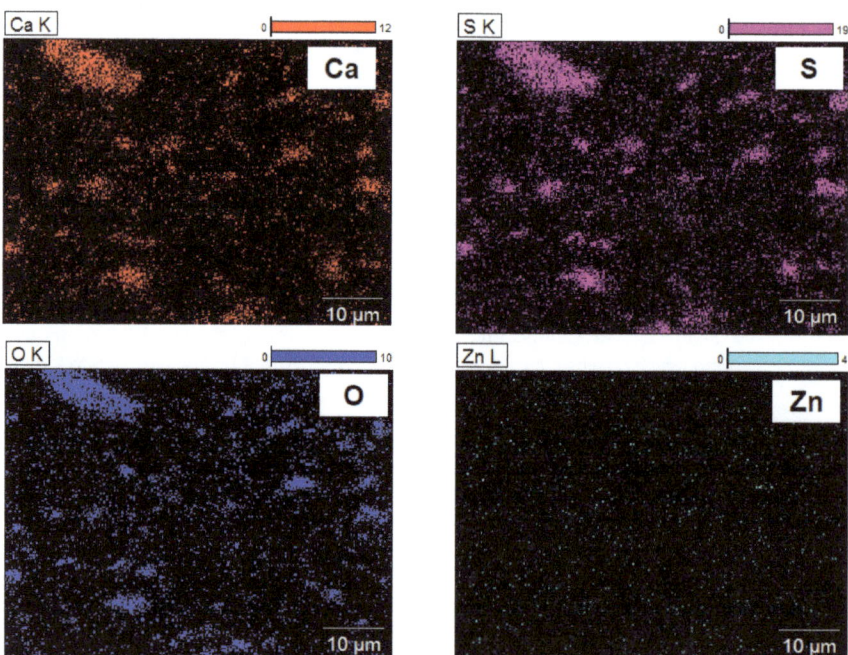

Figure 8. SEM-EDX mapping on the surface of cryo-fractured PP–AII–ZnDA composites to highlight the distribution within the PP matrix of main elements from filler (Ca, S, and O) and of Zn from ZnDA.

It is important to note that the use of ZnDA together with peroxides to cure and improve the rubber's properties, has been applied in industry for decades [68,74,75]. On the other hand, the addition of ionomeric additives, such as ZnDA, to conventional PP is reported to improve its melt strength, and also other properties (hardness, impact strength, HDT, chemical resistance, gas barrier, etc.) [69,76–78]. High melt-strength PP-based homopolymers and copolymers are required for IM, thermoforming, and foaming applications. Moreover, ionic crosslinking (e.g., with Zn acrylates) is of interest in the recycling process to strengthen, toughen, and compatibilize polymer blends [79].

Zn acrylates are reported to form ionomeric crosslinks with the polyolefin (PP) chains. These crosslinks are "thermo-reversible", allowing the material to be processed using traditional equipment (IM machines, extruders, etc.) while providing significant enhancements of properties at temperatures below 150 °C, that are of interest in the case of engineering materials. Regarding the mechanism of the reaction, it is considered that the addition of zinc acrylate monomers during melt mixing/extrusion at temperatures above 210 °C allows acrylate double bonds to graft on PP chains. Then, the polar zinc salts (Zn ionomer) could associate to form ionomeric crosslinks during cooling [80]. Hence, in the frame of this study, it was assumed that the conditions of reactive melt blending using an internal mixer (i.e., melt blending for 10 min, at 220 °C) represent a good experimental choice to allow the grafting of ZnDA on PP chains and to lead to high density crosslinked structures through ionic bonds. Moreover, the reaction between PP and ZnDA is feasible even in the absence of peroxides [80], which are commonly used as initiators for the grafting on PP chains, or for the long-chain branching PP [81,82]. In fact, PP is very sensitive to thermo-mechanical/oxidative degradation at temperature; therefore, the easy generation of radicals and formation of degradation products by β-scission [83,84] during melt mixing can be assumed, even in the absence of peroxides.

Scheme 2a suggests, based on the information from the state of the art, the formation of zinc salt clusters following the grafting of ZnDA onto PP chains (formation of Zn ionomer), ionic domains deemed as thermo-reversible "crosslinks" at temperatures ≤ 190 °C [80,85]. Furthermore, to explain the enhancements of properties obtained by the co-addition of AII and ZnDA, Scheme 2b proposes an alternative structure, based on the hypothesis that the Zn ionomer and AII filler can also interact via ionic bounds (e.g., ion pairs, metal–ion coordination) [75,86], leading to the formation of ionic clusters/crosslinks, physical entanglements/aggregates, and supramolecular networks. Indeed, from the category of multivalent ions, calcium is considered available for ionic crosslinking and chelation with carboxylate groups (−COO$^-$) [87,88].

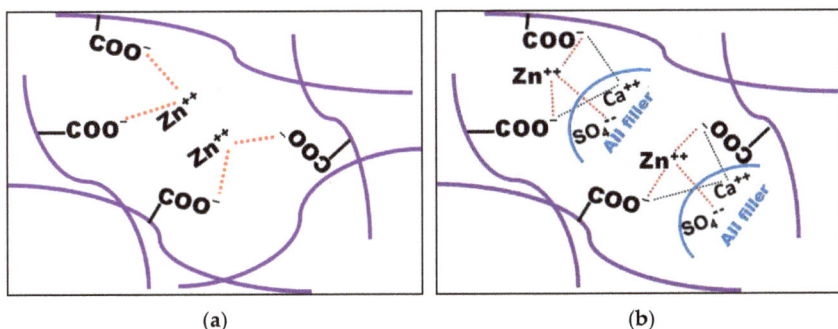

Scheme 2. (a,b) Suggested structures of ionomeric domains and clusters formed following the reactive modification of PP with ZnDA: (a) without filler; (b) presumed structures/ionic interactions (dash lines) in presence of AII.

The ionic interactions and/or the creation of new ionic bonds/interactions, i.e., between Zn^{2+} and Ca^{2+} as cations, with SO_4^{2-} and $-COO^-$ anions, respectively, could be (hypothetically) responsible for the good compatibility/interfacial properties between PP and filler, and finally, can explain the thermo-mechanical performances of these new custom composites, i.e., PP–AII–ZnDA. It is also important to be precise that for additional information, an unfilled PP composition modified with 2% ZnDA was prepared and characterized under similar conditions. Compared to the neat PP, the mechanical properties remain almost comparable, except for the impact resistance, which has increased from 2.3 kJ/m^2 (neat PP) to about 3 kJ/m^2 for the reactive modified PP (results shown in Supplementary Material, Table S1).

For potential utilization in engineering applications, it emerges that the PP–AII–ZnDA composites are characterized by the most interesting performances. Obviously, the formation of ionic crosslinks via the addition of ZnDA (formation of Zn ionomer) and the strong interactions between components play a key role to strengthen (σ_t, 32–34 MPa) and toughen the PP–AII blends. The dramatic decrease in impact resistance by high filling (only 1.4 kJ/m^2 for PP–40AII composites), is clearly corrected to higher values using ZnDA (>3 kJ/m^2), which is significantly better than that of pristine PP (2.3 kJ/m^2). Accordingly, these tailored PP–AII–ZnDA composites have been proposed for upscaling by REx using TSE (details hereinafter).

3.3. Current Prospects: Production of PP–AII Composites by REx

For the supplementary confirmation of the results obtained with laboratory mixers (Section 3.2), and in the perspective of upscaling on pilot plants, it was decided to use twin-screw extruders (TSE) to produce higher quantities of PP–AII–ZnDA composites, and to process them by IM. Nevertheless, regarding the presumed differences between the reactive melt mixing with internal mixers and TSE, it is believed that the temperatures of melt compounding, the shear and residence time, and the conditions of processing used to obtain final products (e.g., IM vs. CM) are among the factors that are requiring increased attention.

3.3.1. Characterization of PP–AII Composites Produced by REx

For a simplified reading and more comprehension, the discussion hereinafter is focused on two key compositions produced in larger quantities, to allow their processing by IM, i.e., PP–20AII–ZnDA (TSE) and PP–40AII–ZnDA (TSE). These PP composites are filled with 20% and 40% AII, respectively, and both are modified with 2% ZnDA (details in the experimental section). For comparative reasons, the unfilled PP (processed on TSE under similar conditions) is used as a reference. As was mentioned in the experimental part, the realization of PP–AII composites modified with PPMA2 (i.e., PP containing 0.8–1.2% grafted MA) was also considered, but unfortunately, the mechanical properties (i.e., the tensile strength and impact resistance) were less promising than using ZnDA (results shown in Supplementary Material, Table S1).

Hereinafter are summarized the key results obtained following the production and analyses of these novel composites (PP–AII–ZnDA), mostly to confirm the good reproducibility of results obtained using different reactive melt-mixing procedures:

(a) Morphology of composites: Figure 9 shows representative SEM images of the cryofractured surfaces of PP–AII–ZnDA composites produced by REx. From the micrographs obtained at low and higher magnification, it emerges that following the melt blending/REx at shorter residence time and higher shear with TSE, once more, AII has a good distribution within the PP matrix, since there is not any evidence for the presence of agglomerates of microparticles, even at high filing.

(b) Thermal properties: TGAs attest again that, by the progressive increase in filler amounts up to 40 wt.%, the thermal stability of PP is remarkably improved (Figure 10a), e.g., the onset of thermal degradation ($T_{5\%}$) is increased by more than 50 °C for PP–40AII–ZnDA (TSE) composites with respect to the neat polymer. These enhancements are ascribed to the effects of the filler and its outstanding distribution/dispersion, and, to a stabilizing effect linked to the presence of Zn ionomer, following the grafting by REx of ZnDA onto PP chains. Moreover, the DSC analyses reconfirm the previous findings obtained with laboratory internal mixers. From the analysis of DSC traces obtained during non-isothermal crystallization (DSC cooling, Figure 10b), it is seen that the samples containing AII and Zn ionomer show higher T_c compared to the neat PP (T_c of 129 °C and only 116 °C, respectively). By considering overall DSC results, it can be assumed that both components (AII and ZnDA) act as effective nucleating agents for PP, allowing improved kinetics of crystallization, a propriety of interest for better processing. On the other hand, regarding the second DSC heating, there is only a slightly noticeable increase in T_m for PP–AII–ZnDA (TSE) composites, from 163 °C to 165 °C.

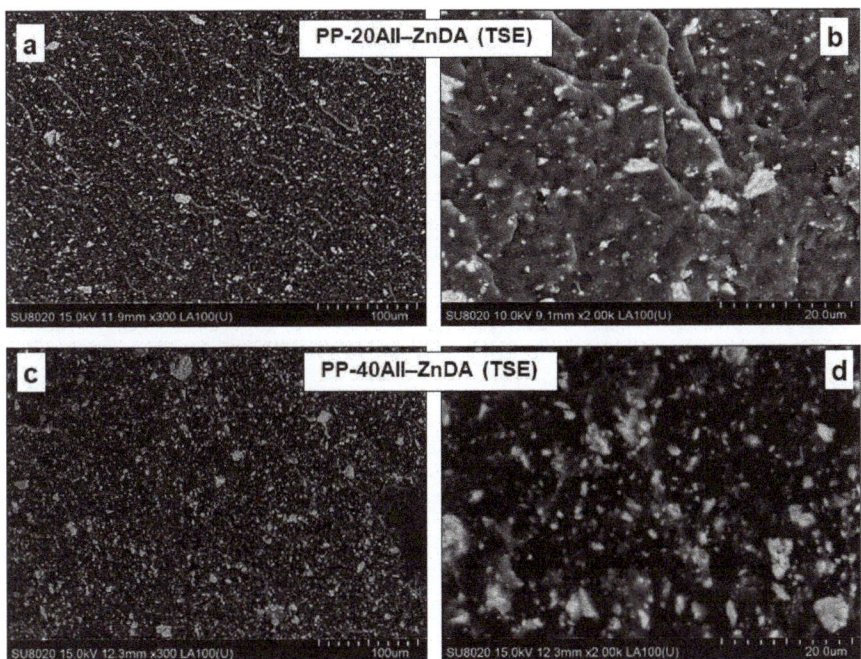

Figure 9. (**a**–**d**). SEM (LA-BSE) pictures at different magnifications on cryo-fractured surfaces of PP–AII–ZnDA composites produced by REx: (**a**,**b**) PP–20AII–ZnDA (TSE); (**c**,**d**) PP–20AII–ZnDA (TSE).

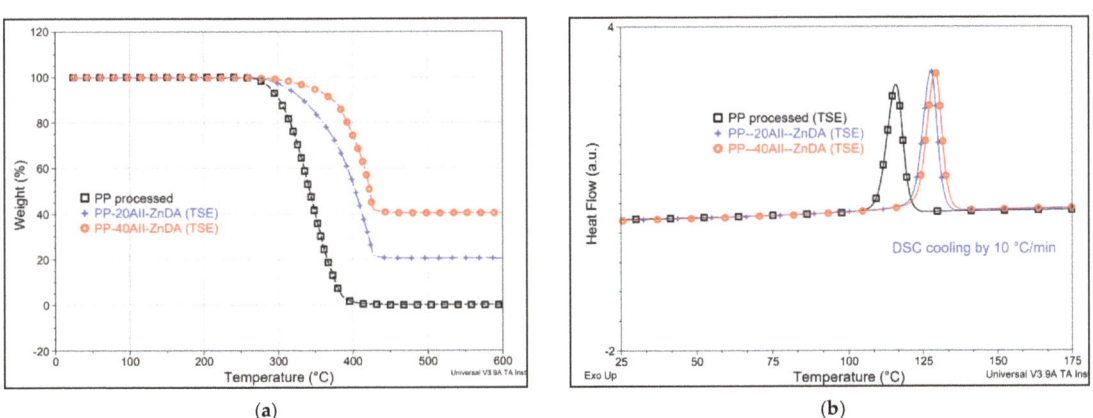

Figure 10. (**a**,**b**) Comparative thermal analyses of PP and PP–AII–ZnDA composites produced by REx: (**a**) TG traces; (**b**) crystallization from the molten state (DSC method, cooling by 10 °C/min).

(c) Mechanical properties: Because the materials have been produced in larger quantities by REx, the specimens required for mechanical characterizations were obtained by IM processing. Table 4 shows the comparison of the main properties of PP and PP–AII–ZnDA (TSE) composites, including their mechanical properties. Obviously, the PP–AII composites modified by REx with ZnDA show excellent tensile strength (σ_t of 34–35 MPa), whereas their rigidity/Young's modulus is progressively increased in correlation with the amounts of AII, e.g., by about 70%, by filling PP with 40% AII. These results are in good agreement with those obtained in the selection phase of key compositions, following the experimental tests with internal mixers.

Table 4. Comparative properties of neat PP and PP filled with 20–40% AII and modified by REx.

Properties	PP (TSE)	PP–20AII–ZnDA (TSE)	PP–40AII–ZnDA (TSE)
Tensile properties *			
Maximum tensile/yield strength, MPa	29 (±1)/yield	34 (±1)	35 (±1)
Tensile strength at break, MPa	34 (±1)	33 (±1)	34 (±1)
Young's modulus, MPa	1300 (±50)	1800 (±10)	2200 (±50)
Nominal strain at max strength, %	8.4 (±0.5)/yield	6.1 (±0.2)	4.8 (±0.1)
Nominal elongation at break, %	476 (±30)	8.1 (±0.5)	5.1 (±0.1)
Flexural properties *			
Max. flexural strength, MPa	37 (±2)	48 (±2)	52 (±1)
Flexural modulus, MPa	1160 (±70)	1850 (±40)	2550 (±90)
Max. deflection, mm	>20	>20	14 (±1)
Izod impact resistance, kJ/m^2 *	1.9 (±0.4)	2.7 (±0.1)	2.8 (±0.1)
Rheological properties			
Melt volume rate (MVR) **, cm^3/10 min	15.3	8.7	5.5

* Mechanical properties determined on specimens produced by IM. ** Melt volume rate (190 °C, 2.16 kg).

In the context of engineering applications, it is important to note that the high-filled composites (40% AII) exhibit a higher flexural strength (65% greater than that of neat PP), while the flexural modulus is almost twice as high for composites compared to the neat PP (2550 MPa vs. 1160 MPa), mechanical properties which may be relevant for the realization of automotive and mechanical components. Then again, the PP–AII–ZnDA (TSE) composites show better impact resistance than the neat PP (2.7–2.8 kJ/m^2, compared to 1.9 kJ/m^2, respectively). IM specimens are, however, slightly less impact-resistant than the CM specimens, highlighting the significance of processing conditions. Additionally, it was found that ε_b of PP was higher after IM than following the processing by CM, the difference was assigned to the better orientation of PP chains by IM.

As was mentioned elsewhere, the good mechanical properties of PP–AII composites can be ascribed not only to the effects of crosslinking, due to the formation of clusters of Zn ionomer, but also to the existence of good compatibility between the PP matrix and filler. Additionally, it is not excluded that the pendant ionic groups in the ionomer might increase the polarity of PP [66,67,89], and this plays a key role in improving the interactions with AII. As in the case of other ionomers, Zn ionomer can act as a compatibilizer, allowing a finer filler dispersion and better adhesion between the PP and AII microparticles [66,67,90].

Nevertheless, for better evidence of the interfacial properties (PP–filler), additional SEM investigations have been performed on the fractured surfaces of composites, obtained after impact and tensile testing (selected images shown in Figure 11a,b, respectively). First, it is important to note that during tensile testing the composites are stretched at low speed (10 mm/min) until they break. In impact tests, a sudden shock is given to the material at a high energy/deformation rate (respectively, 3.9 J and 3.46 m/s), and this leads to the fracture of PP specimens. By observing the SEM images of the fractured surfaces following the impact solicitation (Figure 11a), it is seen that the filler has quite good adhesion to the PP matrix, as the pull-out of particles from the matrix and/or signs of debonding at the interface polymer filler are not observed. Moreover, the distribution/dispersion of AII microparticles and their good wetting/surrounding by the polymer matrix is noticeable (Supplementary Information, Figure S3).

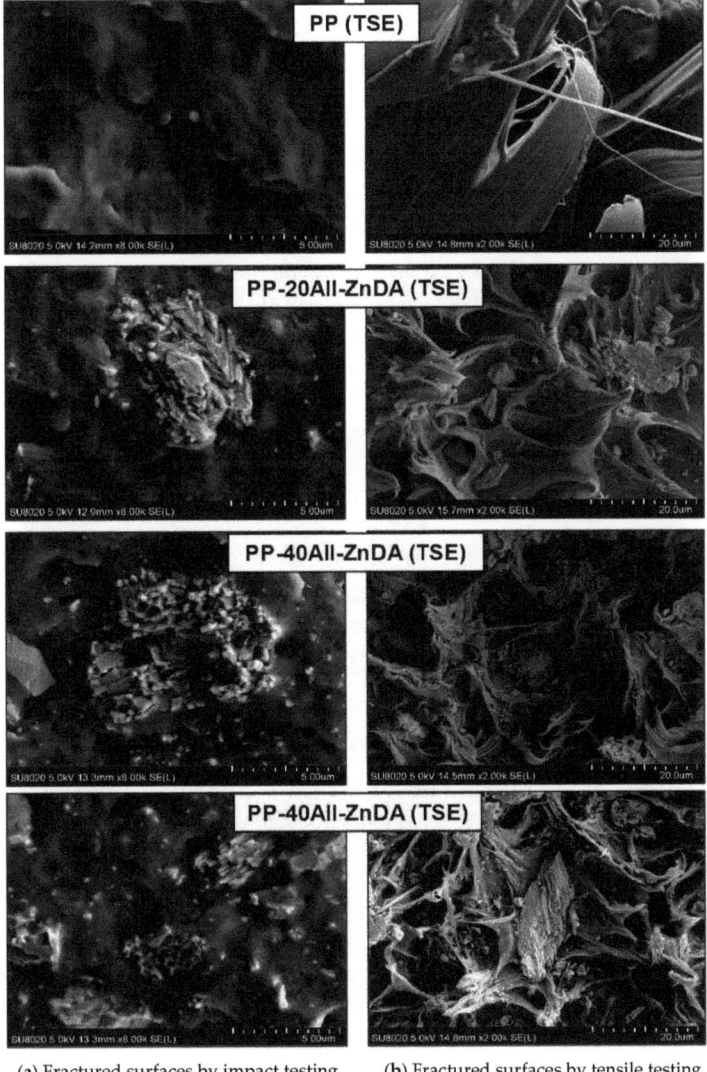

(a) Fractured surfaces by impact testing   (b) Fractured surfaces by tensile testing

Figure 11. (**a**,**b**) SEM images (SE mode) of PP and PP–AII–ZnDA (TSE) composites performed on the fractured surfaces obtained by: (**a**) impact (left side) and (**b**) tensile testing.

On the other hand, regarding the specimens fractured at lower speed by tensile testing (Figure 11b), together with the evidence of ductile/plastic deformation specific to PP, the presence of numerous zones of contact at the interface between the polymeric matrix and AII microparticles, as fibrils or elongated/deformed plastic regions, is reasonably ascribed to the existence of good adhesion/interfacial properties, which finally reflect the level of mechanical performances.

(d) Rheological properties: Following the evolution of MVR (data shown in Table 4), it emerges that the melt viscosity of PP composites is progressively rising (NB: MVR is decreased) in correlation with the AII amounts. Nevertheless, under the specific conditions of testing (190 °C), an additional effect assigned to the presence of Zn ionomer, linked to the formation of ionic crosslinks that are able to increase the melt strength of PP, is not excluded. However, more comprehensive rheological studies can highlight other features of interest, to allow the optimal processing of these composites using different techniques: injection molding (IM), which requires high fluidity and increased IM temperatures; extrusion, which needs increased viscosity and good melt strength.

3.3.2. Other Properties of Interest for Engineering Applications

Figure 12 shows selected images to illustrate the aesthetic aspect/whiteness of samples as granules and as specimens produced by IM, together with the results of color analysis. The color measurements (CIELab) confirm the increase in the lightness (L^*) of the composites in direct correlation with the amounts of filler (N.B. the AII used in this study has the $L^* = 98$; therefore, it is also used as the TiO_2 extender).

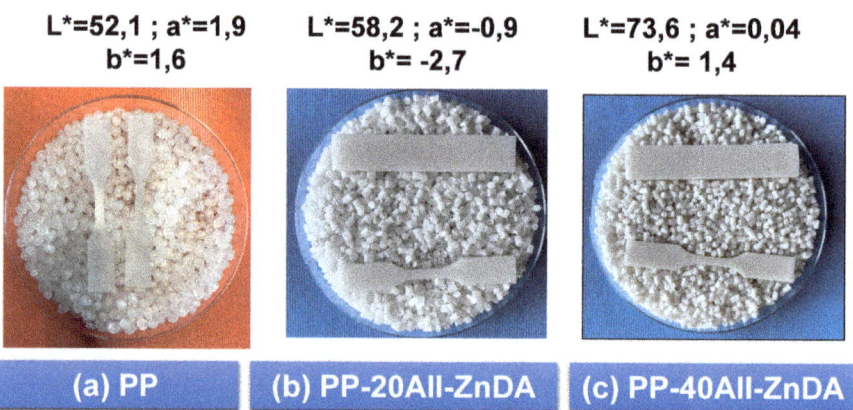

Figure 12. (a–c) Images of PP and PP–AII–ZnDA composites as granules and as specimens obtained by IM, together with the results of color measurements (CIELab): (a) neat PP is translucent, whereas the IM specimens show some warpage; (b,c) PP–AII–ZnDA composites showing high whiteness and good surface quality for IM specimens.

From the perspective of engineering applications, the high rigidity of modified PP–AII composites was also proved under dynamic mechanical solicitations by DMA (Figure 13a,b). The tests performed in the range of temperatures −75 °C–+150 °C, highlight the possibility to use these composites at higher mechanical stress and temperature, than the neat PP. For illustration purposes only, Table 5 shows comparative values of storage modulus (E', Figure 13a) at different temperatures.

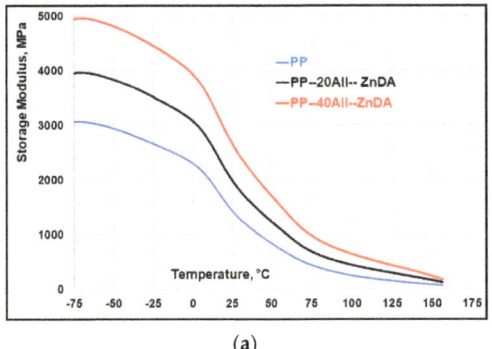
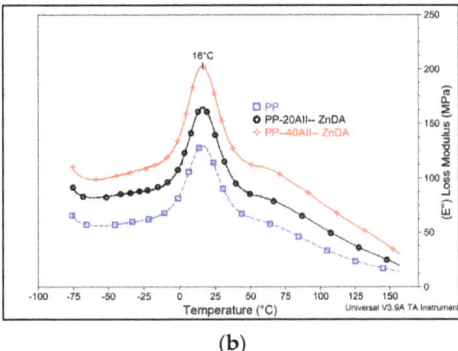

(a) (b)

Figure 13. (**a**,**b**) DMA evaluations of PP vs. PP–AII–ZnDA composites obtained by REx. Evolution of (**a**) storage modulus (E') and of (**b**) loss modulus (E").

Table 5. Comparative values of storage modulus (E') at different temperatures for the neat PP and PP–AII composites modified by REx.

Sample	Temperature, °C			
	−50	0	50	100
	↓ Storage Modulus, MPa			
PP (TSE)	2950	2300	850	250
PP–20AII–ZnDA (TSE)	3850	3100	1250	450
PP–40AII–ZnDA (TSE)	4800	3950	1750	650

On the other hand, the loss modulus (E", Figure 13b) displays a main peak at about 16 °C for all samples (assigned to a β-relaxation), whereas the rising of amounts of filler to 40% leads to the progressive increasing of E" in all range of temperatures, and this is mainly ascribed to the contribution of the mechanical loss generated in the interfacial regions [91]. Furthermore, the DMA results agree with those of HDT testing (Figure 14), which evidence properties that are relevant for engineering applications, since the HDT of neat PP (about 70 °C) is significantly increased to 100–110 °C in the case of filled PP composites.

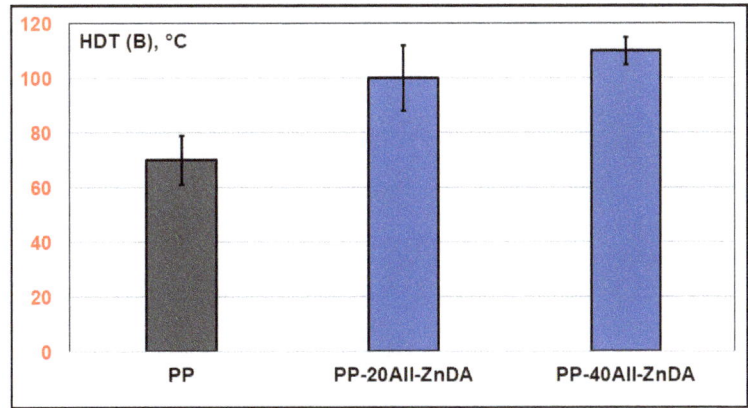

Figure 14. Evaluation of heat deflection temperature (HDT- method B, 0.45 MPa) of PP and PP–AII composites produced by REx.

Before concluding, it is important to point out that the results of REx are highly dependent on the type of equipment and the conditions of processing. Experimental fine-tuning is highly recommended to find the optimal conditions/compositions. Nevertheless, the distinct conditions of production (use of internal mixer or TSE, differences in temperatures and residence time) and processing (CM or IM), could lead to some changes in the final properties of composites. Nonetheless, the overall experimental results obtained in this study are in good agreement. However, in the frame of further prospects, it would be important to reconfirm the performances of these novel composites and to obtain additional validations regarding their behavior under different conditions/temperatures of utilization, or on the crystallization mechanisms, via alternative techniques of investigation (polarized light microscopy, X-ray diffraction, etc.). Nevertheless, starting from the requirements of the application, it is considered that some more specific characterizations will be more necessary (e.g., evaluation of aging at elevated temperature or under UV conditions, tests of permeability, abrasion, scratch resistance, and so on).

4. Conclusions

The study addresses current requests regarding the utilization of natural products (i.e., gypsum derivatives) to produce new high-performance mineral-filled PP composites designed for engineering applications. First, using melt blending internal mixers, the effect of filling PP with up to 40% $CaSO_4$ β-anhydrite II (AII) has been evaluated in terms of morphology, mechanical and thermal properties. The PP–AII composites (without any modifier) showed quite good filler distribution within the PP matrix, enhanced thermal stability, and stiffness. However, especially at high filling (40% AII), the impact resistance and tensile strength were drastically reduced. Therefore, for compatibilization and enhanced properties, the addition of reactive modifiers (PP-g-MA and ZnDA) was considered. Finer filler dispersion and surprising thermal and mechanical properties were obtained following the co-addition of AII and ZnDA into PP, rather than using PP-g-MA.

PP–AII–ZnDA composites show remarkable performances: improved kinetics of crystallization, high thermal stability, and excellent mechanical properties (tensile strength, rigidity, improved/surprising impact resistance). It was presumed that the formation of clusters/ionic crosslinks of Zn ionomer, the favorable interactions with the filler, and the good interfacial properties allowed us to compatibilize, strengthen, and toughen the composites. Moreover, experimental trials by REx have been performed to produce larger quantities of materials for processing by injection molding and to reconfirm their properties. Due to high thermal stability and notable mechanical performances, the PP–AII–ZnDA composites proved attractive for use in technical applications: tensile strength of about 35 MPa, improved stiffness/Young's modulus, impact resistance higher than that of PP, advanced rigidity under dynamic solicitations (DMA) in the range of temperatures from −75 °C to + 150 °C, and the increasing of HDT from 70 °C (neat PP) to 110 °C (for composites). Based on the performances of these new composites, it is anticipated that due to their specific features, they will be highly sought after as potential new candidates for technical/engineering applications, requiring rigidity, heat resistance, dimensional stability, and improved processing.

Supplementary Materials: The following supporting information can be downloaded at: https://www.mdpi.com/article/10.3390/polym15040799/s1, as Supplementary Material: Figure S1. Selected SEM pictures at different magnifications on the cryo-fractured surfaces of PP composites filled with 20% AII; Figure S2. Representative SEM images (SE mode) of PP and PP–AII composites performed on the fractured surfaces obtained by tensile testing; Figure S3. SEM images (SE mode) of PP and PP–AII (TSE) composites performed on the fractured surfaces obtained by impact testing; Table S1. Comparative properties of neat PP, PP modified with 2% ZnDA, and PP–40% AII composites modified by REx with 4% PPMA2.

Author Contributions: Data curation, M.M. and O.M.; Formal analysis, M.M., Y.P., and O.M.; Funding acquisition, M.M. and F.L.; Investigation, M.M., Y.P., and O.M.; Supervision, P.D.; Writing—

original draft, M.M.; Writing—review and editing, M.M., F.L., and P.D. All authors have read and agreed to the published version of the manuscript.

Funding: This research was funded by Toro Gips S.L. (Barcelona, Spain), grant "TOROGYPS," and by the Materia Nova R&D Center using internal funding.

Institutional Review Board Statement: Not applicable.

Informed Consent Statement: Not applicable.

Data Availability Statement: Not applicable.

Acknowledgments: The authors thank Toro Gips S.L. (Barcelona, Spain) for supplying ToroWhite (CS) samples and for accepting the publication of the scientific results obtained in the framework of the TOROGYPS project. Special thanks are addressed to Erdem Kütükçü and Ludwig Soukup from Toro Gips S.L. for the kind collaboration and fruitful discussions regarding ToroWhite products. M.M. is also grateful to Materia Nova Materials R&D Center for the great support received for the development of the topics and results linked to this study.

Conflicts of Interest: The authors declare no conflict of interest. The funders gave their agreement for the publication of scientific results, and they had no role in the design of the study; in the collection, analyses, or interpretation of data; or in the writing of the manuscript.

References

1. Greene, J.P. Chapter 7-Commodity plastics. In *Automotive Plastics and Composites*; Greene, J.P., Ed.; William Andrew Publishing: Oxford, UK, 2021; pp. 83–105.
2. Klein, J.; Wiese, J. High performance engineered polypropylene compounds for high temperature automotive under-the-hood applications. In Proceedings of the Society of Plastics Engineers-11th-Annual Automotive Composites Conference and Exhibition (ACCE 2011), Troy, MI, USA, 13–15 September 2011.
3. Alsabri, A.; Tahir, F.; Al-Ghamdi, S.G. Environmental impacts of polypropylene (PP) production and prospects of its recycling in the GCC region. *Mater. Today Proc.* **2022**, *56*, 2245–2251. [CrossRef]
4. Dennis, B.; Malpass, E.I.B. (Eds.) The future of polypropylene. In *Introduction to Industrial Polypropylene*; John Wiley & Sons, Inc.: Hoboken, NJ, USA; Scrivener Publishing LLC: Salem, MA, USA, 2012; pp. 269–278.
5. Elaheh, G. Materials in automotive application, state of the art and prospects. In *New Trends and Developments in Automotive Industry*; Marcello, C., Ed.; IntechOpen: Rijeka, Croatia, 2011; pp. 365–394. [CrossRef]
6. Wang, S.; Muiruri, J.K.; Soo, X.Y.D.; Liu, S.; Thitsartarn, W.; Tan, B.H.; Suwardi, A.; Li, Z.; Zhu, Q.; Loh, X.J. Bio-PP and PP-based biocomposites: Solutions for a sustainable future. *Chem.-Asian J.* **2023**, *18*, e202200972. [CrossRef] [PubMed]
7. Jansz, J. Polypropylene in automotive applications. In *Polypropylene: An A-Z Reference*; Karger-Kocsis, J., Ed.; Springer: Dordrecht, The Netherlands, 1999; pp. 643–651.
8. Thenepalli, T.; Jun, A.Y.; Han, C.; Ramakrishna, C.; Ahn, J.W. A strategy of precipitated calcium carbonate ($CaCO_3$) fillers for enhancing the mechanical properties of polypropylene polymers. *Korean J. Chem. Eng.* **2015**, *32*, 1009–1022. [CrossRef]
9. Sadiku, R.; Ibrahim, D.; Agboola, O.; Owonubi, S.J.; Fasiku, V.O.; Kupolati, W.K.; Jamiru, T.; Eze, A.A.; Adekomaya, O.S.; Varaprasad, K.; et al. 15-Automotive components composed of polyolefins. In *Polyolefin Fibres (Second Edition)*; Ugbolue, S.C.O., Ed.; Woodhead Publishing (imprint of Elsevier): Duxford, UK, 2017; pp. 449–496.
10. Patil, A.; Patel, A.; Purohit, R. An overview of polymeric materials for automotive applications. *Mater. Today Proc.* **2017**, *4*, 3807–3815. [CrossRef]
11. Delli, E.; Gkiliopoulos, D.; Bikiaris, D.; Chrissafis, K. Thermomechanical characterization of E-glass fiber reinforced random polypropylene. *Macromol. Symp.* **2022**, *405*, 2100226. [CrossRef]
12. Doddalli Rudrappa, S.; Yellampalli Srinivasachar, V. Significance of the type of reinforcement on the physicomechanical behavior of short glass fiber and short carbon fiber-reinforced polypropylene composites. *Eng. Rep.* **2020**, *2*, e12098. [CrossRef]
13. Mihalic, M.; Sobczak, L.; Pretschuh, C.; Unterweger, C. Increasing the impact toughness of cellulose fiber reinforced polypropylene composites—Influence of different impact modifiers and production scales. *J. Compos. Sci.* **2019**, *3*, 82. [CrossRef]
14. Shirvanimoghaddam, K.; Balaji, K.V.; Yadav, R.; Zabihi, O.; Ahmadi, M.; Adetunji, P.; Naebe, M. Balancing the toughness and strength in polypropylene composites. *Compos. Part B Eng.* **2021**, *223*, 109121. [CrossRef]
15. Abreu, F.O.M.S.; Forte, M.M.C.; Liberman, S.A. Sbs and SEBS block copolymers as impact modifiers for polypropylene compounds. *J. Appl. Polym. Sci.* **2005**, *95*, 254–263. [CrossRef]
16. Zhao, W.; Kumar Kundu, C.; Li, Z.; Li, X.; Zhang, Z. Flame retardant treatments for polypropylene: Strategies and recent advances. *Compos. Part A: Appl. Sci. Manuf.* **2021**, *145*, 106382. [CrossRef]
17. Zhang, C.; Jiang, Y.; Li, S.; Huang, Z.; Zhan, X.-Q.; Ma, N.; Tsai, F.-C. Recent trends of phosphorus-containing flame retardants modified polypropylene composites processing. *Heliyon* **2022**, *8*, e11225. [CrossRef] [PubMed]

18. Móczó, J.; Pukánszky, B. Particulate fillers in thermoplastics. In *Fillers for Polymer Applications*; Rothon, R., Ed.; Springer International Publishing: Cham, Switzerland, 2017; pp. 51–93.
19. Bertini, F.; Canetti, M.; Audisio, G.; Costa, G.; Falqui, L. Characterization and Thermal Degradation of Polypropylene-Montmorillonite Nanocomposites. *Polym. Degrad. Stab.* **2006**, *91*, 600–605. [CrossRef]
20. Morajane, D.; Sinha Ray, S.; Bandyopadhyay, J.; Ojijo, V. Impact of melt-processing strategy on structural and mechanical properties: Clay-containing polypropylene nanocomposites. In *Processing of Polymer-based Nanocomposites*; Sinha Ray, S., Ed.; Springer Series in Materials Science; Springer: Cham, Switzerland, 2018; Volume 277, pp. 127–154.
21. Delva, L.; Ragaert, K.; Allaer, K.; Gaspar-Cunha, A.; Degrieck, J.; Cardon, L. Influence of twin-screw configuration on the mechanical and morphological properties of polypropylene-clay composites. *Int. J. Mater. Prod. Technol.* **2016**, *52*, 176–192. [CrossRef]
22. Al-Samhan, M.; Al-Attar, F. Comparative analysis of the mechanical, thermal and barrier properties of polypropylene incorporated with $CaCO_3$ and nano $CaCO_3$. *Surf. Interfaces* **2022**, *31*, 102055. [CrossRef]
23. Goldman, A.Y.; Copsey, C.J. Polypropylene toughened with calcium carbonate mineral filler. *Mater. Res. Innov.* **2000**, *3*, 302–307. [CrossRef]
24. Jing, Y.; Nai, X.; Dang, L.; Zhu, D.; Wang, Y.; Dong, Y.; Li, W. Reinforcing polypropylene with calcium carbonate of different morphologies and polymorphs. *Sci. Eng. Compos. Mater.* **2018**, *25*, 745–751. [CrossRef]
25. Peng, Y.; Musah, M.; Via, B.; Wang, X. Calcium Carbonate Particles Filled Homopolymer Polypropylene at Different Loading Levels: Mechanical Properties Characterization and Materials Failure Analysis. *J. Compos. Sci.* **2021**, *5*, 302. [CrossRef]
26. de Oliveira, C.I.R.; Rocha, M.C.G.; de Assis, J.T.; da Silva, A.L.N. Morphological, mechanical, and thermal properties of PP/SEBS/talc composites. *J. Thermoplast. Compos. Mater.* **2022**, *35*, 281–299. [CrossRef]
27. Putfak, N.; Larpkasemsuk, A. Wollastonite and talc reinforced polypropylene hybrid composites: Mechanical, morphological and thermal properties. *J. Met. Mater. Miner.* **2021**, *31*, 92–99. [CrossRef]
28. Leong, Y.W.; Abu Bakar, M.B.; Ishak, Z.A.M.; Ariffin, A.; Pukanszky, B. Comparison of the mechanical properties and interfacial interactions between talc, kaolin, and calcium carbonate filled polypropylene composites. *J. Appl. Polym. Sci.* **2004**, *91*, 3315–3326. [CrossRef]
29. Várdai, R.; Schäffer, Á.; Ferdinánd, M.; Lummerstorfer, T.; Jerabek, M.; Gahleitner, M.; Faludi, G.; Móczó, J.; Pukánszky, B. Crystalline structure and reinforcement in hybrid PP composites. *J. Therm. Anal. Calorim.* **2022**, *147*, 145–154. [CrossRef]
30. Chen, X.; Zhang, T.; Sun, P.; Yu, F.; Li, B.; Dun, L. Study on the performance and mechanism of modified mica for improving polypropylene composites. *Int. J. Low-Carbon Technol.* **2022**, *17*, 176–184. [CrossRef]
31. García-Martínez, J.-M.; Collar, E.P. The Variance of the Polypropylene α Relaxation Temperature in iPP/a-PP-pPBMA/Mica Composites. *J. Compos. Sci.* **2022**, *6*, 57. [CrossRef]
32. Yang, N.; Zhang, Z.C.; Ma, N.; Liu, H.L.; Zhan, X.Q.; Li, B.; Gao, W.; Tsai, F.C.; Jiang, T.; Chang, C.J.; et al. Effect of surface modified kaolin on properties of polypropylene grafted maleic anhydride. *Results Phys.* **2017**, *7*, 969–974. [CrossRef]
33. Liu, G.; Xu, H.; Song, S.; Wang, H.; Fan, B.; Lu, H.; Liu, Q. Preparation and properties of PP/modified kaolin composites. *Gaofenzi Cailiao Kexue Yu Gongcheng/Polym. Mater. Sci. Eng.* **2012**, *28*, 121–124.
34. Zhang, S.; Jiang, P.; Liu, X.; Gu, X.; Zhao, Q.; Hu, Z.; Tang, W. Effects of kaolin on the thermal stability and flame retardancy of polypropylene composite. *Polym. Adv. Technol.* **2014**, *25*, 912–919. [CrossRef]
35. Wang, K.; Wu, J.; Ye, L.; Zeng, H. Mechanical properties and toughening mechanisms of polypropylene/barium sulfate composites. *Compos. Part A: Appl. Sci. Manuf.* **2003**, *34*, 1199–1205. [CrossRef]
36. Móczó, J.; Pukánszky, B. Particulate filled polypropylene: Structure and properties. In *Polypropylene Handbook: Morphology, Blends and Composites*; Karger-Kocsis, J., Bárány, T., Eds.; Springer International Publishing: Cham, Switzerland, 2019; pp. 357–417.
37. Nofar, M.; Ozgen, E.; Girginer, B. Injection-molded PP composites reinforced with talc and nanoclay for automotive applications. *J. Thermoplast. Compos. Mater.* **2020**, *33*, 1478–1498. [CrossRef]
38. de Medeiros, E.S.; Tocchetto, R.S.; de Carvalho, L.H.; Santos, I.M.G.; Souza, A.G. Nucleating effect and dynamic crystallization of a poly(propylene)/talc system. *J. Therm. Anal. Calorim.* **2001**, *66*, 523–531. [CrossRef]
39. Ammar, O.; Bouaziz, Y.; Haddar, N.; Mnif, N. Talc as reinforcing filler in polypropylene compounds: Effect on morphology and mechanical properties. *Polym. Sci.* **2017**, *3*, 1–7. [CrossRef]
40. Leong, Y.W.; Abu Bakar, M.B.; Ishak, Z.A.M.; Ariffin, A. Filler treatment effects on the weathering of talc-, $CaCO_3$- and kaolin-filled polypropylene hybrid composites. *Compos. Interfaces* **2006**, *13*, 659–684. [CrossRef]
41. Leong, Y.W.; Bakar, M.B.A.; Ishak, Z.A.M.; Ariffin, A. Effects of filler treatments on the mechanical, flow, thermal, and morphological properties of talc and calcium carbonate filled polypropylene hybrid composites. *J. Appl. Polym. Sci.* **2005**, *98*, 413–426. [CrossRef]
42. Balköse, D. Influence of filler surface modification on the properties of PP composites. In *Surface Modification of Nanoparticle and Natural Fiber Fillers*; Wiley-VCH Verlag GmbH & Co. KGaA: Weinheim, Germany, 2015; pp. 83–108.
43. Othman, N.; Ismail, H.; Mariatti, M. Effect of compatibilisers on mechanical and thermal properties of bentonite filled polypropylene composites. *Polym. Degrad. Stab.* **2006**, *91*, 1761–1774. [CrossRef]
44. Bischoff, E.; Simon, D.A.; Liberman, S.A.; Mauler, R.S. Compounding sequence as a critical factor in the dispersion of OMMT/hydrocarbon resin/PP-g-MA/PP nanocomposites. *Polym. Bull.* **2019**, *76*, 849–863. [CrossRef]

45. Morlat, S.; Mailhot, B.; Gonzalez, D.; Gardette, J.-L. Photo-oxidation of polypropylene/montmorillonite nanocomposites. 1. Influence of nanoclay and compatibilizing agent. *Chem. Mater.* **2004**, *16*, 377–383. [CrossRef]
46. Toro, P.; Quijada, R.; Peralta, R.; Yazdani-Pedram, M. Influence of grafted polypropylene on the mechanical properties of mineral-filled polypropylene composites. *J. Appl. Polym. Sci.* **2007**, *103*, 2343–2350. [CrossRef]
47. Hejna, A.; Przybysz-Romatowska, M.; Kosmela, P.; Zedler, Ł.; Korol, J.; Formela, K. Recent advances in compatibilization strategies of wood-polymer composites by isocyanates. *Wood Sci. Technol.* **2020**, *54*, 1091–1119. [CrossRef]
48. DeArmitt, C.; Rothon, R. Surface modifiers for use with particulate fillers. In *Fillers for Polymer Applications*; Rothon, R., Ed.; Springer International Publishing: Cham, Switzerland, 2017; pp. 29–49.
49. Leong, Y.W.; Ishak, Z.A.M.; Ariffin, A. Mechanical and thermal properties of talc and calcium carbonate filled polypropylene hybrid composites. *J. Appl. Polym. Sci.* **2004**, *91*, 3327–3336. [CrossRef]
50. Chen, D.; Yang, H. Polypropylene/combinational inorganic filler micro-/nanocomposites: Synergistic effects of micro-/nanoscale combinational inorganic fillers on their mechanical properties. *J. Appl. Polym. Sci.* **2010**, *115*, 624–634. [CrossRef]
51. Gahleitner, M.; Grein, C.; Bernreitner, K. Synergistic mechanical effects of calcite micro- and nanoparticles and β-nucleation in polypropylene copolymers. *Eur. Polym. J.* **2012**, *48*, 49–59. [CrossRef]
52. Mittal, P.; Naresh, S.; Luthra, P.; Singh, A.; Dhaliwal, J.S.; Kapur, G.S. Polypropylene composites reinforced with hybrid inorganic fillers: Morphological, mechanical, and rheological properties. *J. Thermoplast. Compos. Mater.* **2019**, *32*, 848–864. [CrossRef]
53. Saujanya, C.; Radhakrishnan, S. Structure development in PP/CaSO$_4$ composites: Part I Preparation of the filler by an in situ technique. *J. Mater. Sci.* **1998**, *33*, 1063–1068. [CrossRef]
54. Dong, F.; Liu, J.; Tan, H.; Wu, C.; He, X.; He, P. Preparation of calcium sulfate hemihydrate and application in polypropylene composites. *J. Nanosci. Nanotechnol.* **2017**, *17*, 6970–6975. [CrossRef]
55. Xiang, G.; Liu, T.; Zhang, Y.; Xue, N. Synthesis of polypropylene composites with modified calcium sulfate whisker prepared from shale vanadium neutralization slag. *Results Phys.* **2018**, *10*, 28–35. [CrossRef]
56. Dou, Q.; Duan, J. Melting and crystallization behaviors, morphology, and mechanical properties of β-polypropylene/polypropylene-graft-maleic anhydride/calcium sulfate whisker composites. *Polym. Compos.* **2016**, *37*, 2121–2132. [CrossRef]
57. Murariu, M.; Paint, Y.; Murariu, O.; Laoutid, F.; Dubois, P. Recent advances in production of ecofriendly polylactide (PLA)-calcium sulfate (anhydrite II) composites: From the evidence of filler stability to the effects of PLA matrix and filling on key properties. *Polymers* **2022**, *14*, 2360. [CrossRef]
58. Murariu, M.; Dubois, P. PLA composites: From production to properties. *Adv. Drug Deliv. Rev.* **2016**, *107*, 17–46. [CrossRef]
59. Wirsching, F. Calcium sulfate. In *Ullmann's Encyclopedia of Industrial Chemistry*; Wiley-VCH Verlag GmbH & Co. KGaA: Weinheim, Germany, 2012; Volume 6, pp. 519–550.
60. Murariu, M.; Paint, Y.; Murariu, O.; Laoutid, F.; Dubois, P. Tailoring and long-term preservation of the properties of PLA composites with green plasticizers. *Polymers* **2022**, *14*, 4836. [CrossRef]
61. Maddah, H. Polypropylene as a promising plastic: A review. *Am. J. Polym. Sci.* **2016**, *6*, 1–11. [CrossRef]
62. Liang, J.-Z. Toughening and reinforcing in rigid inorganic particulate filled poly(propylene): A review. *J. Appl. Polym. Sci.* **2002**, *83*, 1547–1555. [CrossRef]
63. Zuiderduin, W.C.J.; Westzaan, C.; Huetink, J.; Gaymans, R.J. Toughening of polypropylene with calcium carbonate particles. *Polymer* **2003**, *44*, 261–275. [CrossRef]
64. Shubhra, Q.T.H.; Alam, A.K.M.M.; Quaiyyum, M.A. Mechanical properties of polypropylene composites: A review. *J. Compos.* **2011**, *26*, 362–391. [CrossRef]
65. Tucker, J.D.; Lear, P.L.; Atkinson, G.S.; Lee, S.; Lee, S.J. Use of polymeric compatibilizers in polypropylene/calcium carbonate composites. *Korean J. Chem. Eng.* **2000**, *17*, 506–509. [CrossRef]
66. Liu, H.; Tag Lim, H.; Hyun Ahn, K.; Jong Lee, S. Effect of ionomer on clay dispersions in polypropylene-layered silicate nanocomposites. *J. Appl. Polym. Sci.* **2007**, *104*, 4024–4034. [CrossRef]
67. Santamaría, P.; Eguiazábal, J.I.; Nazábal, J. Structure and properties of polyethylene ionomer based nanocomposites. *J. Appl. Polym. Sci.* **2010**, *116*, 2374–2383. [CrossRef]
68. Giordano, G. Finetuning process performance. *Plast. Eng.* **2021**, *77*, 18–21. [CrossRef]
69. Technical data sheet Dymalink®9200 (rev aug/2022). Cray Valley TotalEnergies. 2022. Available online: https://www.crayvalley.com/products/metallic-monomers/9200-series/ (accessed on 12 December 2022).
70. Rivera-Armenta, J.L.; Salazar-Cruz, B.A.; Espindola-Flores, A.C.; Villarreal-Lucio, D.S.; De León-Almazán, C.M.; Estrada-Martinez, J. Thermal and thermomechanical characterization of polypropylene-seed shell particles composites. *Appl. Sci.* **2022**, *12*, 8336. [CrossRef]
71. Lushnikova, N.; Dvorkin, L. Sustainability of gypsum products as a construction material. In *Sustainability of Construction Materials*; Elsevier: Amsterdam, The Netherlands, 2016; pp. 643–681.
72. Murariu, M.; Da Silva Ferreira, A.; Degée, P.; Alexandre, M.; Dubois, P. Polylactide compositions. Part 1: Effect of filler content and size on mechanical properties of PLA/calcium sulfate composites. *Polymer* **2007**, *48*, 2613–2618. [CrossRef]
73. Ma, Y.; Yang, G.; Xie, L. Morphology, nonisothermal crystallization behavior and mechanical properties of polypropylene modified by ionomers. *J. Macromol. Sci. Part B* **2014**, *53*, 1829–1845. [CrossRef]

74. Naebpetch, W.; Junhasavasdikul, B.; Saetung, A.; Tulyapitak, T.; Nithi-Uthai, N. The influence of zinc dimethacrylate on crosslink density, physical properties and heat aging resistance of sulfur vulcanized styrene butadiene rubber. *Adv. Mater. Res.* **2014**, *844*, 45–48. [CrossRef]
75. Chen, Y.; Xu, C.; Liang, X.; Cao, L. In situ reactive compatibilization of polypropylene/ethylene–propylene–diene monomer thermoplastic vulcanizate by zinc dimethacrylate via peroxide-induced dynamic vulcanization. *J. Phys. Chem. B* **2013**, *117*, 10619–10628. [CrossRef]
76. Total Cray Valley develops ionomeric additives to improve polyolefin performance. *Addit. Polym.* **2015**, *2015*, 1–2. [CrossRef]
77. Li, Y.; Yao, Z.; Chen, Z.-h.; Qiu, S.-l.; Zeng, C.; Cao, K. High melt strength polypropylene by ionic modification: Preparation, rheological properties and foaming behaviors. *Polymer* **2015**, *70*, 207–214. [CrossRef]
78. Wu, M.-H.; Wang, C.-C.; Chen, C.-Y. Preparation of high melt strength polypropylene by addition of an ionically modified polypropylene. *Polymer* **2020**, *202*, 122743. [CrossRef]
79. Cai, D.; Li, Y.; Wang, W.; Ma, Y.; Cao, N.; Zhang, J.; Pan, D.; Naik, N.; Wei, S.; Huang, M.; et al. Reinforcing and toughening blends of recycled acrylonitrile-butadiene-styrene/recycled high-impact polystyrene through ionic crosslinking. *Surf. Interfaces* **2022**, *28*, 101607. [CrossRef]
80. Robb, B. An additive approach to tailored melt strength in PP and TPO. In Proceedings of the 21st SPE Automotive TPO Engineered Polyolefins Global Conference 2019 (SPE Automotive TPO 2019), Troy, MI, USA, 6–9 October 2019.
81. Zhou, S.; Zhou, J.; Li, L.; Zhao, S.; Shi, Y.; Xin, Z. Relationship between peroxide initiators and properties of styrene grafted polypropylene via reactive extrusion. *J. Macromol. Sci. Part B* **2018**, *57*, 377–394. [CrossRef]
82. Stanic, S.; Gottlieb, G.; Koch, T.; Göpperl, L.; Schmid, K.; Knaus, S.; Archodoulaki, V.-M. Influence of different types of peroxides on the long-chain branching of PP via reactive extrusion. *Polymers* **2020**, *12*, 886. [CrossRef]
83. Saikrishnan, S.; Jubinville, D.; Tzoganakis, C.; Mekonnen, T.H. Thermo-mechanical degradation of polypropylene (PP) and low-density polyethylene (LDPE) blends exposed to simulated recycling. *Polym. Degrad. Stab.* **2020**, *182*, 109390. [CrossRef]
84. Rätzsch, M.; Arnold, M.; Borsig, E.; Bucka, H.; Reichelt, N. Radical reactions on polypropylene in the solid state. *Prog. Polym. Sci.* **2002**, *27*, 1195–1282. [CrossRef]
85. Marozsan, A.; Lodefier, P.; Robb, B.; Chabrol, V. A New Method to Modify PP for Improved Melt Strength. In Proceedings of the SPE International Polyolefin Conference 2017, Houston, TX, USA, 26 February 2017.
86. Capek, I. Dispersions of polymer ionomers: I. *Adv. Colloid Interface Sci.* **2004**, *112*, 1–29. [CrossRef]
87. Wurm, F.; Rietzler, B.; Pham, T. Multivalent ions as reactive crosslinkers for biopolymers-a review. *Molecules* **2020**, *25*, 1840. [CrossRef]
88. Najeeb, S.; Khurshid, Z.; Zafar, M.S.; Khan, A.S.; Zohaib, S.; Martí, J.M.N.; Sauro, S.; Matinlinna, J.P.; Rehman, I.U. Modifications in glass ionomer cements: Nano-sized fillers and bioactive nanoceramics. *Int. J. Mol. Sci.* **2016**, *17*, 1134. [CrossRef]
89. Li, J.; Li, J.; Feng, D.; Zhao, J.; Sun, J.; Li, D. Excellent rheological performance and impact toughness of cellulose nanofibers/PLA/ionomer composite. *RSC Adv.* **2017**, *7*, 28889–28897. [CrossRef]
90. Villanueva, M.P.; Cabedo, L.; Lagarón, J.M.; Giménez, E. Comparative study of nanocomposites of polyolefin compatibilizers containing kaolinite and montmorillonite organoclays. *J. Appl. Polym. Sci.* **2010**, *115*, 1325–1335. [CrossRef]
91. Pluta, M.; Murariu, M.; Da Silva Ferreira, A.; Alexandre, M.; Galeski, A.; Dubois, P. Polylactide compositions. II. Correlation between morphology and main properties of PLA/calcium sulfate composites. *J. Polym. Sci. Part B Polym. Phys.* **2007**, *45*, 2770–2780. [CrossRef]

Disclaimer/Publisher's Note: The statements, opinions and data contained in all publications are solely those of the individual author(s) and contributor(s) and not of MDPI and/or the editor(s). MDPI and/or the editor(s) disclaim responsibility for any injury to people or property resulting from any ideas, methods, instructions or products referred to in the content.

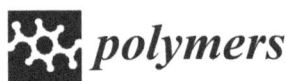

Article

Versatile Polypropylene Composite Containing Post-Printing Waste

Krzysztof Moraczewski [1,*], Tomasz Karasiewicz [1], Alicja Suwała [1], Bartosz Bolewski [1], Krzysztof Szabliński [1] and Magdalena Zaborowska [2]

[1] Faculty of Materials Engineering, Kazimierz Wielki University, Chodkiewicza 30 Str., 85-064 Bydgoszcz, Poland
[2] Blue System Sp. z o.o., Rynkowska 17D Str., 85-503 Bydgoszcz, Poland
* Correspondence: kmm@ukw.edu.pl; Tel.: +48-52-3419331

Abstract: The paper presents the results of the research on the possibility of using waste after the printing process as a filler for polymeric materials. Remains of the label backing were used, consisting mainly of cellulose with glue and polymer label residue. The properly prepared filler (washed, dried, pressed and cut) was added to the polypropylene in a volume ratio of 2:1; 1:1; 1:2; and 1:3 which corresponded to approximately 10, 5, 2.5 and 2 wt % filler. The selected processing properties (mass flow rate), mechanical properties (tensile strength, impact strength, dynamic mechanical analysis) and thermal properties (thermogravimetric analysis, differential scanning calorimetry) were determined. The use of even the largest amount of filler did not cause disqualifying changes in the determined properties. The characteristics of the obtained materials allow them to be used in various applications while reducing costs due to the high content of cheap filler.

Keywords: circular economy; polypropylene; post-printing waste

1. Introduction

Wastes are substances or objects resulting from human activity, as well as residues from their production, which are intended for disposal. The proper management of waste and by-products from industrial and natural production and activities (agriculture, horticulture and others) is the key to sustainable development, reduction in pollution, increase in storage space, minimization of landfill, reduction in energy consumption and facilitation of the circular economy [1].

The circular economy is a regenerative economic system in which the consumption of raw materials and the amount of waste as well as the emission and loss of energy are minimized by creating a closed loop of processes in which waste from one process is used as raw materials for others, which minimizes the amount of waste production [2,3]. Waste management is a series of processes related to the collection, transport and processing, including the supervision of this type of activity, as well as the subsequent handling of waste disposal sites, as well as activities related to waste trading [4,5].

One of the possibilities of reusing post-production waste is its use for the production of other materials, including the production of composite materials with various matrices. In recent years, composites with a polymer matrix and post-production waste fillers or reinforcements have attracted a lot of interest. Due to the characteristics of polymeric materials, thermoplastics in particular, it is possible to easily add post-production wastes in various forms to the polymer mass and subsequently easily process them through extrusion and injection processes.

In the production of polymer composites, the post-production waste used can be divided into two types. The first consists of organic wastes from the agricultural or food industry. The second consists of inorganic wastes from heavy industry such as the steel industry or petrochemical industry [1].

The most frequently used post-production waste materials in the production of polymer composites are natural wastes. Natural wastes are cheap and easily renewable, and their biodegradability is one of their most important features. Among the organic waste, the most commonly used post-production wastes is wood obtained from the wood and furniture industries. Post-production wastes in the form of flour or chips from the processes of cutting, turning, milling, planing, etc. are used as fillers for the production of wood-polymer composites (WPC), most often with polyolefin or poly(vinyl chloride) matrix, but also biodegradable polymers such as polylactide [6–11].

Other organic post-production wastes used in the production of polymer composites include: shells of nuts and seeds [12–15], bamboo [16–19], coconut shells [20–24], husk of rice and cereals [25–30], fruit and vegetables pomace [31–34], as well as waste of animal origin, such as egg shells or shells of crustaceans [35–38].

In most cases, these organic fillers can be easily integrated into thermoplastic or thermoset matrices to change their thermal, mechanical and tribological properties. However, not always only favorable changes in properties are observed. It is very often necessary to properly modify the waste in order to obtain materials with good performance parameters. Therefore, the deterioration of properties, especially mechanical properties, can most often be observed after adding unmodified waste. However, this deterioration does not have to be disqualifying for these composites, as very often, their properties are still sufficient for many planned applications, and their great advantage is a much lower price and environmental friendliness thanks to the use of cheap waste materials.

Printing houses, publishing houses and printing companies have to tackle the issue of increasing amounts of post-production waste on a daily basis. In the printing industry, the different types of waste vary significantly. The most important of the solid, liquid and gaseous wastes produced in the printing industry before, during and after the printing process are waste ink, ink sludge and solvents emerging after machine washing, wastewater of water-based ink, plate and film developer and fixer solutions, cleaning solvents and volatile organic compounds (VOCs) [39]. The printing industry generates large amounts of scrap paper, catalogs, posters, cardboard boxes and plastic foil. The loose waste piles up and is stored in warehouses, taking up space and making it difficult to move around [40]. Some of the produced wastes even fall into the hazardous waste category due to their processing characteristics during the production process. The effective and regular extermination of these wastes is necessary to protect the environment. This can be provided only by the application of waste management. Some wastes are recycled and reused at printing industry, but in some cases, that recycling is impossible, and these wastes should be eliminated without harming human health and the environment [41]. In particular, the materials whose disposal is compulsory should be classified at the source and sent to licensed disposal companies. One of the solutions may be the reuse of generated waste in the production of other materials while being part of the circular economy.

The paper presents the results of research on the possibility of using unmodified post-production waste from the food label printing process. In the product labeling process, the liner is often overlooked by many brands as part of the waste stream. Despite the various recycling programs available on the market, such as UPM Raflatac's RafCycle®, they can still be a major environmental problem. The process of cutting the underlay to the production dimensions still produces from several dozen to several hundred kilograms of cut waste, which does not qualify for the recycling program and requires proper disposal, which is a problem for the company. Due to the lack of modification of the waste, it was expected that the properties of the obtained composites would deteriorate in relation to pure polymer. However, it is purposeful to check whether the properties of the new composites will still be sufficient for the use of the obtained composites.

2. Materials and Methods

The matrix of the tested materials was polypropylene (PP) Moplen EP548U (Lyondell-Basell, Rotterdam, The Netherlands). The basic properties of the polymer according to the data sheet are:

- Melt Flow Rate (230 °C/2.16 kg): 70 g/10 min.
- Density: 0.90 g/cm^3.
- Tensile Stress at Yield: 28 MPa.
- Tensile Strain at Break: 30%.
- Tensile Strain at Yield: 5%.
- Charpy Impact Strength (Notched): 4 kJ/m^2.
- Tensile Modulus: 1450 MPa.
- Melt temperature: 160 °C.

The filler of new materials is post- production waste after the printing process in the form of cut off yellow transparent glassine backing cellulose paper with the trade name HONEY GLASSINE 65 (UPM Raflatac, Tempere, Finland) with possible residues of hotmelt rubber permanent adhesive and a polymer (polypropylene or polyethylene) label. Glassine is a smooth and glossy paper that is air, water, and grease resistant. It is usually available in densities between 50 and 90 g/m^2. It is translucent unless dyes are added to color it or make it opaque. Product is designed for general purpose high-quality multicolor-printed labels to packaged food and homecare applications in ambient conditions. It is intended for all reelstock applications, and it is suitable for automatic dispensing. The basic properties of the backing paper according to the data sheet are:

- Substance: 55 g/m^2.
- Caliper: 49 µm.
- Tensile strength MD: 6.0 kN/m.
- Tensile strength CD: 2.3 kN/m.
- Transparency: 49%.

The input material from the production of the filler was in the form of long strands, approx. 4 mm wide. Initially, the waste was soaked in water, and then, it was formed with a hydraulic press into discs with a diameter of about 10 cm. Then, the discs were dried in a laboratory dryer and comminuted using a laboratory grinder. Ultimately, the filler was in the form of short fragments with a width and thickness similar to the original strips (Figure 1).

Figure 1. Initial and final form of the filler.

The prepared filler was added to the polymer matrix in a 1:3; 1:2; 1:1 and 2:1 volume ratio, which corresponded to mass concentrations at the level of 1.9; 2.5; 5.1 and 10.3 wt %. The test samples are marked with the symbols P_x, where x is the volume ratio of the filler. All test results were compared to the results of a pure polypropylene sample, which was abbreviated as PP. Sample designations with corresponding volume ratio of filler and calculated mass concentration are presented in Table 1.

Table 1. Samples designations.

Sample	Filler: Matrix Volume Ratio	Mass Concentration of Filler [wt.%]
PP	-	-
P_1_3	1:3	1.9
P_1_2	1:2	2.5
P_1_1	1:1	5.1
P_2_1	2:1	10.3

The granules of individual composites were obtained from the prepared composite masterbatches by extrusion. Extrusion was carried out on a W25-30D single screw extruder (Metalchem, Toruń, Poland). In order to obtain a very thorough mixing of both components, intensive mixing screws were used for this purpose, which additionally included kneading and retracting segments. The rotational speed of the screw was constant at 200 rpm. The temperatures of individual zones of the extruder were: 175, 185, 195 and 195 °C, and the temperature of the head was 120 °C. The material coming out of the head was cooled in a bath with water and then cut with a knife granulator.

From the obtained granulate by injection method, test specimens were obtained in the form of standardized bone-shaped samples (Figure 2) and bars. Injection molding was carried out on a TRX 80 Eco (Tederic, Zhejiang, China) injection molding machine. The injection molding process for all compositions was conducted under the following conditions: 170 °C, 170 °C and 175 °C with head temperature—180 °C. Other parameters follow: mold temperature—35 °C, injection pressure—35 bar, cooling time—30 s.

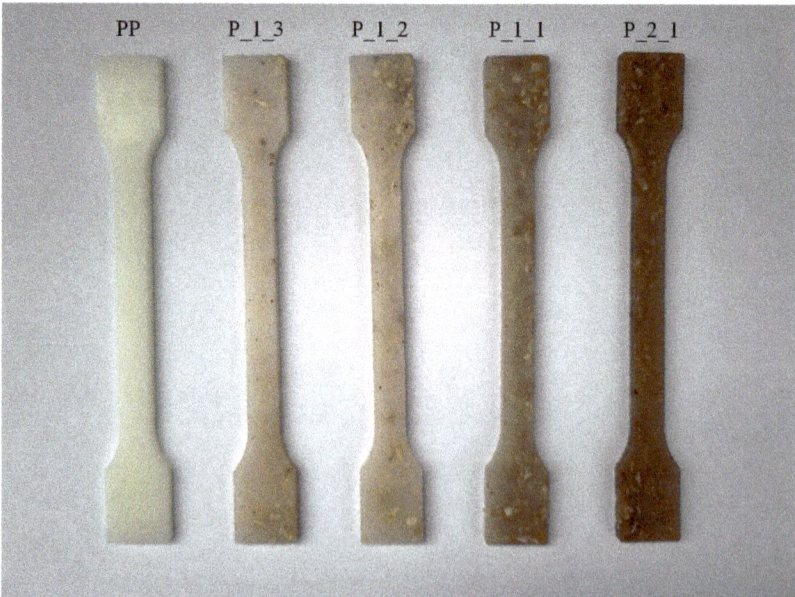

Figure 2. Test specimens obtained by injection method.

Melt flow rate (MFR) studies were performed using an MP600 plastometer (Tinius Olsen, Horsham, PA, USA). The tests were carried out at a temperature of 190 °C with a piston load of 2.16 kg. For each tested material, 12 measurement sections were obtained of which 10 values were taken for the calculations (two extreme ones were rejected).

The mechanical properties of the tested materials were determined by the tensile strength test and the un-notch impact test. For the mechanical tests, twelve samples of each composition were used for each test. The result of mechanical tests are the values obtained as the arithmetic mean of individual parameters together with the calculated values of the standard deviation.

Static tensile tests on PP and filler-containing polymer samples were performed on an Instron 3367 (Instron, Norwood, MA, USA) universal testing machine. The tensile speed was 50 mm/min. As part of the test, the tensile strength (σ_M), stress at break (σ_B), strain at maximum stress (ε_M) and strain at break (ε_B) were determined.

Charpy impact tests were carried out with an XJ 5Z impact hammer (Liangong, Shandong, China) using a 2 J hammer with a fall velocity of 2.9 m/s. Samples in the form of bars with dimensions of 80 mm × 10 mm × 4 mm were tested. As part of the study, the value of the impact strength without notch ($_{ua}$) was determined.

Thermomechanical (DMA) tests were carried out using a Q800 (TA Instruments, New Castle, DE, USA) dynamic mechanical analyzer. The tests were carried out in the temperature range from 30 to 150 °C with a heating rate of 3 °C/min. The samples were bars with dimensions of 80 mm × 10 mm × 4 mm. The strain was 15 μm, and the strain frequency was 1 Hz.

The Q200 (TA Instruments, New Castle, DE, USA) calorimeter was used in the differential scanning calorimetry (DSC) studies. Samples weighing about 4 mg were heated in the temperature range from 0 to 700 °C with the temperature change rate of 10 °C min^{-1}. The test was conducted in a nitrogen atmosphere. Based on the cooling and heating curves, the glass transition temperature (T_g), the cold crystallization temperature (T_{cc}), the change of the cold crystallization enthalpy (ΔH_{cc}), the melting point (T_m), the change of the melting enthalpy (ΔH_{cc}) and the degree of crystallinity (X_c) were determined. X_c values were calculated from equation:

$$X_c = \left(\frac{\Delta H_m - \Delta H_{cc}}{\Delta H_{m100\%} \cdot (1-x)} \right) \cdot 100\% \quad (1)$$

where $\Delta H_{m100\%}$—enthalpy change of 100% crystalline PP; 207 J/g [42]. x—share of post-printing waste.

Thermogravimetric analysis (TGA) studies were performed under nitrogen atmosphere using a Q500 thermobalance (TA Instruments, New Castle, DE, USA). The samples weighing about 21 mg were tested in the temperature range from 25 to 700 °C with the temperature change rate of 10 °C/min. Based on the thermogravimetric curves, the values of $T_{5\%}$, $T_{50\%}$ and $T_{95\%}$ were determined, corresponding to the loss temperature of 5%, 50% and 95% of the initial mass of the sample. The value of $T_{5\%}$ was adopted as the parameter defining the thermal resistance of the material. From the differential thermogravimetric curve (DTG) (the first derivative of the TG curve), the T_{max} values were also determined, defining the temperatures of the fastest mass loss in the individual degradation stages.

3. Results

The determined melt flow rate (MFR) of pure PP was 11.2 g/10 min and was consistent with the literature data (Figure 3). The addition of the filler into the polymer matrix did not cause major changes in the MFR values. Although the applied post-print waste limited the flow of the polymer, which is typical for this type of filler, the obtained decrease in MFR was acceptable. At lower filler contents, i.e., samples P_1_3 and P_1_2, the MFR decreased to the value of about 10 g/10 min. Even a further increase in the amount of filler did not cause a large decrease in MFR, and the recorded values for the samples P_1_1 and P_2_1 were 9.0 g/10 min. Thus, the total decrease in MFR after adding the greatest amount of filler, i.e., twice the volume excess of filler, was 2.2 g/10 min, which is 20% of the value of pure PP.

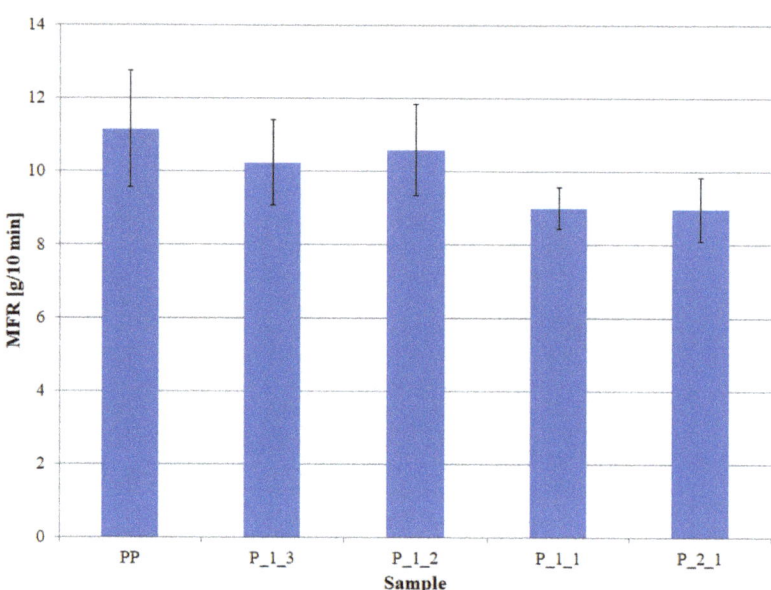

Figure 3. Melt flow rate (MFR) of individual samples.

The determined tensile strength (σ_M) and tensile stress (ε_B) of pure PP samples were slightly lower than the values given in the data sheet for the tested polymer but in line with the literature data for PP. The addition of post-printing waste to the matrix resulted in a slight decrease in the strength of the tested materials (Figure 4) as well as equating the values of σ_M and ε_B, which is related to the change in the elasticity of PP after introducing the filler particles (stress aggregation sites). As for MFR, one observes a clear two-stage decrease in the tensile strength of the tested materials. A smaller drop in strength by approx. 2.5 MPa compared to pure PP was observed for P_1_3 and P_1_2 materials, i.e., materials with a predominance of polymer in the composition. A greater decrease in strength by about 4.5 MPa in relation to pure PP was observed for higher waste content, i.e., P_1_1 and P_2_1 materials, where the initial volume fraction of waste was equal to or higher than the polymer fraction. The overall decrease in the tensile strength of PP after adding the post-printing waste was 4.5 MPa, which is approx. 19% of the value of pure polymer.

The introduction of the printing waste into the PP matrix was also accompanied by a large decrease in elongation at break (ε_B), which was additionally equal to the values of elongation at maximum stress (ε_M) (Figure 5). Due to the characteristics of PP and the occurrence of the phenomenon of necking during the tensile test, where the stretching and ordering of macromolecules occurs, this polymer is characterized by high ε_B values, which significantly exceed the ε_M values. The applied printing waste reduces the values of ε_M and ε_B to the level of approx. 5%, while the values for pure PP were, respectively, 8.1 and 35.9%. The obtained strain drop was the same regardless of the volumetric content of the filler in the matrix. The lack of differences in the deformation between individual materials is probably because, regardless of the amount of waste in the cross-section of the sample, there will always be a filler particle, on which stress aggregation and sample rupture will occur before the necking phenomenon occurs.

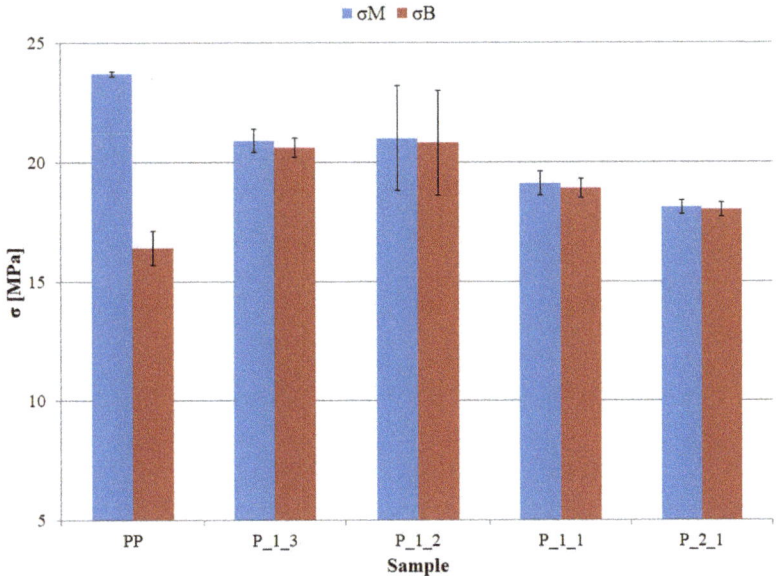

Figure 4. Tensile strength (σ_M) and stress at break (σ_B) of individual samples.

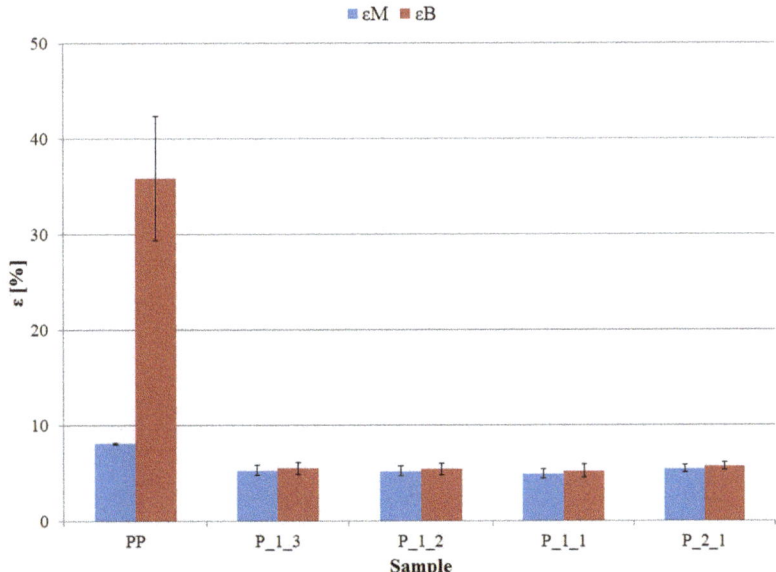

Figure 5. Elongation at maximum stress (ε_M) and elongation at break (ε_B) of individual samples.

The PP sample subjected to impact tests does not break, which is often observed for pure polypropylene due to its high flexibility and hence high impact resistance. Even the lowest content of post-printing waste in the material caused the samples to break as a result of the impact, and the value of the impact strength was recorded (u_a) (Figure 6).

The obtained u_a value for the P_1_3 sample was 28.3 kJ/m^2. With an increase in the filler content, the impact strength decreased, reaching the lowest value of 21.9 kJ/m^2 for the P_2_1 sample, i.e., a material with a waste content twice as high as PP. The total decrease in

u_a between the lowest and the highest content of post-printing waste was 6.4 kJ/m^2, i.e., approx. 22%. The observed decrease was caused by an increase in the amount of filler in the matrix and thus also in the cross-section of the sample (Figure 7), which translated into lower polymer content and lowered the impact strength of the entire system. The large scatter of the obtained results suggests, however, that the distribution of the filler particles was heterogeneous.

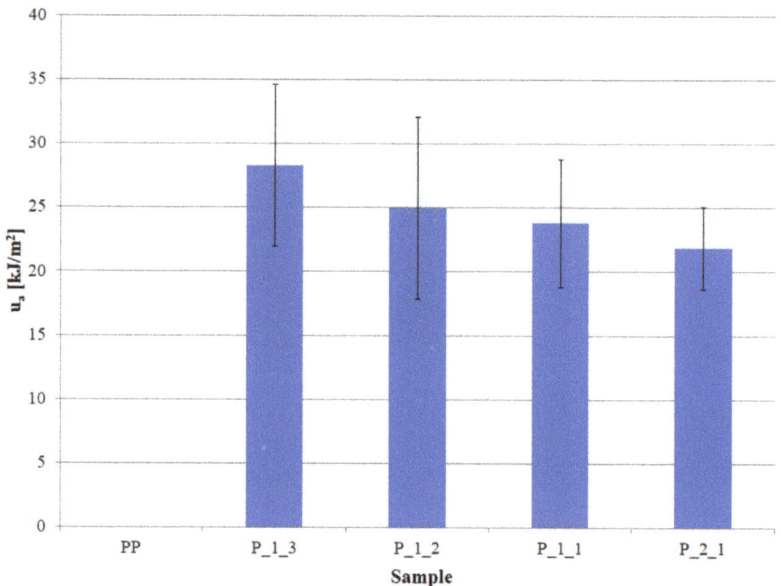

Figure 6. The impact strength (u_a) of individual samples.

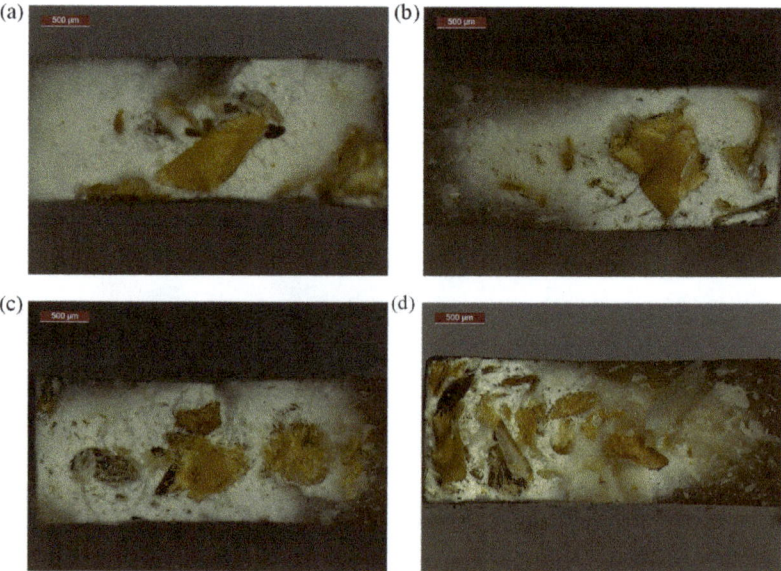

Figure 7. Microscopic photos of the cross-sections of (**a**) P_1_3, (**b**) P_1_2, (**c**) P_1_1, (**d**) P_2_1 sample.

The post-printing waste did not change the thermomechanical characteristics of the polymer. The PP modulus of elasticity at 30 °C (E'_{30}) determined during the three-point bending DMA test was 1225 MPa and decreased with increasing temperature, reaching 132 MPa at 150 °C, i.e., just before the material melting process. The thermomechanical curves with the recorded E' values for materials containing post-printing waste were similar to the values of pure PP, regardless of the amount of filler in the polymer matrix. Thus, the obtained materials retained the PP elasticity even with the highest content of post-printing waste. All results of E' measurements at different temperatures are presented in Table 2.

Table 2. Results of dynamic mechanical analysis (DMA) tests of individual samples.

Sample	E'_{30} (MPa)	E'_{60} (MPa)	E'_{90} (MPa)	E'_{150} (MPa)
PP	1225	759	416	132
P_1_3	1207	782	432	142
P_1_2	1247	779	431	142
P_1_1	1326	791	443	143
P_2_1	1194	744	426	148

Post-printing waste influenced the course of phase transformations in the tested materials, i.e., changed their temperatures and intensity (Figure 8). With the lowest waste content, the thermal characteristics of the P_1_3 sample are still close to the thermal characteristics of pure PP. Only a slight decrease in the intensity of the melting process by 8 J/g is visible, which suggests a slightly lower content of the crystalline phase, which results in a reduction in the calculated degree of crystallinity by 4.7 units. On the other hand, the introduction of a larger amount of filler caused a clear decrease in the crystallization temperature (T_c) by a maximum of approx. 6 °C for the P_2_1 sample, i.e., by approx. 5% compared to the value obtained for the PP sample.

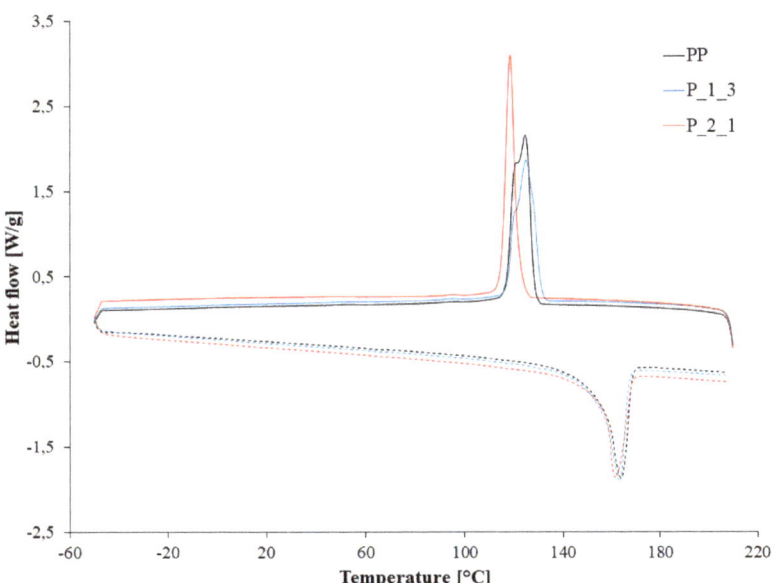

Figure 8. DSC curves of cooling (solid line) and heating (dashed line) of selected samples.

The nature of the crystallization process also changed. At lower waste contents, the recorded peak of the crystallization process was much lower but wider than the peak recorded at the highest filler content. The enthalpy change of the crystallization process

(ΔH_c) was also higher. With the increase in the content of waste, the ΔH_c value decreased by 14.3 J/g, from 95.1 J/g for the P_1_3 sample to 80.8 J/g for the P_2_1 sample. Thus, the waste content affects the structure and amount of the crystalline phase formed in PP. With a lower waste content, the formed crystallites are probably larger, and there are more of them. Changes in the crystallization process are influencing the structure and quantity of the crystalline phase and obviously translate into the recorded melting process and the calculated degree of crystallinity. As the filler content increases, the melting point (T_m) of the materials slightly decreases. The Tm decrease is approx. 2 °C when comparing the PP sample and the P_2_1 sample. The decrease in the melting enthalpy (ΔH_m), and thus the degree of calculated crystallinity (X_c), was much more pronounced. The ΔH_m value decreased by 15.6 J/g comparing the PP and P_2_1 sample. Thus, the degree of crystallinity of the tested materials decreased from 47.7 to 36.2%. All results of DSC are presented in Table 3.

Table 3. The results of the differential scanning calorimetry (DSC) for individual samples.

Sample	Cooling		Heating		
	T_c [°C]	ΔH_c (J/g)	T_m (°C)	ΔH_m (J/g)	X_c (%)
PP	124.5	97.4	163.7	98.8	47.7
P_1_3	124.8	95.1	163.1	90.8	43.0
P_1_2	119.0	80.1	161.5	85.6	40.3
P_1_1	119.0	81.8	161.5	83.1	38.1
P_2_1	118.4	80.8	161.4	83.2	36.2

After introducing the post-printing waste into the polymer matrix, changes in the thermal resistance of the obtained materials were observed. Figure 9 shows the thermogravimetric curves of selected samples, while Table 4 summarizes the results of the determined thermal parameters.

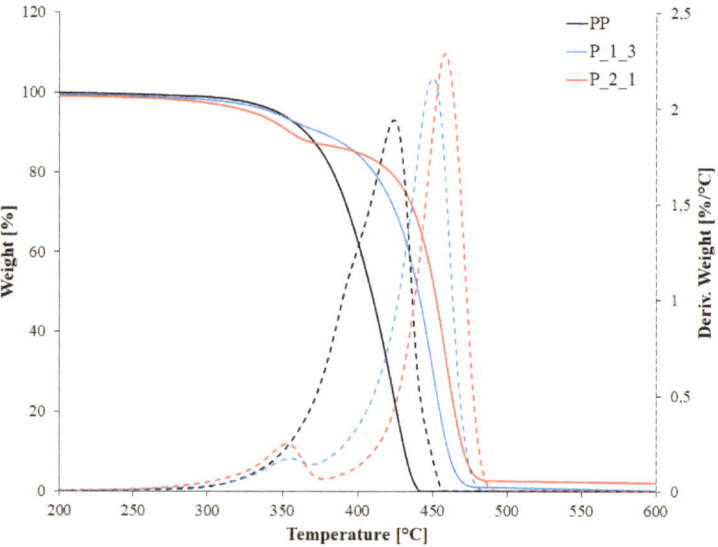

Figure 9. TG (solid line) and DTG (dashed line) curves of selected samples.

As shown in Figure 9, the introduction of post-printing waste into the polymer matrix changes the nature of the degradation process from a one-stage (PP sample) to a two-stage (samples with waste). After introducing the filler, an additional degradation step appears in the range from 320 to 380 °C with the maximum decomposition rate (T_{max1})

at approx. 355 °C, which becomes more significant the higher the content of waste in the polymer matrix.

Table 4. The results of the thermogravimetric analysis (TG) of individual samples.

Sample	$T_{5\%}$ (°C)	$T_{50\%}$ (°C)	$T_{95\%}$ (°C)	T_{max1} (°C)	T_{max2} (°C)
PP	347.5	409.1	434.8	-	423.9
P_1_3	344.8	440.5	465.4	353.6	450.1
P_1_2	343.9	450.1	473.6	354.0	455.4
P_1_1	340.0	440.7	467.9	356.5	450.8
P_2_1	330.6	451.5	476.6	353.2	458.3

The occurrence of an additional degradation stage indirectly contributes to the reduction in the thermal resistance of materials containing waste, which is determined based on the 5% loss temperature of the sample mass ($T_{5\%}$). The thermal resistance of materials dropped from 347.5 °C for a PP sample to 344.8 °C for the P_1_3 sample and 330.6 °C for the P_2_1 sample. Therefore, the decrease in thermal resistance depended on the amount of filler in the polymer matrix. With the lowest post-printing waste content, the decrease in thermal resistance was insignificant: less than 3 °C. Increasing the waste content causes a significant reduction in thermal resistance, as the recorded value of $T_{5\%}$ decreased by 17 °C in relation to the value of pure polymer. The reduction in the recorded $T_{5\%}$ value is caused by the overlapping of the maximum waste degradation rate at the beginning of the matrix degradation, so the higher the content of the filler, the sooner the loss of 5% of the sample weight was achieved and the contractual thermal resistance of the material changed. Regardless of the thermal resistance of the tested materials, their degradation temperature significantly exceeded the typical temperatures of the use of polymer materials, and therefore, developed materials can be successfully used in most typical applications.

4. Conclusions

Based on the conducted research, the following can be concluded:
- The total decrease in the melt flow rate after adding the greatest amount of filler, i.e., twice the volume excess of filler, was 2.2 g/10 min, which is 20% of the value of pure polypropylene.
- The greatest decrease in the tensile strength of polypropylene after adding the post-printing waste was 4.5 MPa, i.e., approx. 19% of the value of pure polymer. This result was obtained with the highest degree of filling of the composite.
- The obtained deformation drop was the same regardless of the volumetric content of the filler in the matrix. The applied post-printing waste reduces the elongation at maximum stress and elongation at break values to the level of approx. 5%, while the values for pure polypropylene were 8.1 and 35.9%, respectively.
- The total decrease in unnotched impact strength between the lowest and the highest content of post-printing waste was 6.4 kJ/m^2, i.e., approx. 22%.
- The introduction of post-printing waste into the polypropylene matrix did not change the thermomechanical characteristics of the polymer.
- The degree of crystallinity of the tested materials decreased from 47.7% for pure polypropylene to 36.2% for the material containing a double excess of printing waste.
- The introduction of the post-printing waste into the matrix causes a reduction in thermal resistance, as the registered value of the 5% weight loss temperature decreased by 17 °C in relation to the value of pure polypropylene.

Thus, the introduction of post-printing waste causes the deterioration of the characteristics of the obtained materials, but the decrease is fully acceptable in terms of the planned applications of the new composites. It should be remembered that the main purpose of the research described in the article was to eliminate post-production waste generated during the production of labels, which was achieved by using them as a filler for the polymer. Better properties of composites containing printing waste could probably be obtained after prior modification of the waste, so this will be the subject of further research.

Author Contributions: Conceptualization, K.M. and M.Z.; methodology, K.M. and T.K.; validation, K.M.; formal analysis, K.M.; investigation, T.K., A.S., B.B. and K.S.; resources, K.M. and M.Z.; data curation, K.M., T.K., A.S. and B.B.; writing—original draft preparation, K.M.; writing—review and editing, K.M. and K.S.; visualization, K.M., A.S. and B.B.; supervision, K.M. All authors have read and agreed to the published version of the manuscript.

Funding: This research received no external funding.

Institutional Review Board Statement: Not applicable.

Data Availability Statement: Not applicable.

Conflicts of Interest: The authors declare no conflict of interest.

References

1. Das, O.; Babu, K.; Shanmugam, V.; Sykam, K.; Tebyetekerwa, M.; Neisiany, R.E.; Försth, M.; Sas, G.; Gonzalez-Libreros, J.; Capezza, A.J.; et al. Natural and Industrial Wastes for Sustainable and Renewable Polymer Composites. *Renew. Sustain. Energy Rev.* **2022**, *158*, 112054. [CrossRef]
2. Bilitewski, B. The Circular Economy and Its Risks. *Waste Manag.* **2012**, *32*, 1–2. [CrossRef] [PubMed]
3. Zhou, Y.; Stanchev, P.; Katsou, E.; Awad, S.; Fan, M. A Circular Economy Use of Recovered Sludge Cellulose in Wood Plastic Composite Production: Recycling and Eco-Efficiency Assessment. *Waste Manag.* **2019**, *99*, 42–48. [CrossRef]
4. Demirbas, A. Waste Management, Waste Resource Facilities and Waste Conversion Processes. *Energy Convers. Manag.* **2011**, *52*, 1280–1287. [CrossRef]
5. Amasuomo, E.; Baird, J. The Concept of Waste and Waste Management. *J. Manag. Sustain.* **2016**, *6*, 88. [CrossRef]
6. Basalp, D.; Tihminlioglu, F.; Sofuoglu, S.C.; Inal, F.; Sofuoglu, A. Utilization of Municipal Plastic and Wood Waste in Industrial Manufacturing of Wood Plastic Composites. *Waste Biomass Valorization* **2020**, *11*, 5419–5430. [CrossRef]
7. Quitadamo, A.; Massardier, V.; Valente, M. Eco-Friendly Approach and Potential Biodegradable Polymer Matrix for WPC Composite Materials in Outdoor Application. *Int. J. Polym. Sci.* **2019**, *2019*, 3894370. [CrossRef]
8. Teuber, L.; Osburg, V.S.; Toporowski, W.; Militz, H.; Krause, A. Wood Polymer Composites and Their Contribution to Cascading Utilisation. *J. Clean. Prod.* **2016**, *110*, 9–15. [CrossRef]
9. Lewandowski, K.; Piszczek, K.; Zajchowski, S.; Mirowski, J. Rheological Properties of Wood Polymer Composites at High Shear Rates. *Polym. Test.* **2016**, *51*, 58–62. [CrossRef]
10. Wei, L.; McDonald, A.G.; Freitag, C.; Morrell, J.J. Effects of Wood Fiber Esterification on Properties, Weatherability and Biodurability of Wood Plastic Composites. *Polym. Degrad. Stab.* **2013**, *98*, 1348–1361. [CrossRef]
11. Friedrich, D. Thermoplastic Moulding of Wood-Polymer Composites (WPC): A Review on Physical and Mechanical Behaviour under Hot-Pressing Technique. *Compos. Struct.* **2021**, *262*, 113649. [CrossRef]
12. Essabir, H.; Hilali, E.; Elgharad, A.; El Minor, H.; Imad, A.; Elamraoui, A.; Al Gaoudi, O. Mechanical and Thermal Properties of Bio-Composites Based on Polypropylene Reinforced with Nut-Shells of Argan Particles. *Mater. Des.* **2013**, *49*, 442–448. [CrossRef]
13. Laaziz, S.A.; Raji, M.; Hilali, E.; Essabir, H.; Rodrigue, D.; Bouhfid, R.; Qaiss, A.E.K. Bio-Composites Based on Polylactic Acid and Argan Nut Shell: Production and Properties. *Int. J. Biol. Macromol.* **2017**, *104*, 30–42. [CrossRef] [PubMed]
14. Leszczyńska, M.; Ryszkowska, J.; Szczepkowski, L. Rigid Polyurethane Foam Composites with Nut Shells. *Polimery* **2020**, *65*, 728–737. [CrossRef]
15. Okonkwo, E.G.; Anabaraonye, C.N.; Daniel-Mkpume, C.C.; Egoigwe, S.V.; Okeke, P.E.; Whyte, F.G.; Okoani, A.O. Mechanical and Thermomechanical Properties of Clay-Bambara Nut Shell Polyester Bio-Composite. *Int. J. Adv. Manuf. Technol.* **2020**, *108*, 2483–2496. [CrossRef]
16. Sun, X.; He, M.; Li, Z. Novel Engineered Wood and Bamboo Composites for Structural Applications: State-of-Art of Manufacturing Technology and Mechanical Performance Evaluation. *Constr. Build. Mater.* **2020**, *249*, 118751. [CrossRef]
17. Lokesh, P.; Surya Kumari, T.S.A.; Gopi, R.; Loganathan, G.B. A Study on Mechanical Properties of Bamboo Fiber Reinforced Polymer Composite. *Mater. Today Proc.* **2020**, *22*, 897–903. [CrossRef]
18. Adediran, A.A.; Akinwande, A.A.; Balogun, O.A.; Olasoju, O.S.; Adesina, O.S. Experimental Evaluation of Bamboo Fiber/Particulate Coconut Shell Hybrid PVC Composite. *Sci. Rep.* **2021**, *11*, 5765. [CrossRef]
19. Mousavi, S.R.; Zamani, M.H.; Estaji, S.; Tayouri, M.I.; Arjmand, M.; Jafari, S.H.; Nouranian, S.; Khonakdar, H.A. Mechanical Properties of Bamboo Fiber-Reinforced Polymer Composites: A Review of Recent Case Studies. *J. Mater. Sci.* **2022**, *57*, 3143–3167. [CrossRef]
20. Agunsoye, J.O.; Isaac, T.S.; Samuel, S.O. Study of Mechanical Behaviour of Coconut Shell Reinforced Polymer Matrix Composite. *J. Miner. Mater. Charact. Eng.* **2012**, *11*, 774–779.
21. Agunsoye, J.O.; Odumosu, A.K.; Dada, O. Novel Epoxy-Carbonized Coconut Shell Nanoparticles Composites for Car Bumper Application. *Int. J. Adv. Manuf. Technol.* **2019**, *102*, 893–899. [CrossRef]
22. Nadzri, S.N.I.H.A.; Sultan, M.T.H.; Shah, A.U.M.; Safri, S.N.A.; Talib, A.R.A.; Jawaid, M.; Basri, A.A. A Comprehensive Review of Coconut Shell Powder Composites: Preparation, Processing, and Characterization. *J. Thermoplast. Compos. Mater.* **2020**, *35*, 2641–2664. [CrossRef]

23. Obasi, H.C.; Mark, U.C.; Mark, U. Improving the Mechanical Properties of Polypropylene Composites with Coconut Shell Particles. *Compos. Adv. Mater.* **2021**, *30*, 263498332110074. [CrossRef]
24. Sundarababu, J.; Anandan, S.S.; Griskevicius, P. Evaluation of Mechanical Properties of Biodegradable Coconut Shell/Rice Husk Powder Polymer Composites for Light Weight Applications. *Mater. Today Proc.* **2021**, *39*, 1241–1247. [CrossRef]
25. Bledzki, A.K.; Mamun, A.A.; Volk, J. Physical, Chemical and Surface Properties of Wheat Husk, Rye Husk and Soft Wood and Their Polypropylene Composites. *Compos. Part A Appl. Sci. Manuf.* **2010**, *41*, 480–488. [CrossRef]
26. Arjmandi, R.; Hassan, A.; Majeed, K.; Zakaria, Z. Rice Husk Filled Polymer Composites. *Int. J. Polym. Sci.* **2015**, *2015*, 501471. [CrossRef]
27. Muthuraj, R.; Lacoste, C.; Lacroix, P.; Bergeret, A. Sustainable Thermal Insulation Biocomposites from Rice Husk, Wheat Husk, Wood Fibers and Textile Waste Fibers: Elaboration and Performances Evaluation. *Ind. Crops Prod.* **2019**, *135*, 238–245. [CrossRef]
28. Bisht, N.; Gope, P.C.; Rani, N. Rice Husk as a Fibre in Composites: A Review. *J. Mech. Behav. Mater.* **2020**, *29*, 147–162. [CrossRef]
29. Suhot, M.A.; Hassan, M.Z.; Aziz, S.A.; Md Daud, M.Y. Recent Progress of Rice Husk Reinforced Polymer Composites: A Review. *Polymer* **2021**, *13*, 2391. [CrossRef]
30. Wilpiszewska, K.; Antosik, A.K. Effect of Grain Husk Microfibers on Physicochemical Properties of Carboxymethyl Polysaccharides-Based Composite. *J. Polym. Environ.* **2022**, *30*, 3129–3138. [CrossRef]
31. Berger, C.; Mattos, B.D.; Amico, S.C.; de Farias, J.A.; Coldebella, R.; Gatto, D.A.; Missio, A.L. Production of Sustainable Polymeric Composites Using Grape Pomace Biomass. *Biomass Convers. Biorefinery* **2020**, *12*, 5869–5880. [CrossRef]
32. Morinaga, H.; Haibara, S.; Ashizawa, S. Reinforcement of Bio-Based Network Polymer with Wine Pomace. *Polym. Compos.* **2021**, *42*, 2973–2981. [CrossRef]
33. Aljnaid, M.; Banat, R. Effect of Coupling Agents on the Olive Pomace-Filled Polypropylene Composite. *E Polym.* **2021**, *21*, 377–390. [CrossRef]
34. Mirowski, J.; Oliwa, R.; Oleksy, M.; Tomaszewska, J.; Ryszkowska, J.; Budzik, G. Poly(Vinyl Chloride) Composites with Raspberry Pomace Filler. *Polymer* **2021**, *13*, 1079. [CrossRef]
35. Hiremath, P.; Shettar, M.; Shankar, M.C.G.; Mohan, N.S. Investigation on Effect of Egg Shell Powder on Mechanical Properties of GFRP Composites. *Mater. Today Proc.* **2018**, *5*, 3014–3018. [CrossRef]
36. Owuamanam, S.; Cree, D. Progress of Bio-Calcium Carbonate Waste Eggshell and Seashell Fillers in Polymer Composites: A Review. *J. Compos. Sci.* **2020**, *4*, 70. [CrossRef]
37. Sakthi Balan, G.; Santhosh Kumar, V.; Rajaram, S.; Ravichandran Ramakrishnan, M.K. Investigation on Water Absorption and Wear Characteristics of Waste Plastics and Seashell Powder Reinforced Polymer Composite. *J. Tribol.* **2020**, *27*, 57–70.
38. Vasanthkumar, P.; Balasundaram, R.; Senthilkumar, N.; Palanikumar, K.; Lenin, K.; Deepanraj, B. Thermal and Thermo-Mechanical Studies on Seashell Incorporated Nylon-6 Polymer Composites. *J. Mater. Res. Technol.* **2022**, *21*, 3154–3168. [CrossRef]
39. Hayta, P.; Oktav, M. The Importance of Waste and Environment Management in Printing Industry. *Eur. J. Eng. Nat. Sci.* **2019**, *3*, 18–26.
40. Carlos Alberto, P.J.; Sonia Karina, P.J.; Francisca Irene, S.A.; Adrielly Nahomee, R.Á. Waste Reduction in Printing Process by Implementing a Video Inspection System as a Human Machine Interface. *Procedia Comput. Sci.* **2021**, *180*, 79–85. [CrossRef]
41. Medeiros, D.L.; Braghirolli, F.L.; Ramlow, H.; Ferri, G.N.; Kiperstok, A. Environmental Improvement in the Printing Industry: The Case Study of Self-Adhesive Labels. *Environ. Sci. Pollut. Res.* **2019**, *26*, 13195–13209. [CrossRef] [PubMed]
42. Lanyi, F.J.; Wenzke, N.; Kaschta, J.; Schubert, D.W. On the Determination of the Enthalpy of Fusion of α-Crystalline Isotactic Polypropylene Using Differential Scanning Calorimetry, X-Ray Diffraction, and Fourier-Transform Infrared Spectroscopy: An Old Story Revisited. *Adv. Eng. Mater.* **2020**, *22*, 1900796. [CrossRef]

MDPI
St. Alban-Anlage 66
4052 Basel
Switzerland
www.mdpi.com

Polymers Editorial Office
E-mail: polymers@mdpi.com
www.mdpi.com/journal/polymers

Disclaimer/Publisher's Note: The statements, opinions and data contained in all publications are solely those of the individual author(s) and contributor(s) and not of MDPI and/or the editor(s). MDPI and/or the editor(s) disclaim responsibility for any injury to people or property resulting from any ideas, methods, instructions or products referred to in the content.

www.ingramcontent.com/pod-product-compliance
Lightning Source LLC
LaVergne TN
LVHW070458100526
838202LV00014B/1747